PRACTICAL
NONPARAMETRIC
STATISTICS

PRACTICAL NONPARAMETRIC STATISTICS

W. J. CONOVER

ASSOCIATE PROFESSOR
OF STATISTICS AND COMPUTER SCIENCE
KANSAS STATE UNIVERSITY

JOHN WILEY & SONS INC.

NEW YORK · LONDON · SYDNEY · TORONTO

Preface

This book is intended as an introductory textbook and as a reference book for applied research workers. As an introductory text, it requires only algebra as a prerequisite. While it is expected that most courses employing this as a text will require a previous course in elementary statistics, such a requirement may be deleted if one is willing to ignore the occasional references to "usual parametric counterparts." In fact, because of the simple nature and general applicability of nonparametric statistics, it may be more practical to introduce the student first to nonparametric statistics, and then to the usual parametric statistics as a special area. Although this book could serve as a text for such a course, I have used it (in preliminary form) as a text for several years in a graduate-level introductory course for nonstatisticians who have had a previous course in statistics. The entire text may be covered in three semester hours by covering approximately one section, including the problems, in each lecture. This allows ample time for discussion of the more interesting problems, and for examinations. For a short course on probability, Chapter 1 may be studied by itself. Or, for a short course on nonparametric statistics, the methods and examples of Chapters 3 to 7 may be studied without the accompanying theory, and without Chapters 1 and 2.

To use this book as a "book of recipes," the chart on page xii may guide in the selection of an appropriate test. Each method is described in a self-contained, clear-cut format. Examples using actual numbers are given to assist in clearing up any ambiguities in the written description. Applications are drawn from the fields of psychology, biology, statistics, engineering business, education, economics, medicine, agriculture and jurisprudence.

I am grateful to the many people who assisted in this work. More than 100 students read the text in various forms and contributed to its clarity. About a dozen professional men read the manuscript and contributed to its validity. Editorial and financial assistance were provided by the publisher. Kansas State University, its Department of Statistics and Computer Science, and its Agricultural Experiment Station supported much of the research that is reported in various places in the book. I was also aided by the National Science Foundation Grant GP-7667 and, in the final stages of the manuscript, by the National Institutes of Health Research Career Development Award, Grant Number 1-KO4-GM42351-01.

Many people contributed to the improvement of the manuscript. I would appreciate personal communications concerning strong points and weak areas, since these may affect the form of any possible later editions.

W. J. Conover

Contents

Tables

PRACTICAL
NONPARAMETRIC
STATISTICS

A CHART FOR QUICK LOCATION OF THE APPROPRIATE NONPARAMETRIC TEST

Type of Sample Obtained	Hypothesis Test Involving . . .	Type of Measurement Involved (See Section 2.1)		
		NOMINAL (observations may be separated according to categories)	ORDINAL* observations map be arranged from smallest to largest)	INTERVAL† (the numerical value of the observation has meaning)
One random sample $X_1, \ldots, X_n$	Means (medians)	Binomial test (p. 96)	Quantile test (p. 105)	Wilcoxon test (p. 206)
	(Conf. int. for means)	Conf. int. for p (p. 99)	Conf. int. for z_p (p. 110)	Conf. int. for mean (p. 216)
	Nonrandomness	Wald-Wolfowitz test (p. 350)	Cox and Stuart test (p. 130) Spearman's rho (p. 251)	See Section 7.4
	Goodness of fit	Chi-square test (p. 186)	Kolmogorov test (p. 295) Cramér-von Mises test (p. 306) Lilliefors test (p. 302)	
	[Conf. band for $F(x)$]		Conf. band for $F(x)$ (p. 299)	
Paired observations, or two matched samples $(X_1, Y_1), \ldots, (X_n, Y_n)$	Means (medians)	McNemar test (p. 127)	Sign test (p. 121)	Wilcoxon test (p. 206) Randomization test (p. 358)
	(Conf. int. for diff. between means)	Conf. int. for p (p. 99)	Conf. int. for z_p (p. 110)	Conf. int. for diff. (p. 216)
	Independence	Chi-square test (p. 154) Fisher's test (p. 158)	Sign test (p. 133) Bell-Doksum test (p. 283) Olmstead-Tukey test (p. 336)	See Section 7.4
	Measures of dependence	See Section 4.4	See Section 5.5	
Multivariate observations, or the randomized complete block design	Means	Cochran test (p. 196)	Friedman test (p. 265) Bell-Doksum test (p. 287)	
	Measures of dependence and tests of independence	See Section 4.4	See Section 5.5	
Two random samples $X_1, \ldots, X_n$ and $Y_1, \ldots, Y_m$. (See also tests for several random samples)	Means (medians)	Chi-square test (p. 141)	Mann-Whitney test (p. 224) Tukey's quick test (p. 328)	Randomization test (p. 360)
	(Conf. int. for diff. between means)		Mann-Whitney conf. int. (p. 238) Tukey's conf. int. (p. 330)	
	Variances		Siegel-Tukey test (p. 230)	
	Identical distributions		Smirnov test (p. 309) Cramér-von Mises test (p. 314) Wald-Wolfowitz test (p. 350)	
Several random samples	Means (medians)	Chi-square test (p. 150)	Median test (p. 167) Kruskal-Wallis test (p. 256) Bell-Doksum test (p. 284) Slippage test (p. 342)	
	Identical distributions		Birnbaum-Hall test (p. 317) Smirnov test (pp. 320, 322)	
Other types	Means (medians)	Many-way contingency table (p. 164)	Median test extended (p. 172) Durbin test for BIBD (p. 276)	

Practical Nonparametric Statistics

One of the dictionary definitions of the word "science" is given as "truth ascertained by observation, experiment, and induction." A vast amount of time, money, and energy is being spent by society today in the pursuit of science. This pursuit is quite often frustrating because, as any scientist knows, the processes of observation, experiment, and induction do not always lay bare the "truth." One experiment, with one set of observations, may lead two scientists to two different conclusions.

For example, a scientist places a rat into a pen with two doors, both closed. One door is painted red and the other blue. The rat is then subjected to twenty minutes of music of the type popular with today's teenagers. After this experience, both doors are opened and the rat runs out of the pen. The scientist notes which color door the rat chose. This experiment is repeated ten times, each time using a different rat.

At the end of the composite experiment, the experimenter notes that the rats chose the red door seven out of ten times. He concludes the "truth" as being that the treatment used causes rats to prefer the red door to the blue door. However, a colleague overhears this conclusion, and jokingly tests the scientist, "If I tossed a coin ten times getting seven heads and, before each toss, I whistled 'Yankee Doodle', would you conclude that my whistling caused the coin to prefer heads?" Seeing the analogy between a rat choosing one of two doors, and a coin landing on one of its two sides, the scientist realized his error and decided that the outcome of his experiment could easily have been the result of chance.

Later the scientist conducts a second experiment. He injects a certain drug into the bloodstream of each of ten rats. Five minutes later he examines the rats and finds that seven are dead, and the other three are apparently healthy. However, since only seven are dead, he recalls the previous experiment and concludes that such a result could easily have occurred by chance and therefore there is no proof that the drug injections are dangerous.

His colleague again interrupts, saying, "With your first experiment each rat had a fifty-fifty chance of choosing the red door, without the music, and therefore we can compare that experiment to tossing a coin. In this experiment, the chances of a rat dying within five minutes are quite slim indeed, if the drug has no effect. Since your experiment resulted in seven of these rare events out of only ten possibilities, it seems safe to conclude that the drug injections caused the deaths."

And so goes research. It soon becomes apparent to most scientists that the ideal way of expressing results of experiments, such as the above, is to be able to say something like, "Without the treatment I administered, experimental results as extreme as the ones I obtained would occur only about three times in a thousand. Therefore I conclude that my treatment has a definite effect." In this way every scientist who reads of this experiment knows just how much subjectivity, or opinion, entered into the stated conclusion.

The purpose of that field of science known as "statistics" is to provide the means for measuring the amount of subjectivity that goes into the scientists' conclusions, and thus to separate "science" from "opinion." This is accomplished by setting up a theoretical "model" for the experiment, such as the model called "tossing a coin" which was set up for the first experiment above. Then laws of probability are applied to this model in order to determine what the "chances" (probabilities) are for the various possible outcomes of the experiment under the assumption that chance alone, and not music or drug injections, determines the outcome of the experiment. Then the experimenter has an objective basis for deciding whether his results were a result of the treatments he applied, or whether the same results could have easily occurred by chance alone with no treatment.

Although it is sometimes difficult to describe an appropriate theoretical model for the experiment, the real difficulty often comes after the model has been defined, in the form of finding the probabilities associated with the model. Many reasonable models have been invented for which no probability solutions have ever been found. For this reason statisticians have often changed the model slightly in order to be able to solve for the desired probabilities, with the hope that the change in the model was slight enough so that the changed model was still fairly realistic. Then they are able to obtain exact solutions for these "approximate problems." This body of statistics is sometimes called "parametric statistics," and embodies such well-known tests as the "t test," the "F test," and others.

In the late 1930's a different approach to the problem of finding probabilities began to gather some momentum. This approach involved making few, if any, changes in the model, and using rather simple and unsophisticated methods to find the desired probabilities, or at least a good approximation to those probabilities. Thus approximate solutions to exact problems were found, as opposed to the exact solution to approximate problems furnished by parametric statistics. This new package of statistical procedures became known as "nonparametric statistics."

Besides the advantage of using a simpler model, nonparametric statistical methods often involve less computational work, and therefore are easier and quicker to apply than other statistical methods. A third advantage of nonparametric statistical techniques is that much of the theory behind the nonparametric methods may be developed rigorously using no mathematics beyond high school algebra. A scientist who understands the theory behind the statistical method is less apt to use that method in a situation where such usage would be incorrect, and he is better able to develop his own statistical methods if his model is one that has not yet been considered by statisticians.

Those parts of nonparametric statistics requiring the use of more advanced mathematics will be presented without deriving them, but whenever convenient there will be a reference to a source where the proof may be found.

The above formulation of parametric statistics versus nonparametric statistics is merely an attempt to give a rough idea concerning the subject of this book. A more precise distinction between the two branches of statistics will be given in Chapter 2, where the philosophy of scientific experimentation is discussed in greater detail. In order to present examples and illustrations in Chapter 2, a preliminary knowledge of some elementary aspects of probability is needed, and so this is the concern of Chapter 1.

From Chapter 3 onward, there is a heavy reliance on the concepts introduced in Chapters 1 and 2. These later chapters present various nonparametric procedures, organized according to the type of model that is being analyzed rather than according to the type of experiment being conducted. For convenience to the experimenter who wants to examine the body of techniques which may be used in his analysis, a cross-referencing table is presented inside the back cover, listing the techniques given in the book according to the type of problem they are intended to solve.

This book attempts to present nonparametric techniques that are already popular among experimenters, in a clearer way than is now available in other books and journals. Also, it presents statistical methods that are not widely known because of their recent development. Some nonparametric methods presented in this book have not yet appeared in the literature, but are included because it is felt that they will be useful to experimenters.

A word about the numbering of examples, equations, and figures would not be out of place at this time. Example 4.2.3 refers to the third example in Section 4.2. When referring to an example within the same section, only the last number is used. For instance, within Section 4.2, Example 4.2.3 is referred to simply as Example 3. The same is true for equations, figures, and problems. No such economy is used with regard to section numbers, so that Section 4.2 is always called Section 4.2, even within Chapter 4.

For those who wish to obtain more information about nonparametric procedures, many references are included at the end of each appropriate section. Most of these references are recent, and earlier, sometimes more important, papers are usually not mentioned. This is because the references given generally refer in turn to the earlier papers on that topic, and so there was no need to repeat them here. The bibliography by Savage (1962) is quite useful for obtaining additional references on each topic.

CHAPTER I

Probability Theory

1.1. PRELIMINARY REMARKS

In order to use statistical methods to analyze data from an experiment, it is essential that some sort of an "idealized" experiment be formulated. Even in such a simple experiment as noting which of two doors a rat chooses after listening to twenty minutes of music, many simplifying assumptions need to be made before the statement "the rat has a fifty-fifty chance of choosing the red door" can be considered to be valid. One assumption might be, "Both doors are of exactly the same size." Another might be, "The rat is not going to prefer the same door the previous rat chose, on the basis of being able to detect the scent of the previous rat's path." Or a broader assumption might be, "The choice of each rat does not depend in any way on the choices of the other rats." The statement, "The music has no effect on the rat's choice of doors" is the statement that is being tested by the experiment, and this will be called the "null hypothesis," which will be discussed in Chapter 2.

The actual experiment is never performed under ideal conditions. However, the experimenter assumes that the idealized experiment contains all of the aspects of the actual experiment, except for those aspects that have negligible effects on the experimental results. Thus, while the actual experiment is concerned with the size and type of pens, the ventilation, the types of rats used, and a multitude of other factors, the idealized experiment may be described quite simply as, "Without the effect of the music, a rat chooses one of two doors, each choice being equally likely. If the two choices are not equally likely, then the playing of the music must have been the reason for the preference of one door to the other."

Probability statements are made concerning the idealized experiment, called "the model." If the model has been realistically formulated, then the probability statements concerning the model are reasonably valid when applied to the actual experiment.

The purpose of this chapter is to introduce some notions of probability theory. Selection has been made so that emphasis is on that portion of probability theory that is most useful in the study of nonparametric statistics.

The reader having difficulty understanding this chapter is referred to other statistics texts such as Mosteller et al. (1961) or Feller (1968).

1.2. COUNTING

The process of computing probabilities often depends on being able to count, in the usual sense of counting, "1, 2, 3," and so on. The usual way of counting becomes quite tedious in some complicated situations, and so some sophisticated methods of counting are developed in this section to handle those complicated situations.

When we speak of tossing a coin, we shall consider only two possible outcomes: either a head (H) appears, or a tail (T) appears. If a coin is tossed once there are two possible outcomes: H or T. If a coin is tossed twice there are $2^2 = 4$ possible outcomes: HH, HT, TH, TT, where HT means a head occurs on the first toss and a tail on the second. Each time we consider one additional toss of the coin, the number of possible outcomes is doubled, since the last toss may result in either of two outcomes. Thus if a coin is tossed n times there are 2^n possible outcomes.

Generalizing this discussion somewhat, we may refer to the tossing of a coin as one example of an *experiment*. Whether the coin is tossed once, twice or, in general, n times, the procedure may be considered to be an experiment. Since tossing a coin three times may be considered to be an experiment, and is a composite of three separate experiments where the coin is tossed only once each time, we may refer to the shorter experiments as *trials* and the collection of trials as "the experiment."

Few scientists seriously consider coin tossing as an experiment worthy of merit by itself. The value of coin tossing is that it serves as a prototype for many different models in many different situations. If an unbiased coin is being considered, one in which each face is equally likely to result, the experiment is not unlike experiments involving rats who have two choices of doors, consumers choosing between two products, educators determining which of two teaching methods is more effective, market analysts deciding whether the market tends to be higher or lower on Mondays, and many other situations.

If we allow the coin to be biased, where one face is more likely to turn up than the other, then a much broader class of experiments is included under the same model. Examples include experiments where a drug is injected into the bloodstream of rats to see if the drug is lethal, a new cure is tested on sick patients, a consumer is given several choices of a product and asked to choose one where only one of the products is manufactured by Company X, and other situations. In each case there are two outcomes of interest, such as "life" versus "death," "cure" versus "no cure," "our brand" versus "other brands," and the two outcomes might not be equally likely to occur.

Throughout this chapter and the next, models involving coin tossing, dice rolling, drawing chips from a jar, placing balls into boxes, and so on, will be discussed as if they were experiments worthy of merit, while actually the value of these models lies mainly in the fact that they serve as useful and simple prototypes of many more complicated models arising from experimentation in such diverse areas as electron physics, psychology, sociology, education, biology, economics, chemistry, etc. An excellent study of the diversity of such models is given by Feller (1968). Some justification for the study of these models will be presented in this chapter, but for the most part the justification will be deferred until later chapters, where the various nonparametric procedures are introduced.

Thus we may refer to coin tossing as an experiment and each individual toss of the coin as a trial. The possible outcomes of one trial, several trials, or the entire experiment will be called *events*. The coin tossing experiment described above consists of n trials, where each trial may result in either the event H or the event T. A combination of events may ifself be an event. Therefore, it is permissible to consider each of the 2^n possible outcomes of the experiment as an event. Examples of other events would include the event "at least one head," the event "a tail on the fourth toss," and the event "at least twice as many heads as tails."

Further generalization leads to the following rule.

RULE 1. If an experiment consists of n trials, where each trial may result in one of k outcomes, then there are k^n possible outcomes of the experiment.

Example 1. Suppose an experiment is composed of n trials, where each trial consists of throwing a ball into one of k boxes. The first throw may result in one of k different outcomes. For each outcome, the second throw also may result in one of k different outcomes. Thus there are k^2 outcomes associated with the first two trials combined. This reasoning extends to the n throws comprising the experiment, resulting in k^n different outcomes of the experiment.

Now consider a box containing n plastic chips numbered 1 to n. One chip is selected from the box, the number on the chip is recorded, and the chip is put aside. The number recorded may be any of the n numbers. A second chip is selected, the number on the chip is recorded, and it also is put aside. The second number may be any one of $n - 1$ numbers. That is, the second number may be any one of the numbers from 1 to n except the number that appeared on the first chip, since the first chip was not replaced. Thus there are $n(n - 1)$ different ways that the first two numbers may be recorded. The process is continued with a third chip, a fourth chip, and so on, until the last chip has been drawn, and its number recorded. There are

$$n(n - 1)(n - 2) \ldots (3)(2)(1) = n!$$

different ways the n numbers may have been recorded, and therefore $n!$ different orders in which the n chips may have been selected.

Instead of considering the number of ways of selecting n objects, we could have considered the number of ways of arranging n objects into a row. This leads to the following.

RULE 2.   There are $n!$ ways or arranging n objects into a row.

If the n objects are distinguishable one from another, then each of the $n!$ arrangements is unique. But suppose two of the objects are identical. Then for each arrangement of the n objects, there is a second arrangement that is indistinguishable from the first, namely the arrangement in which $n - 2$ of the objects are in the same position as in the first arrangement, but the two identical objects are interchanged. Each of the $n!$ arrangements may be paired in this manner with another identical arrangement. The number of different arrangements is thus $n!/2$, or $n!/2!$.

Suppose three of the objects are identical, and $n - 3$ are distinguishable from each other. If we divide the $n!$ arrangements into groups of identical arrangements, we find there are $3!$ arrangements in each group. This is because the three identical objects may be placed $3!$ different but indistinguishable ways into their 3 positions, using Rule 2. Then the number of different arrangements, equal to the number of groups of identical arrangements, is $n!/3!$. If exactly n_1 objects are identical, the $n!$ arrangements may be divided into groups of identical arrangements, each group being of size $n_1!$. If there are n_1 identical objects of type 1, and n_2 identical objects of a different type 2, then for each arrangement of the objects of type 1 there are $n_2!$ identical arrangements of type 2. So there are, in all, $n_1! \, n_2!$ arrangements in each group of identical arrangements. Therefore, the number of groups is $n!/(n_1! \, n_2!)$. This leads to another counting rule.

RULE 3.   If a group of n objects is composed of n_1 identical objects of type 1, n_2 identical objects of type 2, $\ldots$, n_r identical objects of type r, then the number of distinguishable arrangements in a row, denoted by $\begin{bmatrix} n \\ n_i \end{bmatrix}$, is given by

(1)
$$\begin{bmatrix} n \\ n_i \end{bmatrix} = \frac{n!}{n_1!\,n_2!\,\ldots\,n_r!}$$

To justify the use of Rule 3, let us divide the $n!$ arrangements into groups of identical arrangements. Each group then has $n_1!\,n_2!\,\ldots\,n_r!$ arrangements in it. Since no arrangement may appear in two different groups, the number of groups is $n!/(n_1!\,n_2!\,\ldots\,n_r!)$. We may assume without loss of generality that $n_1 + n_2 + \ldots + n_r$ equals n, because some of the n_i may equal 1, representing objects that are similar only to themselves. Since 1! equals 1, and since dividing (1) by 1 does not affect the numerical value, Rule 3 remains unaffected by the above assumption. It is also apparent now that Rule 2 is a special case of Rule 3, where all of the n_i equal 1.

Example 2. In a coin tossing experiment, n trials result in k heads and $n - k$ tails. The number of different sequences of k heads and $n - k$ tails equals the number of distinguishable arrangements of k objects of one kind and $n - k$ objects of another, which is $n!/[k!\,(n - k)!]$. When only two types of objects are considered, the abbreviation

(2)
$$\begin{bmatrix} n \\ n_i \end{bmatrix} = \binom{n}{k}$$

is used. Therefore

(3)
$$\binom{n}{k} = \frac{n!}{k!\,(n - k)!} = \binom{n}{n - k}$$

Throughout this book we shall use the convention that $\binom{n}{k}$ equals zero if k is greater than n, where n and k are positive integers. This is natural, because there is no way of considering arrangements of n objects, where more than n of them are alike.

Example 3. How many different groups of k objects may be formed from n objects? Suppose that the n objects are lined up in a row, and we have k tags to place on k of the n objects. Then the number of ways of selecting k objects from n objects equals the number of different ways of arranging the k tags in the n possible positions, which in turn equals the number of distinguishable ways of arranging k objects of one kind (tags) with $n - k$ objects of another kind (no tags), which is $\binom{n}{k}$. In this situation, $\binom{n}{k}$ is often read "the number of ways of taking n things k at a time."

Example 4. R distinguishable red balls and B distinguishable black balls are to be placed in four boxes as shown in Figure 1, so that r red and b black balls will be in the upper boxes, with the red balls occupying the boxes on the left and the black balls occupying the boxes on the right. The number of ways of

r red balls	b black balls
R — r red balls	B — b black balls

selecting r red balls out of R is $\binom{R}{r}$. For each selection of r red balls, there are $\binom{B}{b}$ ways of dividing the B black balls so that b are in the upper box and $B - b$ are in the lower box. Thus there are $\binom{R}{r}\binom{B}{b}$ different ways of selecting from the $R + B$ balls, so that exactly r red and b black balls are in the upper boxes.

Example 5. The binomial $(x + y)^n$ may be considered as the product of n binomials $(x + y)(x + y) \ldots (x + y)$. The term x^n occurs only when the x term from the first factor is multiplied by the x term from the second factor, and so on for all n factors. The term $x^{n-1}y$ results from multiplying the x term from $n - 1$ of the factors times the y term from one factor. Since the y term may be selected from any one of the n factors, expansion of $(x + y)^n$ results in n terms involving $x^{n-1}y$. Similarly the term $x^k y^{n-k}$ results from the selection of k x's from the n terms, and y from the remaining $n - k$ terms. Therefore the term $x^k y^{n-k}$ appears $\binom{n}{k}$ times in the expansion of $(x + y)^n$. Since all terms in the expansion are added together, we may write

$$(4) \qquad (x + y)^n = x^n + \binom{n}{n-1}x^{n-1}y^1 + \binom{n}{n-2}x^{n-2}y^2 + \cdots$$
$$+ \binom{n}{2}x^2 y^{n-2} + \binom{n}{1}x^1 y^{n-1} + y^n$$

If we define 0! as 1, and if we use the notation

$$\sum_{i=a}^{b} c_i = c_a + c_{a+1} + c_{a+2} + \ldots + c_{b-1} + c_b$$

which is read as "the sum of the terms c_i as i goes from a to b," then we may write

$$(5) \qquad (x + y)^n = \sum_{i=0}^{n} \binom{n}{i} x^i y^{n-i}$$

which is known as the "binomial expansion," and is found in most high school algebra textbooks.

Example 5 illustrates why the term "binomial coefficient" is often used to describe the symbol $\binom{n}{i}$. Similarly it may be shown that the coefficient of $x_1^{n_1} x_2^{n_2} \ldots x_r^{n_r}$ in the expansion of $(x_1 + x_2 + \ldots + x_r)^n$ is given by the "multinomial coefficient" $\begin{bmatrix} n \\ n_i \end{bmatrix}$.

PROBLEMS

1. How many arrangements are there of ten distinguishable marbles in three cups?
2. How many four-digit numbers (from 0000 to 9999) may be formed using the ten digits, where each digit may be repeated any number of times?
3. How many four-digit numbers may be formed using only the odd digits 1, 3, 5, 7, 9, where no digit may be used more than once per number?
4. How many different four letter arrangements are there, using the twenty-six letters in the alphabet, where each letter may be used repeatedly.
5. How many four letter arrangements are there if no letter may be used twice?
6. How many four letter arrangements are there if no letter may appear next to itself?
7. In how many ways may a committee of three be chosen from a club with twelve members?
8. What is the coefficient of $x^3 y^3$ in the expansion of $(x + y)^6$?
9. What is the coefficient of $x^2 y^4 z$ in the expansion of $(x + y + z)^7$?
10. What is the coefficient of $x^2 y^5$ in the expansion of $(w + x + y + z)^7$?
11. In how many ways may a committee of two men and three women be chosen from a group of seven men and six women?
12. Evaluate $\displaystyle\sum_{i=1}^{3} \binom{4}{i}$.
13. Evaluate $\displaystyle\sum_{i=0}^{3} \binom{4}{i}\left(\frac{1}{2}\right)^2$.
14. Evaluate $\displaystyle\sum_{i=3}^{5} \binom{6}{i}\left(\frac{1}{3}\right)^i\left(\frac{2}{3}\right)^{6-i}$.
15. In how many ways may a group of two men, two women, and four children be formed from a larger group of five men, six women, and twelve children?
16. How many ways are there of choosing n_1 objects of the first kind, n_2 objects of the second kind, and so on, to n_k objects of the kth kind, where there are altogether N_1 objects of the first kind, N_2 objects of the second kind, and so on? How many ways are there if n_i is greater than N_i for some i?

1.3. PROBABILITY

Let us assume that we have a specified experiment in mind, such as, "six independent tosses of an unbiased coin are made," or "two fair dice are rolled." We may just as validly consider more complicated experiments, and the same concepts introduced below are applicable.

Now we shall define the important terms "sample space," and "points in the sample space," in connection with an experiment.

Definition 1. The *sample space* is the collection of all possible different outcomes of an experiment.

Definition 2. A *point in the sample space* is a possible outcome of an experiment.

Each experiment has its own sample space, which consists essentially of a list of the different outcomes of the experiment that are possible. It is tacitly assumed that the sample space is subdivided as finely as reasonably possible with each subdivision being called a point. Also it is tacitly assumed that each possible outcome is represented by one and only one point.

Example 1. If an experiment consists of tossing a coin twice, the sample space consists of the four points *HH*, *HT*, *TH*, and *TT*.

Example 2. An examination consisting of ten "true or false" questions is administered to one student, as an experiment. There are $2^{10} = 1024$ points in the sample space, where each point consists of the sequence of possible answers to the ten successive questions, such as "*TTFTFFTTTT*."

It is now possible to define *event*, in terms of the points in the sample space.

Definition 3. An *event* is any set of points in the sample space.

In Example 1 we may speak of the event "two heads" which consists of the single point *HH*, the event "one head" which consists of the two points *HT* and *TH*, the event "at least one tail" which consists of the points *TH*, *HT*, and *TT*, as well as the event "four heads" which has no points in it. A set with no points in it is sometimes called "the empty set." The event consisting of all points in the sample space is sometimes called "the sure event," because it is certain to occur every time the experiment is performed.

Two different events may have points common to both. The events "at least one tail" and "at least one head" have the two points *TH* and *HT* in common. If two events have no points in common, then they are called *mutually exclusive* events, because the occurrence of one event automatically excludes the possibility of the other event occurring at the same time.

If all of the points in one event are also contained in a second event, then we say that the first event *is contained in* the second event, or that the second event *contains* the first event. The event "at least one head" contains the event "two heads." Each event therefore contains itself.

To each point in the sample space there corresponds a number, called "the probability of the point" or "the probability of the outcome." These probabilities may be any number from 0 to 1. If we can conceive of a long series of repetitions of the experiment under fairly uniform conditions, then the relative frequency of the occurrence of the point or event in mind represents an approximation to the probability of that point or event.

Definition 4.   If A is an event associated with an experiment, and if n_A represents the number of times A occurs in n independent repetitions of the experiment, then the *probability of the event A*, denoted by $P(A)$, is given by

(1)
$$P(A) = \lim_{n \to \infty} \frac{n_A}{n}$$

which is read "the limit of the ratio of the number of times A occurs to the number of times the experiment is repeated, as the number of repetitions approaches infinity."

A formal definition of "independent" is deferred until later. For the present we may think of experiments as independent if the outcome of any one experiment does not influence the outcome of the other experiments.

The definition of the probability of an event includes the definition of the probability of an outcome as a special case, since an event may be considered as consisting of a single outcome. It is apparent from the definition that the probability of an event equals the sum of the probabilities of all outcomes comprising the event, since the number of times the event occurs equals the sum of the numbers of times the mutually exclusive outcomes comprising the event occur.

In practice, the set of probabilities associated with a particular sample space is seldom known, but the probabilities are assigned according to the experimenter's preconceived notions. That is, the experimenter formulates a model as an idealized version of his experiment. Then the sample space of the model experiment is examined, and probabilities are assigned to the various points of the sample space in some manner which the experimenter feels can be justified.

Example 3.   In an experiment consisting of the single toss of an unbiased coin, it is reasonable to assume that the outcome H will occur about half the time. Thus we may assign the probability 1/2 to the outcome H, and the same to the outcome T. We write this as $P(H) = 1/2$, $P(T) = 1/2$.

Example 4.   In an experiment consisting of three tosses of an unbiased coin, it is reasonable to assume that each of the $2^3 = 8$ outcomes *HHH, HHT, HTH, HTT, THH, THT, TTH, TTT* is equally likely. Thus the probability of each outcome is 1/8. Also $P(3 \text{ tails}) = 1/8$, $P(\text{at least one head}) = 7/8$, and $P(\text{more heads than tails}) = P(\text{at least 2 heads}) = 4/8 = 1/2$.

Example 5.   Referring to the experiment described in Example 1.2.4, we wish to find the probability of obtaining exactly r red and b black balls in the upper boxes, under the assumption that any ball is equally likely to fall into either an upper or a lower box. Since each of the $R + B$ balls may fall two ways (upper or lower) there are 2^{R+B} points in the sample space, by Rule 1. By our above assumption, each of these points is equally likely, and therefore each point has a probability $1/2^{R+B}$. Since $\binom{R}{r}\binom{B}{b}$ of the points are in the event "r red and b black balls in the upper boxes," the probability of that event, $P(r, b)$, is given by

$$(2) \quad P(r, b) = \frac{\binom{R}{r}\binom{B}{b}}{2^{R+B}} \quad \text{for} \quad 0 \le r \le R, \quad \text{and} \quad 0 \le b \le B$$

We have been working with *probability functions* in the previous three examples.

Definition 5.   A *probability function* is a function which assigns probabilities to the various events in the sample space.

In Example 3 the probability function was given by $P(H) = 1/2$, $P(T) = 1/2$. It is necessary that the probability function assign a probability to each point in the sample space. Then the probabilities of all events in the sample space are automatically specified by the probabilities of the sample points contained in the events.

Several properties of probability functions become apparent. Let S be a sample space and let A be any event in S. Then if P is a probability function,

(a) $P(S) = 1$, because

$$P(S) = \lim_{n \to \infty} \frac{n}{n} = 1;$$

(b) $P(A) \ge 0$, because $n_A \ge 0$, and therefore

$$\lim_{n \to \infty} \frac{n_A}{n} \ge 0;$$

and

(c) $P(\bar{A}) = 1 - P(A)$, where $\bar{A}$ is the event "the event A does not occur," because $n_{\bar{A}} = n - n_A$, and

$$\lim_{n \to \infty} \frac{n_{\bar{A}}}{n} = \lim_{n \to \infty} \frac{n - n_A}{n} = \lim_{n \to \infty} \left(1 - \frac{n_A}{n}\right) = 1 - \lim_{n \to \infty} \frac{n_A}{n} = 1 - P(A).$$

We mentioned earlier that while the various outcomes of an experiment are mutually exclusive, the various events associated with an experiment do not necessarily have that property. In our experiment of tossing a coin three times the events "three heads" and "at least two heads" may both occur at the same time. Now consider the probability of the event "three heads" if we are given that the event "at least two heads" has occurred. If at least two heads have occurred, then we know that several points in the sample space may be eliminated, namely *TTT*, *TTH*, *THT*, and *HTT*. The possible outcomes of the experiment are reduced to four equally likely points. Therefore the probability of each point is now 1/4, and hence the probability of the event "three heads," or *HHH*, is 1/4, if we are given the fact that at least two heads have occurred. The additional information that we are given has the effect of eliminating some of the outcomes from consideration, and thus artificially reducing the sample space.

In another experiment, consider rolling a die. Let *S* be the sample space, let *A* be the event "a 4, 5, or 6 occurs," and let *B* be the event "an even number (2, 4, or 6) occurs," as depicted by Figure 1. The probability that the event *A*

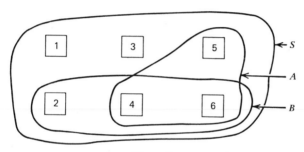

Figure 1

has occurred, given that *B* has occurred, is written $P(A \mid B)$, and is usually read "the probability of *A* given *B*." Since we know that *B* has occurred, we may not only eliminate the points that are in neither *A* nor *B*, namely the points 1 and 3, but we may even eliminate the point in *A* that is not in *B*, or 5. Thus all points not in *B* are eliminated, and the sample space is just the set of points in *B*. The only points in *B* that can result in the event *A* are the points in both *A* and *B*, or 4 and 6. These points represent the event "both *A* and *B* occur."

Definition 6. If *A* and *B* are two events in a sample space *S*, then the event "both *A* and *B* occur," representing those points in the sample space that are in both *A* and *B* at the same time, is called *the joint event A and*

B, and is represented by *AB*. The probability of the joint event is represented by $P(AB)$.

Then the probability of "*A* given *B*" is given by the probability of "*AB*" relative to the reduced sample space "*B*." Or, symbolically,

$$P(A \mid B) = \frac{P(AB)}{P(B)}$$

Looking at it another way, suppose that the above experiment is repeated *n* times. However, only those outcomes resulting in the event *B* are recorded, and the outcomes not resulting in *B* are ignored. Let n_B represent the number of times *B* occurs, and let n_{AB} represent the number of times *A* occurs when *B* occurs. Then

(3) $$P(A \mid B) = \lim_{n \to \infty} \frac{n_{AB}}{n_B} = \lim_{n \to \infty} \frac{n_{AB}/n}{n_B/n} = \frac{P(AB)}{P(B)}$$

We have intuitively justified the following definition.

Definition 7. The *conditional probability* of *A* given *B* is the probability that *A* occurred given that *B* occurred, and is given by

(4) $$P(A \mid B) = \frac{P(AB)}{P(B)}$$

where $P(B) > 0$. If $P(B) = 0$, $P(A \mid B)$ is not defined.

Example 6. Referring to the experiment described by Example 1.2.4, we wish to find the probability of obtaining exactly *r* red and *b* black balls in the upper boxes, given that the total number of balls in the upper boxes is known to be $r + b$. We shall also make the "equally likely" assumption stated in Example 5. Represent the event "*r* red and *b* black balls in the upper boxes" by *A*, and represent the event "$r + b$ balls in the upper boxes" by *C*. Then the desired probability is given by (4) as

(5) $$P(A \mid C) = \frac{P(AC)}{P(C)}$$

If the event *A* occurs, the event *C* automatically occurs. Therefore the joint event *AC* is the same as the event *A*, and $P(AC) = P(A)$, which is given by (2) as

(6) $$P(AC) = P(A) = \frac{\binom{R}{r}\binom{B}{b}}{2^{R+B}}$$

Since $\binom{R + B}{r + b}$ of the 2^{R+B} equally likely outcomes result in the event C, we have

(7)
$$P(C) = \frac{\binom{R + B}{r + b}}{2^{R+B}}$$

Therefore, combining (5), (6), and (7) we have

(8)
$$P(A \mid C) = \frac{\binom{R}{r}\binom{B}{b}}{\binom{R + B}{r + b}}$$

The idea of conditional probability leads quite naturally into the idea of independent events. That is, if the probability of A, given that B occurs, is the same as the probability of A without any information on the occurrence or nonoccurrence of B, then we feel that the occurrence or nonoccurrence of A is independent of whether or not B occurs. That is, we feel that A is independent of B if $P(A \mid B)$ equals $P(A)$. In fact, this may be used as the definition of independence, but it is not clear from this form of the definition whether B then is also independent of A. So it is better to substitute $P(A)$ for $P(A \mid B)$ in (4), the definition of conditional probability. This leads to the following.

Definition 8. Two events A and B are *independent* if

(9)
$$P(AB) = P(A)P(B)$$

Because of the symmetry of (9) it is readily apparent that if A is independent of B, then also B is independent of A, and so it is better to say "A and B are independent," where it is meant that they are independent of each other.

Example 7. In an experiment consisting of two tosses of a balanced coin, the four points in the sample space are assumed to have equal probabilities. Let A be the event "a head occurs on the first toss" and let B be the event "a head occurs on the second toss." Then A has the points HH and HT, B has the points HH and TH, and AB has the point HH. Also $P(A) = 2/4$, $P(B) = 2/4$, and $P(AB) = 1/4$. Therefore (9) is satisfied and A and B are independent.

The following example illustrates the fact that the independence of two events is not always intuitively obvious, and should always be determined directly from the definition and (9).

Example 8. Consider again the experiment consisting of one roll of a balanced die, where the sample space consists of the six equally likely points 1, 2, 3, 4, 5, and 6. Let A be the event "an even number occurs," including the points 2, 4, and 6. Let B be the event "at least a 4 occurs," including the points 4, 5, and 6. Finally, let C be the event "at least a 5 occurs," including the points 5 and 6. Then A and B are not independent, because $P(A)P(B)$ equals $(1/2)(1/2)$, or $1/4$, while $P(AB)$ equals $1/3$. However, A and C are independent, because $P(A)P(C)$ equals $(1/2)(1/3)$, or $1/6$, the same as $P(AC)$.

Sometimes the notions of "independent events" and "mutually exclusive events" are confused with each other, because both notions give the impression that "the two events don't have anything to do with each other." The property of independence depends not only on the two events being considered, but also on the particular probability function defined on the sample space. It is possible for $P(AB)$ and $P(A)P(B)$ to be equal to each other with one set of probabilities, and to be unequal with another set of probabilities. But "mutually exclusive" simply means the two events have no points in common, and no matter what probability function is defined on the sample space, AB is empty, so $P(AB) = 0$. If A and B are mutually exclusive, they will be independent only if either $P(A)$ or $P(B)$ equals zero, since (9) must be satisfied.

Now we shall define the concept of *independent experiments*.

Definition 9. Two experiments are *independent* if for every event A associated with one experiment and every event B associated with the second experiment,

$$P(AB) = P(A)P(B)$$

It is equivalent to define two experiments as independent if every event associated with one experiment is independent of every event associated with the other experiment.

It is quite tedious to examine every pair of events associated with two experiments to see if they satisfy Definition 9. However, it is sufficient to verify the definition only for those events consisting of a single point each. Then the definition is automatically verified for all other events.

In practice, the model is usually set up assuming independence, and the assumption of independence is then used to find $P(AB)$ using $P(A)$ and $P(B)$ in Definition 9. This is the main value of the definition of independence. Thus it is reasonable to extend the definition of independent experiments to cover the eventuality of more than two experiments being involved.

Definition 10. *n experiments are mutually independent* if for every set of n events, formed by considering one event from each of the n experiments, the following equation is true:

(10) $$P(A_1 A_2 \ldots A_n) = P(A_1)P(A_2) \ldots P(A_n)$$

where A_i represents an outcome of the ith experiment, for $i = 1, 2, \ldots, n$.

The word "mutually" may be omitted in the preceding definition if no confusion results.

Example 9. Let an experiment consist of one toss of a biased coin, where the event H has probability p and the event T has probability $q = 1 - p$. Consider three independent repetitions of the experiment, where a subscript will be used to denote the experiment with which the outcome is associated. Thus $H_1 T_2 H_3$ means the first experiment resulted in H, the second in T, and the third in H. Because of our assumption of independence,

$$P(H_1 T_2 H_3) = P(H_1)P(T_2)P(H_3) = pqp$$

If we consider the event "exactly two heads" associated with the combined experiments, then this may occur $\binom{3}{2} = 3$ ways, and hence

$$P(\text{exactly two heads}) = 3p^2 q$$

Obviously the above might just as well have been described as one experiment with three independent trials. The extension to considering an experiment consisting of n independent tosses may be made. The probability of obtaining "exactly k heads" then equals the term $p^k q^{n-k}$ times the number of times that term can appear. Therefore, in n independent tosses of a coin,

$$(11) \qquad\qquad P(\text{exactly } k \text{ heads}) = \binom{n}{k} p^k q^{n-k}$$

where $p = P(H)$ on any one toss.

The four preceding definitions, as is true for all definitions, work both ways. Example 7 presents a situation where the satisfaction of (9) implies that two events are independent. Example 9 presents a situation where the assumption of independence implies that (10) is satisfied. It follows then that if (10) is not satisfied, the experiments are not independent, and conversely if the experiments are not independent, (10) is not satisfied for at least one set of events $A_1 A_2 \ldots A_n$.

PROBLEMS

1. In an experiment consisting of three tosses of a coin, list the points in the sample space.

2. Referring to Problem 1, give
  (a) Two mutually exclusive events.
  (b) Two events that are not mutually exclusive.

3. Referring to Problem 1, how many events (containing at least one point) are there?

4. Show that in a sample space with n points, there are exactly $2^n - 1$ events containing at least one point.

5. What is the maximum number of events that may be obtained from a sample space with n points in it, if each of the events is to contain at least one point, and all of the events are to be mutually exclusive?

6. If a football team has an equal probability of winning or losing each game (assume no tie games occur), what is the probability of the team losing at least seven games in an eight game season?

7. In three independent tosses of an unbiased coin, what is the probability of obtaining three heads?

8. In three independent tosses of an unbiased coin, what is the probability of obtaining at least one tail?

9. In three independent tosses of an unbiased coin, what is the probability of obtaining three heads if we know that at least one head has occurred?

10. In three independent tosses of an unbiased coin, what is the probability of obtaining three heads if we know that the first toss resulted in a head? (*Note.* Problems 9 and 10 have different answers.)

11. What is the probability of obtaining at least one head in nine independent tosses of a balanced coin.

12. If ten red balls and eight black balls are tossed into two boxes, such that each ball is equally likely to fall into box 1 or box 2, what is the probability of exactly four red and three black balls falling into box 1?

13. In Problem 12, what is the probability of exactly four red balls falling into box 1, if altogether seven balls end up in box 1?

14. In Problem 12, what is the probability of an equal number of red and black balls falling into box 1?

15. In Problem 12, what is the probability that, among the balls falling into box 1, exactly 3 of them are red?

1.4. RANDOM VARIABLES

Outcomes associated with an experiment may be numerical in nature, such as the score on an examination, or nonnumerical, such as the choice "red door" by a rat escaping from a pen. In order to analyze the results of an experiment it is necessary to assign numbers to the points in the sample space. Any rule for assigning such numbers is called a random variable.

Definition 1. A *random variable* is a function which assigns real numbers to the points in a sample space.

We shall usually denote random variables by the capital letters W, X, Y, or Z, with or without subscripts. The real numbers assigned by the random variables will be denoted by lowercase (small) letters.

Example 1. In an experiment where a consumer is given a choice of three products, soap, detergent, or Brand A, the sample space consists of the three points representing the three possible choices. Let the random variable assign the number 1 to the choice "Brand A," and the number 0 to the other two possible outcomes. Then $P(X = 1)$ equals the probability that the consumer chooses Brand A.

At times it is convenient to define more than one random variable for a single sample space, such as the following example.

Example 2. R girls and B boys are each asked whether they communicate more easily with their mother or their father. Let X be the number of girls who feel they communicate more easily with their mothers, and let Y be the total number of children who feel they communicate more easily with their mothers. If X equals r, then we know the event "r girls feel they communicate more easily with their mothers" has occurred. If at the same time Y equals k, then we know that the event "r girls and $k - r$ boys feel they communicate more easily with their mothers" has occurred. Therefore, if we assume that each child is equally likely to choose either parent, the probability $P(X = r, Y = k)$ may be found in the same manner that was used in Example 1.3.5. The result is

$$(1) \qquad P(X = r, Y = k) = \frac{\binom{R}{r}\binom{B}{k-r}}{2^{R+B}} \qquad \begin{array}{l} \text{for } 0 \le r \le R \\ \text{and } 0 \le k - r \le B \end{array}$$

If X is a random variable, then "$X = x$" corresponds to some event in the sample space, namely the event consisting of the set of all points to which the random variable has assigned the value "x."

Example 3. In an experiment consisting of two tosses of a coin, let X be the number of heads. Then "$X = 1$" corresponds to the event containing only the points HT and TH.

Thus "$X = x$" is sometimes referred to as "the event $X = x$," when the intended meaning is "the event consisting of all outcomes assigned the number x by the random variable X."

Because of this close correspondence between random variables and events, the definitions of *conditional probability* and *independence* apply equally well to random variables.

Definition 2. The *conditional probability of X given Y*, written $P(X = x \mid Y = y)$, is the probability that the random variable X has assumed the value x, given that the random variable Y has assumed the value y.

The equation for determining conditional probabilities may be obtained from Definition 1.3.7 as

$$(2) \qquad P(X = x \mid Y = y) = \frac{P(X = x, Y = y)}{P(Y = y)} \qquad \text{if } P(Y = y) > 0$$

Example 4. Let X and Y be defined as in Example 2. Then $P(X = r, Y = k)$ is given by (1). Also $P(Y = k)$ is the probability that exactly k of the children feel they communicate more easily with their mothers, and is given in Example 1.3.6 as

$$(3) \qquad P(Y = k) = \frac{\binom{R + B}{k}}{2^{R+B}} \qquad 0 \leq k \leq R + B$$

Substitution of (3) and (1) into (2) gives

$$(4) \quad P(X = r \mid Y = k) = \frac{\binom{R}{r}\binom{B}{k - r}}{\binom{R + B}{k}} \qquad 0 \leq r \leq R, \quad 0 \leq k - r \leq B$$

which is essentially the same result as given by (1.3.8)

Just as the points in a sample space are mutually exclusive, the values that a random variable may assume are mutually exclusive. That is, for a single outcome of an experiment, the random variable defined for that experiment furnishes us with only one number. Thus the entire set of values that a random variable may assume has many of the same properties as a sample space. The individual values assumed by the random variable correspond to the points in a sample space, a set of values corresponds to an event, and the probability of the random variable assuming any value within a set of values equals the sum of the probabilities associated with all values within the set. For example,

$$P(a < X < b) = \sum_{a < x < b} P(X = x)$$

where the summation extends over all values of x between, but not including, the numbers a and b, and

$$P(X = \text{even number}) = \sum_{x \text{ even}} P(X = x)$$

where the summation applies to all values of x that are even numbers. Because of this similarity between the set of possible values of X and a sample space, the description of the set of probabilities associated with the various values X may assume is often called the *probability function of the random variable X*, just as a sample space has a probability function. However, the probability function of a random variable is not an arbitrary assignment of probabilities, as is the probability function for a sample space, because once the probabilities are assigned to the points in a sample space, and once a random variable X is defined on the sample space, the probabilities associated with the various values of X are known, and the probability function of X is thus already determined.

Definition 3. *The probability function of the random variable X*, usually denoted by $f(x)$, is the function which gives the probability of X assuming the value x, for any real number x. In other words,

(5) $$f(x) = P(X = x)$$

The probability function always equals 0 at values of x which X can not assume.

Sometimes it is convenient to represent the probability function as a bar graph, with the values of the random variable as the abscissa (along the horizontal axis), and the probabilities as the ordinate (the height of the bar). For instance, if $P(X = 1)$ equals .3, $P(X = 2)$ equals .4, and $P(X = 4)$ equals .3, the bar graph of the probability function looks like this. The heights of

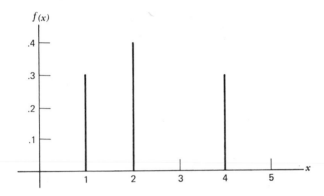

the bars represent the various probabilities associated with the random variable X.

It is not always convenient to use $f(x)$ to denote the probability function of a random variable. Other expressions that may be used include $f_0(x)$, $f_1(x)$, $f_2(x)$, $g(x)$, $h(x)$, and so on. However, the meaning of the various expressions used will always be clear from the context.

We have seen that the distribution of probabilities associated with a random variable may be described by a probability function. Another way of accomplishing the same thing is by means of a *distribution function*, which describes the accumulated probabilities.

Definition 4. The *distribution function of a random variable X*, usually denoted by $F(x)$, is the function which gives the probability of X being less than or equal to any real number x. In other words,

$$(6) \qquad\qquad F(x) = P(X \leq x) = \sum_{t \leq x} f(t)$$

where the summation extends over all values of t that do not exceed x.

Distribution functions also may be represented graphically, with the x as the abscissa and $F(x)$ as the ordinate. As an illustration, suppose, as above, that $P(X = 1)$ equals .3, $P(X = 2) = .4$, and $P(X = 4) = .3$. Then the graph of $F(x)$ looks like this.

The graph actually consists only of the horizontal lines; the vertical lines

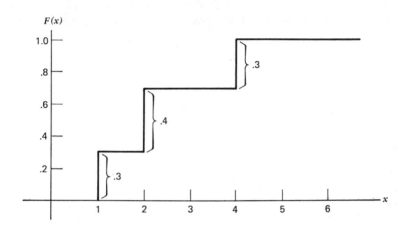

are drawn in merely to give the graph a somewhat "connected" appearance, and to assist in the finding of "quantiles," as explained in the next section. The lengths of the vertical lines are the same as the lengths of the bars in the graph of the probability function.

Some probability distributions are well known, and consequently have been given names.

Definition 5.   Let X be a random variable. The *binomial distribution* is the probability distribution represented by the probability function

$$(7) \qquad f(x) = P(X = x) = \binom{n}{x} p^x q^{n-x} \qquad x = 0, 1, \ldots, n$$

where n is a positive integer, $0 \leq p \leq 1$, and $q = 1 - p$. Note that we are using the usual convention that $0!$ equals 1.

The distribution function is then

$$
(8) \qquad F(x) = P(X \leq x) = \sum_{i \leq x} \binom{n}{i} p^i q^{n-i}
$$

where the summation extends over all possible values of i less than or equal to x. Table 3 at the end of this book gives the values of $F(x)$ for some selected values of the parameters n and p.

Example 5. An experiment consists of n independent trials where each trial may result in one of two outcomes, "success" or "failure," with probabilities p and q, respectively, such as with the tossing of a coin. Let X equal the total number of "successes" in the n trials. Then, as was shown by (1.3.11),

$$
P(X = x) = \binom{n}{x} p^x q^{n-x}
$$

for integer x from 0 to n. Thus X has the binomial distribution.

Another useful probability distribution is the *discrete uniform distribution*.

Definition 6. Let X be a random variable. The *discrete uniform distribution* is the probability distribution represented by the probability function

$$
(9) \qquad f(x) = \frac{1}{N} \qquad x = 1, 2, \ldots, N
$$

Thus X may assume any integer value from 1 to N with equal probability, if X has the discrete uniform probability function.

Example 6. A jar has N plastic chips, numbered 1 to N. An experiment consists of drawing one chip from the jar, where each chip is equally likely to be drawn. The sample space has N points, representing the N chips that may be drawn. Let X equal the number on the drawn chip. Then X has the discrete uniform distribution.

When several random variables are defined on the same sample space, or when several experiments, each with one or more random variables defined for them, are considered as a combined experiment, it becomes useful to consider joint distributions, described by *joint probability functions* and *joint distribution functions*.

Definition 7. The *joint probability function* $f(x_1, x_2, \ldots, x_n)$ of the random variables $X_1, X_2, \ldots, X_n$ is the probability of the joint occurrence of $X_1 = x_1, X_2 = x_2, \ldots,$ and $X_n = x_n$. Stated differently,

$$
(10) \qquad f(x_1, x_2, \ldots, x_n) = P(X_1 = x_1, X_2 = x_2, \ldots, X_n = x_n)
$$

Definition 8. The *joint distribution function* $F(x_1, x_2, \ldots, x_n)$ of the random variables $X_1, X_2, \ldots, X_n$ is the probability of the joint occurrence of $X_1 \leq x_1, X_2 \leq x_2, \ldots,$ and $X_n \leq x_n$.
Stated differently,

$$(11) \qquad F(x_1, x_2, \ldots, x_n) = P(X_1 \leq x_1, X_2 \leq x_2, \ldots, X_n \leq x_n)$$

Example 7. Consider the random variables X and Y as defined in Example 2. Let $f(x, y)$ and $F(x, y)$ be the joint probability function and the joint distribution function respectively. Then, from (1),

$$(12) \qquad f(r, k) = P(X = r, Y = k) = \frac{\binom{R}{r}\binom{B}{k-r}}{2^{R+B}}, \qquad r = 0, 1, \ldots$$

and

$$(13) \qquad F(r, k) = P(X \leq r, Y \leq k) = \sum_{\substack{0 \leq x \leq r \\ x \leq y \leq k}} \frac{\binom{R}{x}\binom{B}{y-x}}{2^{R+B}}$$

where the summation extends over all values of x and y, such that $x \leq r$ and $y \leq k$, with the usual restriction that x and $y - x$ be nonnegative integers.

Definition 9. The *conditional probability function* of X given Y, $f(x \mid y)$, is

$$(14) \qquad \qquad f(x \mid y) = P(X = x \mid Y = y)$$

From (2) we see that

$$(15) \qquad f(x \mid y) = P(X = x \mid Y = y) = \frac{P(X = x, Y = y)}{P(Y = y)} = \frac{f(x, y)}{f(y)}$$

where $f(x, y)$ is the joint probability function of X and Y, and $f(y)$ is the probability function of Y itself.

Example 8. Let X and Y be defined as in Example 2. Then the conditional probability function of X given Y is derived in Example 4 as

$$(16) \quad f(r \mid k) = P(X = r \mid Y = k) = \frac{\binom{R}{r}\binom{B}{k-r}}{\binom{R+B}{k}} \qquad \begin{array}{l} 0 \leq r \leq R \\ 0 \leq k - r \leq B \end{array}$$

We have been working with a probability distribution known as the *hypergeometric distribution*, which has the probability function given by (16).

Definition 10. Let X be a random variable. The *hypergeometric distribution* is the probability distribution represented by the probability function

$$(17) \quad f(x) = P(X = x) = \frac{\binom{A}{x}\binom{B}{k-x}}{\binom{A+B}{k}} \quad \begin{matrix} 0 \le x \le A \\ 0 \le k - x \le B \end{matrix}$$

where A, B, and k are nonnegative integers, and $k \le A + B$.

Example 9. Consider four boxes arranged as shown in Figure 1. Let an experiment consist of placing A objects into the boxes on the left in some

x	$k - x$	k
$A - x$	$B - k + x$	
A	B	

Figure 1

random manner so that each object has an equal probability p of being placed into the upper box, and of placing B objects into the boxes on the right in the same random manner so that each object has an equal probability p of being placed into the upper box. Define the random variables X, Y, and Z as

X = the number of objects in the upper left box,
Y = the number of objects in the upper right box,
Z = the total number of objects in the upper boxes.

It is desired to find $P(X = x \mid Z = k)$. We have, from (15),

$$(18) \qquad P(X = x \mid Z = k) = \frac{P(X = x, Z = k)}{P(Z = k)}$$

and also

$$(19) \qquad P(Z = k \mid X = x) = \frac{P(X = x, Z = k)}{P(X = x)}$$

Rewrite (19) as

$$(20) \qquad P(X = x, Z = k) = P(Z = k \mid X = x) \cdot P(X = x)$$

and substitute (20) into the numerator of (18) to get

$$(21) \qquad P(X = x \mid Z = k) = \frac{P(Z = k \mid X = x) \cdot P(X = x)}{P(Z = k)}$$

We shall now evaluate the terms in (21). Consider first the A objects, acting independently of each other, with a probability p of each object being placed in an upper box. The situation is similar to that described in Example 5, and therefore

$$(22) \qquad P(X = x) = \binom{A}{x} p^x (1 - p)^{A-x}$$

Also the random variable Z behaves in the same way, only with all $A + B$ objects being distributed. Hence

$$(23) \qquad P(Z = k) = \binom{A + B}{k} p^k (1 - p)^{A+B-k}$$

Finally, if we are given that $X = x$, then Z will equal k only when Y equals $k - x$. Therefore

$$(24) \qquad P(Z = k \mid X = x) = P(Y = k - x \mid X = x)$$

But since the placement of the B objects is assumed to be independent of the placement of the A objects, the probability of a particular value of Y does not depend on what X equals. So (24) becomes, again with the assistance of Example 5,

$$(25) \quad P(Z = k \mid X = x) = P(Y = k - x) = \binom{B}{k - x} p^{k-x} (1 - p)^{B-k+x}$$

The substitution of (25), (23), and (22) into (21) gives

$$(26) \qquad P(X = x \mid Z = k) = \frac{\binom{A}{x}\binom{B}{k - x}}{\binom{A + B}{k}} . \qquad \begin{array}{l} 0 \le x \le A \\ 0 \le k - x \le B \end{array}$$

The conditional distribution of X given Z is then recognized as the hypergeometric distribution.

If we set $p = 1/2$, Example 9 reduces to a simpler situation described in Example 2 and elsewhere. It should be noted that the conditional probability distribution given by (26) does not depend at all on the value of p.

Mutually independent random variables may be defined in a manner similar to Definitions 1.3.10 and 1.3.11 of independent experiments.

Definition 11. Let $X_1, X_2, \ldots, X_n$ be random variables with the respective probability functions $f_1(x_1), f_2(x_2), \ldots, f_n(x_n)$, and with the joint probability function $f(x_1, x_2, \ldots, x_n)$. Then $X_1, X_2, \ldots, X_n$ are *mutually independent* if

$$(27) \qquad f(x_1, x_2, \ldots, x_n) = f_1(x_1) f_2(x_2) \ldots f_n(x_n)$$

for all combinations of values of $x_1, x_2, \ldots, x_n$.

Example 10. Consider the experiment described in Example 9. Then the probability function of X is given by (22) as

$$(28) \qquad f_1(x) = P(X = x) = \binom{A}{x} p^x (1 - p)^{A-x}$$

and the probability function of Z is given by (23) as

$$(29) \qquad f_2(z) = P(Z = z) = \binom{A + B}{z} p^z (1 - p)^{A+B-z}$$

Since

$$f(x, z) = P(X = x, Z = z) = P(X = x \mid Z = z)P(Z = z)$$

the use of (26) and (29) results in the joint probability function of X and Z being given by

$$f(x, z) = \frac{\binom{A}{x}\binom{B}{z - x}}{\binom{A + B}{z}} \binom{A + B}{z} p^z (1 - p)^{A+B-z}$$

$$= \binom{A}{x}\binom{B}{z - x} p^z (1 - p)^{A+B-z}$$

But since

$$f_1(x)f_2(z) = \binom{A}{x}\binom{A + B}{z} p^{x+z}(1 - p)^{2A+B-x-z}$$

we see that

$$f(x, z) \neq f_1(x)f_2(z)$$

and therefore X and Z are not independent.

PROBLEMS

1. If $f(x)$ is the binomial probability function with $n = 6$ and $p = 1/3$, find

(a) $f(6)$ (d) $F(2.5)$
(b) $f(0)$ (e) $F(-3)$
(c) $f(2.5)$ (f) $F(7)$
(g) Draw a bar graph of the probability function.
(h) Draw a graph of the distribution function.

2. Suppose $f(x)$ is the discrete uniform probability function with N equal to 12. Find

(a) $f(2)$ (e) $F(0)$
(b) $f(12)$ (f) $F(3.1)$
(c) $f(0)$ (g) $F(1000)$
(d) $f(1.5)$ (h) $F(-1000)$
(i) Draw a bar graph of the probability function.
(j) Draw a graph of the distribution function.

3. Let $f(r, k)$ be defined for integer values of r and k to be

$$f(r, k) = \frac{\binom{3}{r}\binom{4}{k - r}}{2^7} \qquad 0 \leq r \leq 3, \quad 0 \leq k - r \leq 4$$

$$= 0 \qquad\qquad \text{for all other values of } r \text{ and } k$$

Find

(a) $f(0, 0)$ (e) $f(4, 4)$
(b) $f(0, 1)$ (f) $F(0, 0)$
(c) $f(1, 0)$ (g) $F(1, 1)$
(d) $f(3, 4)$ (h) $F(3, 4)$

4. Let $f(r \mid k)$ be the hypergeometric probability function, where $A = 3$ and $B = 4$. Find

(a) $f(0 \mid 0)$ (d) $f(1 \mid 5)$
(b) $f(1 \mid 1)$ (e) $f(1 \mid 6)$
(c) $f(2 \mid 1)$

5. Let $f(x) = 1/6$ for $x = 0, 2, 3, 5, 8$, and let $f(x) = 0$ for all other values of x. Why is this not a probability function?

6. A diner selects one sandwich at random out of six possible sandwich varieties.
   (a) What is the sample space?
   (b) What is the probability function on the sample space?
   (c) Define a random variable on the sample space, such that the random variable has the discrete uniform distribution.

7. Assume that every patient with a particular type of disease has probability .1 of being cured within a week, if he is given no treatment for the disease. Ten patients with that type of disease are given a new type of drug. After one week nine of the ten patients are cured.
   (a) What is the probability of at least nine or more patients being cured if the drug is assumed to have no curative effects?
   (b) In your opinion, would you consider the drug to be beneficial?
   (c) What sample space did you use in this analysis?
   (d) What probability function did you define on the sample space?
   (e) What random variable did you define on the sample space?
   (f) What is the name of the probability distribution of your random variable?

8. Seven boys and ten girls take an examination, and each student has probability .2 of failing the examination.
   (a) What is the sample space for this experiment?
   (b) Given that three students failed the examination what is the probability that all three were boys?
   (c) What is the name of the probability distribution you are using?
   (d) If the probability of each failure is .8 instead of .2, what is the answer to part (b)?

1.5. SOME PROPERTIES OF RANDOM VARIABLES

We have already discussed some of the properties associated with random variables, such as their probability functions and their distribution functions. The probability function describes all of the properties of a random variable that are of interest, because the probability function reveals the possible values the random variable may assume, and the probability associated with each value. A similar statement may be made concerning the distribution function.

At times, however, it is inconvenient or confusing to present the entire probability function to describe a random variable, and some sort of a "summary description" of the random variable is needed. And so we shall now introduce some other properties of random variables which may be used to present a brief, but incomplete, description of the distribution of the random variable.

The most common method used in this book for summarizing the distribution of a random variable is by giving some selected *quantiles* of the random variable. The term "quantile" is not as well known as the terms "median," "quartile," "decile," and "percentile," and yet these latter terms are popular names given to particular quantiles. The median of a random variable, for example, is some number that the random variable will exceed with probability one-half or less and will be smaller than with probability one-half or less. This definition may be extended as follows.

Definition 1. The number x_p, for a given value of p between 0 and 1, is called the pth *quantile of the random variable X*, if $P(X < x_p) \leq p$ and $P(X > x_p) \leq 1 - p$.

If more than one number satisfies the definition of the pth quantile, we shall avoid confusion by adopting the convention that x_p equals the average of the largest and the smallest numbers that satisfy Definition 1.

That is, X is less than x_p with probability p or less, and X exceeds x_p with probability $1 - p$ or less. The *median* is the .5 quantile, the third *decile* is the .3 quantile, the *upper and lower quartiles* are the .75 and .25 quantiles respectively, and the sixty-third *percentile* is the .63 quantile.

Perhaps the easiest method of finding the pth quantile involves using the graph of the distribution function of the random variable. The pth quantile is the abscissa of the point on the graph which has the ordinate value of p, as illustrated in the following example.

Example 1. Let X be a random variable with the following probability distribution

$$P(X = 0) = 1/4$$
$$P(X = 1) = 1/4$$
$$P(X = 2) = 1/3$$
$$P(X = 3) = 1/6$$

Then the distribution function of X may be represented by the following graph. The .75 quantile $x_{.75}$, called the upper quartile, may be found by drawing a horizontal line through .75 on the vertical axis, as indicated by the dotted line in the figure on the following page. The value of x where the dotted line intersects the graph is the upper quartile, which equals 2 in this example. Therefore we may say that $x_{.75} = 2$, which may be verified directly from the definition, since

$$P(X < 2) = 1/2$$

which is less than .75, and since

$$P(X > 2) = 1/6$$

which is less than $1 - .75$.

Similarly the median is found by drawing a line through .5 on the vertical scale. The median is any value from 1 to 2 inclusive, and it is easy to see that any of these values satisfies the definition of the median. By our convention we select 1.5 as the median.

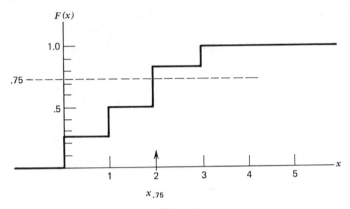

Figure 1

Certain random variables called "test statistics" play an important role in most statistical procedures. These test statistics are useless unless their distribution function is at least partially known. Most of the tables at the back of this book give information concerning the distribution functions of various test statistics used in nonparametric statistics. This information is condensed with the aid of quantiles, otherwise the tables would be inconveniently bulky.

Often we shall define a random variable, and instead of working with that random variable we shall work with a function of the random variable. A real valued function of a random variable X is a rule for assigning new real numbers to the sample space, instead of the usual numbers assumed by X. For example, if Y is $X + 4$, then Y is a real valued function of X, and if X equals x, then Y equals $x + 4$. If X equals 3, then Y equals 7. This is usually written as $Y = u(X)$, where $u(X)$ in this case is $X + 4$. Other functions might include $u(X) = X^2$, $u(X) = X$, and $u(X) = (X - a)^2$ for some constant a. Since Y also assigns real numbers to points in the sample space, even though Y uses X in the process, we see that Y is a random variable. It is true in general that a real valued function of a random variable is also a random variable.

Another very useful property of a random variable is its *expected value*. First we shall present a general definition of expected value, which will be followed by some particular cases.

Definition 2. Let X be a random variable with the probability function $f(x)$, and let $u(X)$ be a real valued function of X.

Then the *expected value of* $u(X)$, written $E[u(X)]$, is

$$(1) \qquad\qquad E[u(X)] = \sum_x u(x)f(x)$$

where the summation extends over all possible values of X. If the sum on the right hand side of (1) is infinite, or does not exist, then we say that the expected value of $u(X)$ does not exist.

Our interest is confined mainly to two special expected values, namely the *mean* and the *variance* of X.

Definition 3. Let X be a random variable with the probability function $f(x)$. The *mean of* X, usually denoted by μ, is

$$(2) \qquad\qquad \mu = E(X)$$

From (1) we have

$$(3) \qquad\qquad \mu = E(X) = \sum_x xf(x)$$

which shows our mean to be the same as the "centroid" in physics. The mean, as does the centroid, marks a central point, a point of balance. If weights, in proportion to the various probabilities, were to be placed on a yardstick at the appropriate values of X, the yardstick would balance right at the mean. Because of this tendency to "locate" the center of the distribution, the mean is sometimes called a "location parameter." The mean and the median, discussed previously, are the two most commonly used location parameters.

Example 2. Consider a simple experiment that results in "success" with probability p, or "failure" with probability q equal to $1 - p$. Let X equal 1 if a "success" occurs, and 0 if a "failure" occurs. Thus X has the binomial distribution with n equal to 1. From (3) the expected value of X is determined as follows.

$$(4) \qquad\qquad E(X) = 1(p) + 0(1 - p) = p$$

The mean of X is equal to p. If the outcomes have equal probability, then p equals 1/2, and the mean of X equals 1/2.

Example 3. A certain businessman always eats lunch at a certain restaurant, which has lunches priced at \$1.00, \$1.50, \$2.00, and \$2.50. The businessman knows from past experience that on any given day he will select the dollar lunch with probability .25, the dollar and a half lunch with probability .35, and the remaining two lunches with probability .20 each. Let X be the price of the lunch, in dollars. Then the probability function of X is

$$P(X = 1) = .25$$
$$P(X = 1.5) = .35$$
$$P(X = 2) = .20$$
$$P(X = 2.5) = .20$$

The mean of X is found using (3).

$$E(X) = (1)(.25) + (1.5)(.35) + (2)(.20) + (2.5)(.20) = 1.675$$

Over a long period of time the businessman may expect the average luncheon expense to be somewhere near $1.67\frac{1}{2}$, even though no single lunch will cost him that amount.

Just as the mean and the median are called location parameters, the name "scale parameter" is given to properties of the random variable that measure the amount of spread, or variability, of the random variable. One scale parameter based on quantiles is the *interquartile range*, the number obtained by subtracting $x_{.25}$ from $x_{.75}$. Another scale parameter, based more directly on the probability function, is the *range*, which equals the largest possible value of the random variable minus its smallest possible value. The most common scale parameter is the *standard deviation*, which equals the square root of the *variance*, defined as follows.

Definition 4.   Let X be a random variable with mean μ and the probability function $f(x)$. The *variance of X*, usually denoted by σ^2 or by Var (X), is

$$(5) \qquad \sigma^2 = E[(X - \mu)^2]$$

Using (1) the variance of X may be written as

$$(6) \qquad \sigma^2 = \sum_x (x - \mu)^2 f(x)$$

$$= \sum_x (x^2 - 2\mu x + \mu^2) f(x)$$

$$= \sum_x x^2 f(x) - 2\mu \sum_x x f(x) + \mu^2 \sum_x f(x)$$

Because $\sum_x f(x)$ equals 1, and because of (3), the above becomes

$$(7) \qquad \sigma^2 = E(X^2) - 2\mu^2 + \mu^2 = E(X^2) - \mu^2$$

which is often a more useful form of the variance for computing purposes.

The positive square root of the variance is called the *standard deviation* of X and is usually denoted by σ.

Example 4. If X has the binomial distribution with n equal to 1, then $P(X = 1)$ equals p, and $P(X = 0)$ equals $1 - p$. In Example 2 the mean of X was found to equal p. Therefore, from (6),

$$\sigma^2 = (1 - p)^2(p) + (0 - p)^2(1 - p)$$
$$= p(1 - p)$$

(8)
$$= pq$$

Alternatively we might use (7) to compute σ^2. Then we would first compute $E(X^2)$ using (1).

$$E(X^2) = (1)^2(p) + (0)^2(1 - p)$$
$$= p$$

The variance of X is then found to be

$$\sigma^2 = E(X^2) - \mu^2$$
$$= p - p^2$$
$$= p(1 - p)$$

as before.

Example 5. There are six identical chips numbered 1 to 6. A monkey has been trained to select one chip and give it to its trainer, whereupon it receives a number of bananas equal to the number on the chip. The sample space is the chip selected by the monkey. Let X be the number on the chip. If each chip has probability 1/6 of being selected, then X has the discrete uniform distribution. For the mean of X we have

$$E(X) = 1(\tfrac{1}{6}) + 2(\tfrac{1}{6}) + 3(\tfrac{1}{6}) + 4(\tfrac{1}{6}) + 5(\tfrac{1}{6}) + 6(\tfrac{1}{6})$$
$$= 3\tfrac{1}{2}$$

The expected value of X^2 is given by

$$E(X^2) = 1(\tfrac{1}{6}) + 4(\tfrac{1}{6}) + 9(\tfrac{1}{6}) + 16(\tfrac{1}{6}) + 25(\tfrac{1}{6}) + 36(\tfrac{1}{6})$$
$$= 15\tfrac{1}{6}$$

The variance of X is computed using (7).

$$\text{Var}(X) = E(X^2) - \mu^2$$
$$= 15\tfrac{1}{6} - (3\tfrac{1}{2})^2$$
$$= 2\tfrac{11}{12}$$

The standard deviation is the square root of the variance and, for this example, equals 1.71.

Definition 2 defined the expected value of a function of a single random variable. An extension of the definition may be made to include functions of several random variables considered jointly. This extended definition leads us into consideration of the *covariance* of two random variables, and enables us to find the mean and variance of the sum of several random variables.

Definition 5. Let $X_1, X_2, \ldots, X_n$ be random variables with the joint probability function $f(x_1, x_2, \ldots, x_n)$, and let $u(X_1, X_2, \ldots, X_n)$ be a real valued function of $X_1, X_2, \ldots, X_n$. Then the *expected value of* $u(X_1, X_2, \ldots, X_n)$ is

$$(9) \quad E[u(X_1, X_2, \ldots, X_n)] = \sum u(x_1, x_2, \ldots, x_n) f(x_1, x_2, \ldots, x_n)$$

where the summation extends over all possible combinations of values of $x_1, x_2, \ldots, x_n$.

To show that Definition 5 is consistent with Definition 2, consider a function of only one random variable, which we may call $u(X_1)$, although any X_i other than X_1 may be considered in the same way. From (9) we have

$$E[u(X_1)] = \sum u(x_1) f(x_1, x_2, \ldots, x_n)$$

where the summation extends over all combinations of values of x_1, $x_2, \ldots, x_n$. Since

$$f(x_1, x_2, \ldots, x_n) = P(X_1 = x_1, X_2 = x_2, \ldots, X_n = x_n)$$
$$= P(X_2 = x_2, \ldots, X_n = x_n \mid X_1 = x_1) \cdot P(X_1 = x_1)$$

we have

$$E(u(X_1)) = \sum_{x_1} \sum_{x_2} \ldots \sum_{x_n} u(x_1) P(X_2 = x_2, \ldots, X_n = x_n \mid X_1 = x_1) \cdot P(X_1 = x_1)$$
$$= \sum_{x_1} u(x_1) P(X_1 = x_1) \sum_{x_2} \ldots \sum_{x_n} P(X_2 = x_2, \ldots, X_n = x_n \mid X_1 = x_1)$$

However, the last $n - 1$ summations equal unity, because they represent the summing of the conditional probabilities over all of the points in the reduced sample space. We are left with

$$E[u(X_1)] = \sum_{x_1} u(x_1) P(X_1 = x_1)$$

which is the same as in Definition 2. Hence we see that Definition 2 is a special case of Definition 5.

One of the simpler functions of $X_1, X_2, \ldots, X_n$ is

$$(10) \qquad\qquad Y = X_1 + X_2 + \ldots + X_n$$

That is, each value of the random variable Y associated with the combined experiment involving the X_i's is obtained simply by adding the values achieved by all the X_i's. Then

$$E(Y) = \sum (x_1 + \ldots + x_n) f(x_1, \ldots, x_n)$$
$$(11) \qquad = \sum x_1 f(x_1, \ldots, x_n) + \ldots + \sum x_n f(x_1, \ldots, x_n)$$

where each summation extends over all possible combinations of the values of $x_1, \ldots, x_n$. Using Definition 5, (11) immediately becomes

$$(12) \qquad\qquad E(Y) = E(X_1) + \ldots + E(X_n)$$

The result of the above may be stated as a theorem.

Theorem 1.   *Let $X_1, X_2, \ldots, X_n$ be random variables and let*

$$Y = X_1 + X_2 + \ldots + X_n.$$

Then $E(Y) = E(X_1) + E(X_2) + \ldots + E(X_n)$.

The statement in the above theorem holds true in all cases, whether the random variables are independent or not. Often the apparently difficult problem of finding the mean of the sum of several random variables reduces to a trivial exercise with the use of the above theorem.

The results of the next two examples will be used in later chapters.

Example 6.   Let Y be the total number of "successes" in n independent trials, where each trial results in either "success" or "failure" with probability p and $q = 1 - p$, respectively. Then Y has the binomial distribution with parameters n and p. However, Y may be regarded as the sum of n independent random variables $X_1, X_2, \ldots, X_n$, where $X_i = 1$ if the ith trial results in "success," and $X_i = 0$ if the ith trial results in failure, for each i from 1 to n. Then

$$Y = X_1 + X_2 + \ldots + X_n$$

and, from Theorem 1,

$$E(Y) = E(X_1) + E(X_2) + \ldots + E(X_n)$$

In Example 2 the mean of X_i was found to equal p. Therefore

$$(13) \qquad\qquad E(Y) = np$$

gives the mean for the binomial distribution.

Note that in the binomial distribution the trials are assumed to be independent, and therefore the X_i are independent. This assumption is not needed in order to find the mean.

The following lemma is needed in Example 7. This lemma presents a convenient equation for expressing the sum of consecutive integers.

Lemma 1

$$\sum_{i=a}^{N} i = \frac{(N + a)(N - a + 1)}{2} \qquad and \qquad \sum_{i=1}^{N} i = \frac{(N + 1)N}{2}$$

Proof.   The desired sum may be written two ways. Let $S = \sum\limits_{i=a}^{N} i$. Then

$$S = a + (a + 1) + (a + 2) + \ldots + (N - 1) + N$$
$$S = N + (N - 1) + (N - 2) + \ldots + (a + 1) + a$$

Adding the two equations together gives

$$2S = (N + a) + (N + a) + (N + a) + \ldots + (N + a) + (N + a)$$
$$= (N + a)(N - a + 1)$$

Therefore

$$S = \sum_{i=a}^{N} i = \frac{(N + a)(N - a + 1)}{2}$$

For $a = 1$, this becomes

$$\sum_{i=1}^{N} i = \frac{(N + 1)N}{2}$$

completing the proof.

Example 7.   There are N chips in a jar, numbered from 1 to N. One by one, n of those chips, where n is less than N, are drawn from the jar, the number noted, and they are put aside. Let Y be the sum of the numbers on the n drawn chips. Assume the drawings are random; that is, each chip is equally likely to be selected.

The mean of Y would be difficult to find, without using Theorem 1. The successive drawings are not independent, because once a number is recorded, no other chip can have that same number. However, we may regard Y as the sum of the random variables $X_1, X_2, \ldots, X_n$, where each X_i is the number on the ith chip drawn.

Now the chip drawn on the ith drawing is just as likely to be any one chip as any other chip. Therefore the probability distribution of X_i, considered by itself, is the discrete uniform distribution, with the probability function

$$P(X_i = k) = \frac{1}{N}, \quad \text{for} \quad k = 1, 2, 3, \ldots, N$$

Therefore, with the assistance of Lemma 1, we have the following.

$$
\begin{aligned}
E(X_i) &= \sum_{k=1}^{N} k\left(\frac{1}{N}\right) \\
&= \frac{1}{N} \sum_{k=1}^{N} k \\
&= \frac{1}{N} \frac{N(N + 1)}{2} \\
&= \frac{N + 1}{2}
\end{aligned}
$$

(14)

Equation (14) furnishes us with the mean of a discrete uniform random variable. Since Y equals $X_1 + X_2 + \ldots + X_n$, we have

$$E(Y) = E(X_1) + E(X_2) + \ldots + E(X_n)$$

(15)
$$= n\frac{N + 1}{2}$$

A particularly useful function of two random variables is $[X_1 - E(X_1)] \times [X_2 - E(X_2)]$, whose expected value is called the *covariance of X_1 and X_2*. In particular, a comparison of Definition 4 with the following reveals that the variance of X_1 may be considered as the covariance of X_1 with itself.

Definition 6. Let X_1 and X_2 be two random variables with means μ_1 and μ_2, probability functions $f_1(x_1)$ and $f_2(x_2)$ respectively, and joint probability function $f(x_1, x_2)$. The *covariance of X_1 and X_2* is

(16)
$$\text{Cov}\,(X_1, X_2) = E[(X_1 - \mu_1)(X_2 - \mu_2)]$$

The definition of expected value, Definition 5, may be used to give

$$\text{Cov}\,(X_1, X_2) = E[(X_1 + \mu_1)(X_2 - \mu_2)]$$
(17)
$$= \sum (x_1 - \mu_1)(x_2 - \mu_2)f(x_1, x_2)$$

where the summation extends over all values of x_1 and x_2. This expands as

$$\text{Cov}\,(X_1, X_2) = \sum (x_1 x_2 - \mu_1 x_2 - \mu_2 x_1 + \mu_1 \mu_2)f(x_1, x_2)$$
$$= \sum x_1 x_2 f(x_1, x_2) - \mu_1 \sum x_2 f(x_1, x_2) - \mu_2 \sum x_1 f(x_1, x_2)$$
$$+ \mu_1 \mu_2 \sum f(x_1, x_2)$$
$$= E(X_1 X_2) - \mu_1 \mu_2 - \mu_2 \mu_1 + \mu_1 \mu_2$$
(18)
$$= E(X_1 X_2) - \mu_1 \mu_2$$

Equation (18) is often easier to use than (17) when calculating a covariance.

Example 8. An insurance company has noticed that the probability of any particular person having an automobile accident within a given year is about .1. However, this probability becomes .3 if it is known that the person had an automobile accident the previous year.

Let X_1 equal 0 or 1 depending on whether a particular person has no accidents or at least one accident, respectively, during the first year of his insurance period. Let X_2 be similarly defined for the second year of that same person's insurance period. The probability function of X_1, and therefore X_2 also, is

$$P(X_1 = 0) = .9$$
$$P(X_1 = 1) = .1$$

From example 2 we obtain

$$E(X_1) = .1$$
$$E(X_2) = .1$$

The joint probability function of X_1 and X_2 at $X_1 = 1$ and $X_2 = 1$ may be found as follows.

$$
\begin{aligned}
f(1, 1) &= P(X_1 = 1, X_2 = 1) \\
&= P(X_2 = 1 \mid X_1 = 1)P(X_1 = 1) \\
&= (.3)(.1) \\
&= .03
\end{aligned}
$$

The computation of $E(X_1X_2)$ follows directly from Definition 5.

$$
\begin{aligned}
E(X_1X_2) &= (1)(1)f(1, 1) \text{ plus "zero" terms} \\
&= .03
\end{aligned}
$$

The covariance of X_1 and X_2 is then obtained by using (18).

$$
\begin{aligned}
\text{Cov}(X_1, X_2) &= E(X_1X_2) - E(X_1)E(X_2) \\
&= .03 - (0.1)(0.1) \\
&= .02
\end{aligned}
$$

We shall now define the *correlation coefficient*, which is used as a measure of linear dependence between two random variables. Although we shall not prove it here, the correlation coefficient is always between -1 and $+1$. It equals zero when the two random variables are independent, although it may equal zero in other cases also.

Definition 7. The *correlation coefficient* between two random variables is their covariance divided by the product of their standard deviations. That is, the correlation coefficient, usually denoted by ρ, between two random variables X_1 and X_2 is given by

(19)
$$\rho = \frac{\text{Cov}(X_1, X_2)}{\sqrt{\text{Var}(X_1)\,\text{Var}(X_2)}}$$

A lemma that will be used in the next example is now presented. This lemma furnishes us with a convenient formula for expressing the sum of the squares of the first N consecutive integers.

Lemma 2

$$\sum_{i=1}^{N} i^2 = \frac{N(N + 1)(2N + 1)}{6}$$

Proof.   Let $S = \sum\limits_{i=1}^{N} i^2$. Then

$$S = 1^2 + 2^2 + 3^2 + 4^2 + \ldots + N^2$$
$$= 1 + 2 + 3 + 4 + \ldots + N$$
$$+ 2 + 3 + 4 + \ldots + N$$
$$+ 3 + 4 + \ldots + N$$
$$+ 4 + \ldots + N$$
$$\cdots \qquad \cdots$$
$$+ N$$

where the sum of the numbers in the *i*th column is i^2. However, instead of adding down the columns, we shall add across the rows. The sum of the numbers in the *j*th row from the top is found by Lemma 1 to be

$$j + (j + 1) + (j + 2) + \ldots + N = \frac{(N + j)(N - j + 1)}{2}$$
$$= \tfrac{1}{2}(N^2 + N + j - j^2)$$

Adding these row sums together gives

$$S = \sum_{j=1}^{N} \tfrac{1}{2}(N^2 + N + j - j^2)$$

$$= \frac{1}{2}\sum_{j=1}^{N} N^2 + \frac{1}{2}\sum_{j=1}^{N} N + \frac{1}{2}\sum_{j=1}^{N} j - \frac{1}{2}\sum_{j=1}^{N} j^2$$

$$= \tfrac{1}{2}(N \cdot N^2) + \tfrac{1}{2}(N \cdot N) + \tfrac{1}{2} \cdot \frac{(N + 1)N}{2} - \tfrac{1}{2}S$$

since the last sum in the middle equation above is denoted by S in this proof. Rearranging gives

$$\tfrac{3}{2}S = \tfrac{1}{4}(2N^3 + 3N^2 + N) = \tfrac{1}{4}N(N + 1)(2N + 1)$$

so that

$$S = \sum_{i=1}^{N} i^2 = \frac{N(N + 1)(2N + 1)}{6}$$

completing the proof.

Example 9.   A jar contains N plastic chips numbered 1 to N, as in Example 7. An experiment consists of drawing n of these chips from the jar, where $n \leq N$. We assume that each chip is equally likely to be selected, and that the drawing is without replacement. Let $X_1, X_2, \ldots, X_n$ be random variables, where X_i equals the number on the *i*th chip drawn from the jar, for $i = 1, 2, \ldots, n$.

Then, from Example 7, we have

$$E(X_i) = \frac{N+1}{2}$$

Also, from (7) and Lemma 2 we have

$$\text{Var}\,(X_i) = E(X_i^2) - [E(X_i)]^2 = \sum_{k=1}^{N} k^2 \frac{1}{N} - \left(\frac{N+1}{2}\right)^2$$

$$= \frac{1}{N} \frac{N(N+1)(2N+1)}{6} - \left(\frac{N+1}{2}\right)^2$$

(20)
$$= \frac{(N+1)(N-1)}{12}$$

Now consider jointly two random variables X_i and X_j, where $i \neq j$. Their joint probability function is

$$f(x_i, x_j) = P(X_i = x_i, X_j = x_j) = P(X_i = x_i \mid X_j = x_j) \cdot P(X_j = x_j)$$

(21)
$$= \frac{1}{N-1} \cdot \frac{1}{N}, \text{ for } x_i, x_j = 1, 2, \ldots, N; x_i \neq x_j$$

because once X_j is known to equal x_j, X_i may equal any integer from 1 to N except the integer x_j. Hence there are $N - 1$ equally likely values for X_i, each with a probability $1/(N - 1)$.

The covariance of X_i and X_j, using Definition 6, is

$$\text{Cov}\,(X_i, X_j) = E\{[X_i - E(X_i)][X_j - E(X_j)]\}$$

Since the mean of both X_i and X_j is $(N + 1)/2$, we have

$$\text{Cov}\,(X_i, X_j) = \sum_{\substack{k=1 \\ k \neq s}}^{N} \sum_{s=1}^{N} \left(k - \frac{N+1}{2}\right)\left(s - \frac{N+1}{2}\right)\frac{1}{(N-1)N}$$

where the summation extends over all k and s from 1 to N, except that k does not equal s because X_i and X_j cannot equal the same number at the same time. If we, at the same time, add and subtract those terms for $k = s$ the covariance becomes

$$\text{Cov}\,(X_i, X_j) = \sum_{k=1}^{N} \sum_{s=1}^{N} \left(k - \frac{N+1}{2}\right)\left(s - \frac{N+1}{2}\right)\frac{1}{(N-1)(N)}$$

$$- \sum_{k=1}^{N} \left(k - \frac{N+1}{2}\right)^2 \frac{1}{(N-1)N}$$

(22)
$$= \frac{1}{(N-1)N} \sum_{k=1}^{N} \left(k - \frac{N+1}{2}\right) \sum_{s=1}^{N} \left(s - \frac{N+1}{2}\right)$$

$$- \frac{1}{N-1} \sum_{k=1}^{N} \left(k - \frac{N+1}{2}\right)^2 \frac{1}{N}$$

To simplify (22) we note that

$$
(23) \quad \sum_{i=1}^{N} \left(i - \frac{N+1}{2} \right) = \sum_{i=1}^{N} i - \sum_{i=1}^{N} \frac{N+1}{2} = \frac{N(N+1)}{2} - \frac{N(N+1)}{2} = 0
$$

and therefore the longer term in (22) equals zero. Also, from the definition of variance, and from (20) we have

$$
(24) \quad \text{Var}\,(X_i) = \sum_{k=1}^{N} \left(k - \frac{N+1}{2} \right)^2 \frac{1}{N} = \frac{(N+1)(N-1)}{12}
$$

Substitution of (23) and (24) into (22) yields

$$
(25) \quad \text{Cov}\,(X_i, X_j) = -\frac{N+1}{12}
$$

The fundamental importance of the covariance is based on what happens to the covariance in the case of two independent random variables. Let X_1 and X_2 be two independent random variables with probability functions $f_1(x_1)$ and $f_2(x_2)$, and means μ_1 and μ_2, respectively. Then the covariance of X_1 and X_2 is

$$
\text{Cov}\,(X_1, X_2) = E(X_1, X_2) - \mu_1 \mu_2
$$
$$
= \sum x_1 x_2 f_1(x_2, x_2) - \mu_1 \mu
$$

where the summation extends over all combinations of values of x_1 and x_2. Since X_1 and X_2 are independent

$$
f(x_1, x_2) = f_1(x_1) f_2(x_2)
$$

and

$$
\text{Cov}\,(X_1, X_2) = \sum_{x_1, x_2} x_1 x_2 f_1(x_1) f_2(x_2) - \mu_1 \mu_2
$$
$$
= \left[\sum_{x_1} x_1 f_1(x_1) \right] \left[\sum_{x_2} x_2 f_2(x_2) \right] - \mu_1 \mu_2
$$
$$
= \mu_1 \mu_2 - \mu_1 \mu_2 = 0
$$

Therefore, independence of two random variables implies that their covariance is zero, which in turn implies that their correlation coefficient equals zero.

Theorem 2. *If X_1 and X_2 are independent random variables, then the covariance of X_1 and X_2 is zero.*

The converse of Theorem 2 is not necessarily true. That is, zero covariance does not necessarily imply that the random variables are independent, even though the implication is an error made often in practice.

Example 10.   Define the joint probability function of two random variables as follows.

$$P(X = 0, Y = 0) = 1/2$$
$$P(X = 1, Y = 1) = 1/4$$
$$P(X = -1, Y = 1) = 1/4$$

The probability function of X is then

$$P(X = 0) = 1/2$$
$$P(X = 1) = 1/4$$
$$P(X = -1) = 1/4$$

and the probability function of Y is

$$P(Y = 0) = 1/2$$
$$P(Y = 1) = 1/2$$

The expected values of X and Y are

$$E(X) = 0$$
$$E(Y) = 1/2$$

The covariance of X and Y is

$$\begin{aligned} \text{Cov }(X, Y) &= E(XY) - E(X)E(Y) \\ &= (1)(\tfrac{1}{4}) + (-1)\tfrac{1}{4} - (0)(\tfrac{1}{2}) \\ &= 0 \end{aligned}$$

However X and Y are not independent, because

$$P(X = 0, Y = 0) = \tfrac{1}{2}$$

which is not equal to

$$P(X = 0)P(Y = 0) = (\tfrac{1}{2})(\tfrac{1}{2})$$
$$= \tfrac{1}{4}$$

Therefore X and Y have zero covariance, even though they are not independent.

We are now equipped to find the variance of the sum of several random variables. Let Y equal $X_1 + X_2 + \ldots + X_n$, where the X_i's may or may not be independent. We wish to find the variance of Y.

$$\begin{aligned} \text{Var }(Y) &= E\{[Y - E(Y)]^2\} \\ &= E\{[X_1 + X_2 + \ldots + X_n - E(X_1) - E(X_2) - \ldots - E(X_n)]^2\} \\ &= E\{[X_1 - E(X_1) + X_2 - E(X_2) + \ldots + X_n - E(X_n)]^2\} \\ &= E\left\{ \sum_{i=1}^{n} [X_i - E(X_i)]^2 + \sum_{\substack{i=1 \\ i \neq j}}^{n} \sum_{j=1}^{n} [X_i - E(X_i)][X_j - E(X_j)] \right\} \end{aligned}$$

But since the expected value of a sum of random variables equals the sum of the expected values of the random variables,

$$\text{Var}(Y) = \sum_{i=1}^{n} E\{[X_i - E(X_i)]^2\} + \sum_{i=1}^{n}\sum_{\substack{j=1 \\ i \neq j}}^{n} E\{[X_i - E(X_i)][X_j - E(X_j)]\}$$

(26)

$$= \sum_{i=1}^{n} \text{Var}(X_i) + \sum_{i=1}^{n}\sum_{\substack{j=1 \\ i \neq j}}^{n} \text{Cov}(X_i, X_j)$$

If $X_1, \ldots, X_n$ are mutually independent, then from Theorem 2 we have $\text{Cov}(X_i, X_j) = 0$, and

(27)
$$\text{Var}(Y) = \sum_{i=1}^{n} \text{Var}(X_i)$$

We may summarize the above as a theorem.

Theorem 3.   *Let $X_1, X_2, \ldots, X_n$ be random variables and let*

$$Y = X_1 + X_2 + \ldots + X_n$$

Then

$$\text{Var}(Y) = \sum_{i=1}^{n} \text{Var}(X_i) + \sum_{i=1}^{n}\sum_{\substack{j=1 \\ i \neq j}}^{n} \text{Cov}(X_i, X_j)$$

Furthermore, if $X_1, X_2, \ldots, X_n$ are mutually independent, then

$$\text{Var}(Y) = \sum_{i=1}^{n} \text{Var}(X_i)$$

Example 11.   In continuation of Example 9, let X_i equal the number on the ith chip drawn, as before, and let Y equal the sum of the X_i's, as in Example 7. Then Theorem 3 gives us

$$\text{Var}(Y) = \sum_{i=1}^{n} \text{Var}(X_i) + \sum_{i=1}^{n}\sum_{\substack{j=1 \\ i \neq j}}^{n} \text{Cov}(X_i, X_j)$$

The various terms in the above equation are given by (20) and (25). The variance term appears n times, and the covariance term appears $n(n - 1)$ times.

$$\text{Var}(Y) = n\frac{(N + 1)(N - 1)}{12} + n(n - 1)\left(-\frac{N + 1}{12}\right)$$

(28)
$$= \frac{n(N + 1)(N - n)}{12}$$

Note that $\text{Var}(X_i)$ is only a special case of $\text{Var}(Y)$, for $n = 1$.

The variance of a random variable which has the binomial distribution is found in the next example.

Example 12.   Consider n independent trials, where each trial may result in "success" with probability p, or "failure" with probability q, where $p + q$ equals 1. As in Examples 4 and 6, let X_i equal 0 or 1, depending on whether the ith trial results in "failure" or "success," respectively, and let Y equal the total number of "successes" in the n trials. Since the X_i are mutually independent, Theorem 3 states that

$$\text{Var } (Y) = \sum_{i=1}^{n} \text{Var } (X_i)$$

From Example 4, Var (X_i) equals pq, so

$$\text{Var } (Y) = npq$$

furnishes us with the variance of Y, which has the binomial distribution.

The results of some of the preceding examples will be used later in this book, and so they are stated separately as theorems, for convenience.

Theorem 4.   *Let X be a random variable with the binomial distribution*

$$P(X = k) = \binom{n}{k} p^k q^{n-k}$$

Then the mean and variance of X are given by

$$E(X) = np$$
$$\text{Var}(X) = npq$$

Theorem 5.   *Let X be the sum of n integers selected at random, without replacement, from the first N integers 1 to N. Then the mean and variance of X are given by*

$$E(X) = \frac{n(N + 1)}{2}$$

$$\text{Var } (X) = \frac{n(N + 1)(N - n)}{12}$$

Example 13.   An advertising agency drew twelve sample magazine ads for one of their customers, and ranked the ads from 1 to 12 on the basis of the agency's opinion of which ads would be the most effective in selling the product. The "most effective" ad was given the rank 1, and so on. The customer, the manufacturer of the product, selected four of the ads for purchase. These ads were ranked 4, 6, 7, and 11 by the agency.

Assuming that the customer's choice and the agency's rankings were independent, the sum of the ranks on the selected ads should be distributed the same as the sum of the numbers on four chips selected at random out of twelve chips numbered one to twelve. Let X equal the sum of the ranks of four ads if they are selected independently of the ranks. Then Theorem 5 states that the mean of X is

$$E(X) = \frac{(4)(12 + 1)}{2} = 26$$

and the variance of X is

$$\text{Var}(X) = \frac{(4)(12 + 1)(12 - 4)}{12} = 34\tfrac{2}{3}$$

The standard deviation of X is

$$\sigma = \sqrt{\text{Var}(X)} = 5.9$$

The observed value of X is

$$X = 4 + 6 + 7 + 11 = 28$$

which is close to the mean of X under the above assumptions.

PROBLEMS

1. If $P(X = 0) = 1/3$, $P(X = 1) = 1/3$, $P(X = 2) = 1/6$, and $P(X = 3) = 1/6$, find

   (a) $E(X)$ (d) the median

   (b) $\text{Var}(X)$ (e) $x^{\frac{1}{3}}$

   (c) $E(X^2 + 2X)$ (f) the fourth decile

2. If $P(X = 0) = 0$, $P(X = 1) = 1/2$, $P(X = 2) = 1/4$, and $P(X = 4) = 1/4$, find

   (a) $E(X)$ (d) the median

   (b) $\text{Var}(X)$ (e) the upper quartile

   (c) $E(-X)$ (f) the thirty-seventh percentile

3. If $P(X = 0, Y = 0) = 1/4$, $P(X = 0, Y = 1) = 1/4$, $P(X = 1, Y = 0) = 1/4$, and $P(X = 1, Y = 1) = 1/4$, find

   (a) $E(X)$ (e) $\text{Cov}(X, Y)$

   (b) $E(Y)$ (f) $P(X = 0)$

   (c) $E(XY)$ (g) $P(X = 1)$

   (d) $E(X + Y)$ (h) Are X and Y independent?

4. Of $P(X = 0, Y = 0) = 1/8$, $P(X = 0, Y = 1) = 3/8$, $P(X = 1, Y = 0) = 3/8$, and $P(X = 1, Y = 1) = 1/8$, find

   (a) $E(X)$ (d) $E(X^2 Y)$

   (b) $E(Y)$ (e) $\text{Cov}(X, Y)$

   (c) $E(XY)$ (f) $P(X = x)$ for all x

   (g) Are X and Y independent?

5. What is the sum of the sixty-six integers from 1 to 66?

6. What is the sum of the thirty integers from 70 to 99?

7. What is the sum of the squares of the thirty integers from 1 to 30?

8. If X equals the number of spots showing on one roll of a balanced die, find

   (a) $E(X)$

   (b) Var (X)

   (c) $E(X^2 + X)$

9. If thirty tickets are numbered consecutively from 1 to 30, and two of those tickets are selected at random without replacement, find

   (a) E (sum of the numbers on the two tickets).

   (b) Var (sum of the numbers on the two tickets).

10. Prove that a vertical line drawn at the mean of a random variable X divides the distribution function of X in such a way that the area under the distribution function, to the left of the mean, and above 0, equals the area above the distribution function, to the right of the mean, and below 1.

1.6. CONTINUOUS RANDOM VARIABLES

All of the random variables that we have introduced so far in this chapter have one property in common; their possible values can be listed. The list of possible values assumed by the binomial random variable is 0, 1, 2, 3, 4, ... , $n - 1, n$. No other values may be assumed by the binomial random variable. The list of values that may be assumed by the discrete uniform random variable could be written as 1, 2, 3, ... , N. Similar lists could be made for each random variable introduced in the previous definitions and examples.

These lists may be infinite in length, such as in an experiment where the random variable X equals the number of times a monkey pushes the "wrong" button before he finally pushes the "right" button and gets a reward. Then X may equal zero if the right button is pushed the first time, or X may equal 1000 if the monkey has difficulty finding the right button. Theoretically there is no limit to the number of times the monkey may choose the wrong button before pushing the correct one. The possible values of X may be listed, even though the list may be infinitely long. The infinitely long list of possible values is a characteristic of the model rather than the actual experiment in this example and in most situations, because such real factors as the eventual death of the monkey, the absence of research funds, or the waning enthusiasm of the experimenter will prevent the actual experiment from being prolonged to the point of absurdity. Nevertheless, the model may be reasonable, and the random variable in the model may have an infinity of possible values.

A more precise way of stating that the possible values of a random variable may be listed, is to say that there exists a *one to one correspondence* between the possible values of the random variable and some or all of the positive integers. This means that to each possible value there corresponds one and only one positive integer, and that positive integer does not correspond to more than one possible value of the random variable. Random variables with this property are called *discrete*. All of the random variables we have considered so far are discrete random variables.

Definition 1. A random variable X is *discrete* if there exists a one to one correspondence between the possible values of X and some or all of the positive integers.

The distribution function of a discrete random variable is always a "step function"; that is, the graph looks like a series of stair steps, although the

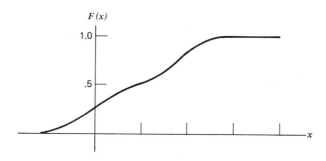

Figure 1

steps may not be very uniform in appearance, and there may be an infinite number of steps. If any portion of the graph appears to rise gradually, instead of rising only in clear-cut steps, then the associated random variable is not discrete.

If the graph of a distribution function has no steps, but rather rises only gradually where it rises, then the distribution function is called "continuous," and the random variable with that distribution function is called a *continuous random variable*. Figure 1 is an example of the graph of a continuous distribution function. Saying that a distribution function has no steps is the same as saying that no two horizontal lines will intersect the graph at the same value as measured along the horizontal axis. That is, if there is a step in the graph of a distribution function, then at least two horizontal lines may be drawn, say at heights p and p', closely enough to each other so that they intersect the graph at the same value, as measured along the horizontal axis. Since this describes the graphical method of finding quantiles, we may say that if there is a step in the distribution function then there are at least two quantiles x_p and $x_{p'}$ that are equal to each other. Conversely, if there are no two quantiles exactly equal to each other, then there are no steps in the distribution function, and the function is continuous. This leads to a method of defining *continuous random variable*.

Definition 2.   A random variable X *is continuous* if no two quantiles x_p and $x_{p'}$ of X are equal to each other, where p is not equal to p'. Equivalently, a random variable X is continuous if $P(X \leq x)$ equals $P(X < x)$ for all numbers x.

Example 1.   If the graph of a distribution function is as follows, then $F(x)$ is a continuous distribution function, and any random variable with the distribution

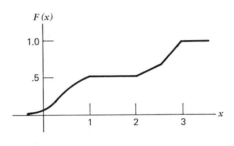

Figure 2

function $F(x)$ is a continuous random variable. For each value of x there is only one value of $F(x)$, and therefore no two quantiles are equal.

In practice, no actual random variable is continuous, because the observed values of actual random variables are always the result of measurements of some sort, and measurements are made with tools that have only a finite capacity for discriminating between two values. Continuous random variables exist only in theory, such as in a model of an actual experiment. At times it is preferable to assume a random variable to be continuous, even though it is known to be discrete, as in the following example.

Example 2.   The time it takes a racehorse to run a mile race is a continuous quantity, because time is in general a continuous quantity. In practice, however, the time is usually measured to the nearest one-fifth of a second. It is not unusual for a horse to run two races in identical lengths of time— measured time, that is. The actual lengths of time will be exactly equal with probability zero, and therefore it is reasonable to assume that the time of a race, measured exactly, is a continuous random variable which is approximately equal to the measured time of the race, a discrete random variable. If two horses run in the same race and cross the finish line ahead of all other horses, with identical measured times, the winner of the race is then determined by examining a photograph taken at the finish line at the moment the horses crossed. Only rarely does this fail to determine the actual winner of the race. This is meant to illustrate that even though the random variable (measured time) is discrete, it is assumed to be continuous because the actual order in which the horses finished the race still may be determined even though two or more measured times appear to be identical.

Another reason for considering continuous random variables is that the distribution function of a discrete random variable sometimes may be approximated by a continuous distribution function, resulting in a convenient method for computing desired probabilities associated with the discrete random variable. Two continuous distribution functions commonly used for this purpose are the *normal distribution* and the *chi-square distribution*.

The distribution function in the following definition might frighten those who are unfamiliar with elementary calculus. There is no cause for alarm, however, because the distribution function is well tabulated. Such tables may be found in most statistics texts, as well as in Table 1 at the end of this book.

Definition 3.   Let X be a random variable. Then X is said to have the *normal distribution* if the distribution function of X is given by

$$(1) \qquad F(x) = P(X \leq x) = \int_{-\infty}^{x} \frac{1}{\sqrt{2\pi}\,\sigma} e^{-\frac{1}{2}[(y-\mu)/\sigma]^2}\, dy$$

where it can be shown (using calculus) that the parameters μ and σ are the mean and standard deviation of X. The *standard normal distribution* is the normal distribution with μ equal to 0 and σ equal to 1.

The normal distribution function cannot be evaluated directly, and so Table 1 may be used to find approximate probabilities associated with normal random variables. Table 1 gives selected quantiles of the standard normal random variable. Quantiles of normal random variables with mean μ and variance σ^2 may be found from Table 1, with the aid of the equations given without proof in the following theorem.

Theorem 1.   *For a given value of p, let x_p be the pth quantile of a normal random variable with mean μ and variance σ^2, and let w_p be the pth quantile of a standard normal random variable. The quantile x_p may be obtained from w_p by using the relationship*

$$(2) \qquad\qquad\qquad x_p = \mu + \sigma w_p$$

Similarly, w_p may be obtained from x_p with the aid of the relationship

$$(3) \qquad\qquad\qquad w_p = \frac{x_p - \mu}{\sigma}$$

Example 3.   Let X be a random variable with the standard normal distribution. To find the probability that X will not exceed 1.42, Table 1 is used. From Table 1 we see that

$$P(X \leq 1.4051) = .92$$

and

$$P(X \leq 1.4758) = .93$$

Therefore, as an approximation we may either interpolate to get an approximate probability

$$P(X \leq 1.42) \cong .922$$

or we may simply use the number closest to 1.42 to obtain

$$P(X \leq 1.42) \cong .92$$

Example 4.   Let X be the "I.Q." of a person selected at random from a large group of people, and assume that X has the normal distribution with mean 100 and standard deviation 15.

Suppose we want to find the probability that X will exceed 125. We have

$$P(X > 125) = 1 - P(X \leq 125)$$

so it suffices to find $P(X \leq 125)$. The quantile w_p, of the standard normal random variable, corresponding to the quantile $x_p = 125$ is found from (3).

$$w_p = \frac{x_p - \mu}{\sigma}$$

$$= \frac{125 - 100}{15}$$

$$= 1.67$$

From Table 1 we see that if w_p equals 1.67, then p equals .95. Therefore 125 is the .95 quantile of X.

$$P(X \leq 125) = .95$$
$$P(X > 125) = .05$$

The desired probability is .05.

To find the upper one percentile, called the ninety-ninth percentile, we want to find the number $x_{.99}$, where

$$P(X \leq x_{.99}) = .99$$

Since, from Table 1, $w_{.99}$ equals 2.3263, $x_{.99}$ may be found from (2) to be

$$x_{.99} = \mu + \sigma w_{.99}$$
$$= 100 + 15(2.3263)$$
$$= 134.9$$

Therefore the probability of the randomly selected person having an I.Q. less than 135 is about .99.

Example 5.   A railroad company has observed over a period of time that the number X of people taking a certain train seems to follow a normal distribution with mean 540 and standard deviation 32. How many seats should the company provide on the train, if the company wants to be 95% certain that everyone will have a seat?

We wish to find the 95th percentile. From (2) we have that

$$x_{.95} = \mu + \sigma w_{.95}$$
$$= 540 + 32(1.6449)$$
$$= 592.6$$

where $w_{.95}$ is obtained from Table 1. The company needs 593 seats on the train so that they can be 95% certain that there will be enough seats for everyone on any one run of that train.

In Example 5 the random variable X is actually a discrete random variable which assumes only the nonnegative integers as values. Therefore, X cannot possibly have a normal distribution. The normal approximation to the distribution of X was used partly for convenience, and partly out of necessity, because a realistic discrete distribution might be difficult to formulate. In other problems a discrete distribution function which agrees well with the data may be known, and yet the normal approximation still might be used for ease in calculations. The validity of using the normal approximation usually depends on the Central Limit Theorem, which we now discuss.

The so-called Central Limit Theorem appears in many different forms. All forms have in common the purpose of stating conditions under which the sum of several random variables may be approximated by a normal random variable. The theorem says that the distribution function of the sum of several random variables approaches the normal distribution function, as the number of random variables being added becomes large (i.e., goes to infinity), and when certain other general conditions are met. These "other general conditions" may be stated many different ways, giving rise to the many forms for the Central Limit Theorem. Although a thorough discussion of this theorem is well beyond the scope of this book, frequent reference in this book to the use of the theorem invites a brief attempt to dispel some of the mystery that might otherwise build.

Theorem 2.   (*Central Limit Theorem*) *Let Y_n be the sum of the n random variables $X_1, X_2, \ldots, X_n$. Let μ_n be the mean of Y_n, and let σ_n^2 be the variance of Y_n. As n, the number of random variables, goes to infinity, the distribution function of the random variable*

$$\frac{Y_n - \mu_n}{\sigma_n}$$

approaches the standard normal distribution function, if one of the following sets of conditions holds.

Set A: *The X_i are independent and identically distributed, with $\infty > \text{Var}(X_i) > 0$ (Fisz, 1963, p. 197).*

Set B: *The X_i are independent but not necessarily identically distributed, but $E(X_i^2)$ exists for all i and satisfies certain conditions as stated on p. 203 of Fisz (1963).*

Set C: *The X_i are neither independent nor identically distributed, but represent the successive drawings, without replacement, of values from a finite population of size N, where N is greater than 2n. Also a condition stated on p. 523 of Fisz (1963) should be satisfied.*

Some of the conditions that must be met, in order for the Central Limit Theorem to apply, are not stated above because they are somewhat mathematical and would not add to the intuitive understanding of the theorem. A convenient reference is given for the benefit of the interested reader.

When the Central Limit Theorem is stated in terms of Set A, it is known as the Lindeberg-Lévy Theorem (Lindeberg, 1922 and Lévy, 1925). When the theorem is stated with Set B conditions it is usually called the Lapunov Theorem (Lapunov, 1901). The theorem with conditions given by Set C was proved by Erdös and Rényi (1959).

In practice, the number of random variables summed never goes to infinity. But the value of the Central Limit Theorem is that in those situations where the theorem holds, the normal approximation is usually considered to be "reasonably good" as long as *n* is "large." The terms "reasonably good" and "large" are subjective terms, and therefore much latitude exists in the practice of using the normal approximation.

A useful illustration of a situation where the Set A conditions hold is presented in the following example.

Example 6.   Let Y_n be a random variable with the binomial distribution (Definition 1.4.5) with mean np and variance npq (Theorem 1.5.4). Then Y_n may be regarded as the sum of n independent random variables, each with the binomial distribution where n equals 1 (Example 1.5.12). Since these other random variables have a variance of pq, which is greater than zero for $0 < p < 1$, the Set A of conditions holds. For large n, the random variable

$$\frac{Y_n - np}{\sqrt{npq}}$$

is distributed approximately the same as a standard normal random variable. This is equivalent to saying that for large n, the distribution function of Y_n may be approximated by the normal distribution with mean np and variance npq (Theorem 1).

The conditions given by Set C are met in the following example, which becomes useful to us in Chapter 5.

Example 7.   Consider the sampling scheme where n integers are selected at random, without replacement, from the first N integers, 1 to N. Let X_i be the ith integer selected, and let

$$Y_n = X_1 + X_2 + \ldots + X_n$$

be the sum of the integers selected. Then for large n and large N, and $n < N/2$, the Set C of conditions hold and the distribution function of

$$\frac{Y_n - \dfrac{n(N+1)}{2}}{\left(\dfrac{n(N+1)(N-n)}{12}\right)^{\frac{1}{2}}}$$

(Theorem 1.5.5) may be approximated by the standard normal distribution function. In other words, the distribution function of Y_n may be approximated by the normal distribution function with mean $n(N+1)/2$ and variance $n(N+1)(N-n)/12$ (Theorem 1).

The widespread applicability of the Central Limit Theorem makes it a very useful theorem. Since it justifies to some extent the use of the normal approximation, the normal distribution is a rather valuable distribution. Other distributions that are related to the normal distribution also become important, such as the *chi-square distribution*.

In the following definition, the chi-square distribution function is given using the "integral" notation of calculus, and using the "gamma function" $\Gamma(k/2)$. This notation needs no explanation, or even understanding, because the tabulated values of the chi-square distribution function, given in Table 2, will be used whenever values of the distribution function are needed. A more extensive table is given by Harter (1964). A convenient nomogram is provided by Boyd (1965).

Definition 4.   A random variable X has the *chi-square distribution with k degrees of freedom* if the distribution function of X is given by

$$(4) \qquad F(x) = P(X \leq x) = \int_0^x \frac{y^{(k/2)-1}e^{(y/2)}}{2^{k/2}\Gamma(k/2)}\, dy \qquad \text{if } x > 0$$

$$= 0 \qquad\qquad \text{if } x \leq 0$$

The distribution function (4) shows that a chi-square random variable may assume only nonnegative values, since $F(x)$ equals zero for negative values of x. The "degrees of freedom," k, is merely a parameter. The values of k are usually restricted to the integers 1, 2, 3, and so on. For different values of the parameter k, the distribution functions are different also. Table 2 gives some selected quantiles of a chi-square random variable, for $k = 1, 2, 3$, up to 30, and for some values of k greater than 30. For k greater than 100 the Central Limit Theorem may be used to obtain approximate quantiles, which will be justified later in this section.

It is shown in most introductory books on mathematical statistics that if X is a random variable with the chi-square distribution with k degrees of freedom, then the mean and variance of X are given by

(5) $$E(X) = k$$
(6) $$\text{Var } (X) = 2k$$

The following theorem is proved on p. 194 of Freund (1962).

Theorem 3.   *Let $X_1, X_2, \ldots, X_k$ be k independent and identically distributed standard normal random variables. Let Y be the sum of the squares of the X_i,*

(7) $$Y = X_1^2 + X_2^2 + \ldots + X_k^2$$

Then Y has the chi-square distribution with k degrees of freedom.

The value of this theorem, for our purposes, is that if the distribution function of X_i is approximately equal to the standard normal distribution function, Theorem 3 justifies using the chi-square distribution function to find approximate probabilities for Y.

Example 8.   A child psychologist asks each of 100 children to tell which of two trucks they would rather play with. The two trucks are identical in all respects, except that one is red and the other is green. The psychologist is interested in knowing whether children have a color preference.

Forty-two children selected the green truck, and the other fifty-eight chose the red truck. In the model "no preference" is assumed, so the random variable X, equal to the number of children who selected the green truck, should have the binomial distribution with mean $np = 50$ and variance $npq = 25$. The normal approximation to the distribution function of X seems appropriate, so

$$\frac{X - 50}{5}$$

is considered to be approximately the same as a standard normal random variable. However, the psychologist is interested in determining differences in either direction; that is, he wants to know whether X is much smaller than 50 as well as whether X is much larger than 50. So he uses the square of the difference essentially, but actually examines the random variable

$$X^* = \left(\frac{X - 50}{5}\right)^2$$

because it may be compared with a chi-square random variable with one degree of freedom. In this experiment X^* equals $[(42 - 50)/5]^2$, or 2.56. The probability of getting a number smaller than 2.56, corresponding to a value of X closer to 50, is found by interpolation in Table 2, $k = 1$, to be about .88. Therefore the psychologist concludes that there is some indication of a color preference among the children. (More will be said concerning this method of drawing a conclusion in later chapters.)

In Example 8 the distribution function of the random variable $(X - 50)/5$ was considered to be approximately equal to the standard normal distribution function. Therefore the chi-square approximation, with one degree of freedom, was used for the distribution of X^*. The desired probability could have been found by using both tails of the normal distribution function. That is,

$$P(X^* \leq (-1.6)^2) = P\left(-1.6 < \frac{X - 50}{5} < +1.6\right)$$

The probability on the left, found from Table 2, should equal the probability on the right, obtained from Table 1. The only difference between the two probabilities results from using interpolation in the two tables. If more than one degree of freedom is involved, then Table 1 may not be used as an alternative to Table 2.

Example 9. In continuation of the experiment described in Example 8, the psychologist obtains two toy telephones, identical except that one is white and the other is blue. He asks each of twenty-five children to choose one to play with. Seventeen children chose the white telephone, and the other eight preferred the blue telephone. Let Y be the random variable equal to the number of children selecting the white telephone. Since

$$\frac{Y - np}{\sqrt{npq}} = \frac{Y - (1/2)(25)}{5/2}$$

is approximately a standard normal random variable under the assumption of no color preference, the random variable

$$Y^* = \left(\frac{Y - (1/2)(25)}{5/2}\right)^2$$

may be compared with a chi-square random variable with one degree of freedom. Since Y equals 17, Y^* equals 3.24. The probability of a chi-square random variable with one degree of freedom being less than 3.24 is found from Table 2 to equal about .92, using interpolation. Therefore, if the assumption that each toy was equally likely to be chosen is in fact true, such a large deviation from the expected value of 12.5 would occur only about 8% of the time.

Since the experiment in Example 8 and this one were designed for the same purpose, it would seem desirable to be able to combine the results in some way. If X^* and Y^* may be considered as independent random variables, a reasonable consideration here, Theorem 3 may be used to combine X^* and Y^* as

$$W = X^* + Y^*$$

and the distribution function of W may be approximated by the chi-square distribution function with two degrees of freedom. Then

$$W = 2.56 + 3.24$$
$$= 5.80$$

The probability of a chi-square random variable with two degrees of freedom being greater than 5.80 is only about .06, which was obtained by interpolation in Table 2.

In this example more information concerning the presence of color preference among children was obtained by combining the information gained in the two studies.

It should be noted that if Y had been defined as the number of children preferring the blue telephone, instead of the way Y was defined in this example, Y^* would still have the same value, because the deviation of Y from the mean was squared, eliminating the directional influence of the difference.

In Example 9, two approximate chi-square random variables were added, and their sum was an approximate chi-square random variable with two degrees of freedom. This method of combining independent chi-square random variables is valid in general. More discussion of this method is given by Radhadrishna (1965) and Nelson (1966).

The following theorem may be found on p. 194 of Freund (1962).

Theorem 4.   *Let $X_1, X_2, \ldots, X_n$ be independent chi-square random variables with $k_1, k_2, \ldots, k_n$ degrees of freedom respectively. Let Y equal the sum of the X_i. Then Y is a chi-square random variable with k degrees of freedom, where*

$$k = k_1 + k_2 + \ldots + k_n$$

Theorem 4 will be used later in this book to approximate the distribution function of the sum of several random variables, where the random variables may be assumed to be independent and to be distributed approximately as chi-square random variables.

Since a chi-square random variable with k degrees of freedom may be considered to be the sum of k independent and identically distributed random variables, each having the chi-square distribution with one degree of freedom, the set A of conditions on the Central Limit Theorem are met. The mean and variance of a chi-square random variable with k degrees of freedom are given by (5) and (6) to be k and $2k$, respectively. Therefore, if Y is a chi-square random variable with k degrees of freedom, the distribution function of

$$(8) \qquad Z = \frac{Y - k}{\sqrt{2k}}$$

may be approximated by the standard normal distribution function. From Theorem 1, if x_p is a quantile from Table 1, the quantile y_p for Table 2 may be approximated, for large k, by

$$(9) \qquad y_p = k + \sqrt{2k}\, x_p$$

This is not as good as the approximations

$$(10) \qquad y_p = \tfrac{1}{2}(x_p + \sqrt{2k - 1})^2$$

or

$$(11) \qquad y_p = k\left(1 - \frac{2}{9k} + x_p\sqrt{\frac{2}{9k}}\right)^3$$

given at the bottom of Table 2.

PROBLEMS

1. Suppose that X is the amount of time (in minutes) it takes a certain high school athlete to run one mile. Assume that X has the normal distribution with mean 4.30 and standard deviation .05. What is the probability that the athlete will break the school record of 4.15 minutes at the annual track meet?

2. Let X be the number of policyholders who make at least one claim to a large insurance company. Assume that there are 2000 policyholders, and each one has probability .2 of making at least one claim during the year. What is the probability that no more than 500 policyholders will make claims during any given year?

3. If the distribution of weights of a certain class of individuals is approximately normal with mean 160 and variance 400, how high should a set of bathroom scales be calibrated so that about 99% of the people will be able to weigh themselves?

4. If Y is a binomial random variable with parameters $n = 60$, $p = .5$, find the probability that the random variable

$$\frac{(Y - np)^2}{npq}$$

will exceed 5.0.

5. If X, Y, and Z are independent chi-square random variables with 3, 2, and 3 degrees of freedom, respectively, find the probability that W will exceed 15, where $W = X + Y + Z$.

6. If X is a chi-square random variable with 100 degrees of freedom,
   (a) Find the exact quantile $x_{.950}$ from Table 2.
   (b) Approximate $x_{.950}$ using (9).
   (c) Approximate $x_{.950}$ using (10).
   (d) Approximate $x_{.950}$ using (11).

7. Let X be a binomial random variable with parameters $n = 100, p = .3$. Estimate

$$P(20 \leq X \leq 40)$$

   (a) using Table 1
   (b) using Table 2

CHAPTER 2

Statistical Inference

2.1. POPULATIONS, SAMPLES, AND STATISTICS

Much of our knowledge concerning the world we live in is the result of samples. We eat at a restaurant once and we form an opinion concerning the quality of the food and service at that restaurant. We know twelve people from England and we feel we know the English people. Quite often the opinions we form from the sample are not accurate. However, in most cases, the opinions are more accurate than if no sample had been observed. And usually the larger the sample, the more accurate the opinion.

Our process of forming opinions may be placed within the framework of an investigation. To investigate the quality of a restaurant we eat there once. To investigate (or study) the English people we recall our experiences with English people.

We shall refer to the collection of all elements under investigation as the *population*. A *sample* is a collection of some elements of a population. Scientific investigations are often concerned with obtaining information about some population. Suppose a psychologist wishes to study the effect of constantly interrupted sleep on the emotional balance of a person. He might consider the population to be all human beings of contemporary times. To conduct his experiment he uses paid volunteers, obtained through an ad in a college newspaper. He can hardly consider his subjects to be representative of the population because they are all college students, at one university, in a rather restricted age group, and possessing an emotional makeup which prompts them to reply to an ad in a newspaper and volunteer for a somewhat personal study. And yet he is forced to use this type of sample for his experiment, for practical reasons such as limited funds and limited time available for research, or to abandon his experiment entirely. Thus it is advisable to speak of two populations, the population under investigation and the population actually sampled. The population about which information is wanted is called the *target population*. The population to be sampled is called the *sampled population*. Our example considered all contemporary human beings as being the target population, and all human beings who responded to the ad as being the sampled population. All experimenters must necessarily work with the sampled population, and the validity of their experiment rests on the assumption that the sampled population is similar to the target population, at least with respect to the properties under investigation.

In order to obtain accurate information about a population it would seem desirable to examine every element in that population. Usually this is impossible or impractical. And so only a sample from that population is observed. The sample may consist of those elements that are easily accessible, such as the citizens of England who are known to the observer. The sample may consist of a haphazard selection of elements from the population, such as the names of people obtained from a mailing list. Perhaps only "typical" elements of the population are selected for study; that is, elements that appear to be average or nearly average. Experimentation that involves discomfort or inconvenience often relies on a sample of volunteers. None of these four methods of obtaining a sample permits the use of statistical techniques to aid in making inferences about the population, because they do not result in a random sample. Usually we assume that the sample is random even if it isn't, but it is much better actually to have a random sample. A random sample may be obtained by numbering all of the elements of the population from 1 to N, and then drawing n numbers in a random manner. The n numbers drawn then correspond to the n elements in the population that are to be included in the sample. The statistical methods presented in this book usually assume that the sample is a random sample, so it is well to discuss the idea of a random sample.

If the population has a finite number of elements, the following definition of random sample is appropriate.

Definition 1.   A sample from a finite population is a *random sample* if each of the possible samples was equally likely to be obtained.

The definition may seem a little strange, in that the term "random" does not really refer to the sample itself, but to the method by which the sample was obtained. In fact, we cannot look at a sample to see if it is a random sample or not. Instead we look at the means by which the sample is obtained. If the finite population has N elements total, then, as seen in Section 1.2, there are $\binom{N}{n}$ possible samples of size n if the sample is obtained without replacement. If the sampling is with replacement there are N^n possible samples. If each of these possible samples is equally likely to be obtained, then the method of sampling is considered to be random, and the resulting sample is a random sample.

The above definition of a random sample seems to be satisfactory for most situations where the population is finite. But suppose we are examining the number of dreams a certain individual has in one night. We think of a "random sample" in this case as the number of dreams he has in one night, and the number he has another night, and so on for, say, seven nights. Even under ideal conditions the sampling method does not fit into the framework of Definition 1, with a concept of "equally likely." What is equally likely? Not the individual, because presumably we are studying only the individual, not a representative of some population (although this may be the ultimate objective in the back of our minds). Are we to select the nights for study in some equally likely fashion out of the remaining nights that the individual can expect to be alive? Clearly this is impossible. We may conclude that at least one more definition of "random sample" is needed.

The definition of "random sample," which is standard among mathematical statisticians, is the following.

Definition 2.   *A random sample of size n is a sequence of n independent and identically distributed random variables $X_1, X_2, \ldots, X_n$.*

This definition requires an explanation. First, what we call a random sample in Definition 2 is called a "simple random sample" by many authors. We shall not make a distinction between the two expressions.

Second, each random variable in Definition 2 may actually be a multivariate random variable. That is, X_i may really represent the k-variate random variable $(Y_{i1}, Y_{i2}, \ldots, Y_{ik})$, where the X_i's are still independent and identically distributed, but where the individual Y_{ij} random variables within each X_i may or may not be independent and/or identically distributed. As an example, consider the "dream" experiment described above. The random variable X_i could be the number of dreams counted during the ith night of observation. Then it may not be too unreasonable to assume that the X_i's are independent, as defined by Definition 1.4.11, and identically distributed (meaning each X_i has the same distribution function). But suppose that each night the experimenter records not only the total number of dreams, but also the total amount of sleep, which we shall call Y_{i1} and Y_{i2} respectively. The number of dreams and the length of sleep during any one night may be related variables, so Y_{i1} and Y_{i2} are probably not independent. However, the sleep pattern on one night may be independent of the sleep pattern on another night. Mathematically, this means that the joint probability function of Y_{i1}, Y_{i2}, Y_{j1}, Y_{j2} may be factored as follows,

$$(1) \qquad f(y_{i1}, y_{i2}, y_{j1}, y_{j2}) = f_1(y_{i1}, y_{i2}) f_2(y_{j1}, y_{j2})$$

where f_1 and f_2 are the joint probability functions of (Y_{i1}, Y_{i2}) and (Y_{j1}, Y_{j2}), respectively. If the joint probability distribution of the sleep patterns does not change from one night to the next, then f_1 is identical with f_2, and we say that (Y_{i1}, Y_{i2}) and (Y_{j1}, Y_{j2}) are identically distributed. A more convenient method of expressing the facts of "between" independence but not necessarily "within" independence, and "between" identical distributions but not necessarily "within" identical distributions, is to let X_i represent both Y_{i1} and Y_{i2} jointly. X_i is called a *bivariate random variable*, and a "value" of X_i actually consists of two numbers, one for Y_{i1} and one for Y_{i2}. Then all of the above statements may be summarized by saying, "The X_i's are independent and identically distributed."

In a similar manner we may consider k measurements being taken each night, and the resulting k random variables $Y_{i1}, Y_{i2}, \ldots, Y_{ik}$ being represented by X_i, which is called a *k-variate random variable*, or also a *multivariate random variable*. Then independence of the X_i's, in the sense of Definition 1.4.11, means that the joint probability distribution of all of the Y_{ij}'s may be factored into the product of n joint probability functions, each being the joint probability function of $Y_{i1}, Y_{i2}, \ldots, Y_{ik}$ for some i. Identically distributed X_i's means that the joint probability functions just mentioned are identical functions.

The third point of explanation concerning Definition 2 is that even though a random variable is a function that assigns real numbers to the outcomes of the experiment (Definition 1.4.1), complete knowledge of those real numbers is not always necessary in order to use nonparametric statistical methods. This is particularly nice when complete knowledge of those real numbers is not available. An experimental subject who has performed four tasks can usually arrange the tasks in order, from "most difficult" to "least difficult." However, it may be unrealistic for the subject to assign a number to the task, where the number represents degree of difficulty. Yet we might speak of the random variable X as measuring degree of difficulty. The random variables which constitute the random sample may be of this type. More will be said concerning various types of measurements later in this section.

We now have two definitions of the expression "random sample." The first definition applies only to samples from a finite population, and may be directly related to the sample space. If each possible sample (of size n) is represented by one point in the sample space, and if each point in the sample space has equal probability of being selected as the sample, then the sampling method is random, and the resulting sample is a random sample. The concepts of sample space and probability function are used in that definition, but there is no mention, explicit or implicit, of a random variable.

Example 1.   A psychologist would like to obtain four subjects for individual training and examination. He advertises and twenty volunteers respond. He has several ways of selecting a sample of four from his sampled population of size twenty.

He might select the first four to volunteer. Thus he may be biasing his selection toward those volunteers who tend to be more prompt or aggressive. This is probably not a random sample.

He might adhere strictly to Definition 1, and consider that there are $\binom{20}{4} =$ 4845 ways in which a sample of size four may be selected. Then he obtains 4845 pieces of paper that are identical and writes four names on each piece of paper, a different combination each time, and puts them in a basket. One slip is randomly drawn, and those four people are used. This is a random sample, but such a psychologist would need a psychiatrist.

Another way of obtaining a random sample would be to write each of the names on a slip of paper, twenty slips in all, and one by one draw four slips in some random manner, such as from a hat. This method also satisfies the definition of a random sample.

The second definition of random sample is concerned directly with random variables, and does not mention the sample space. However, since a random variable is a function defined on a sample space, a sample space is implicitly involved, although it remains in the background. Also, as we mentioned in Section 1.4, the set of possible values of a random variable resembles a sample space. At times it will be necessary to list the points in this pseudo sample space in order to solve statistical problems that may arise. In fact, often no confusion will result if the possible measurements themselves (the values assumed by the random variables) are considered to be the points in the sample space. We usually think of these measurements as being numbers, but sometimes the numerical values of the measurements are obscure, as we mentioned earlier in this section. So it would be well to discuss the various types of measurements.

The types of measurements are usually called *measurement scales*, and are discussed at some length in various publications, including an excellent paper by Stevens (1946). We shall proceed from the "weakest" scale of measurement, the nominal scale, through the ordinal scale and the interval scale to the "strongest" scale, the ratio scale.

The *nominal scale* of measurement uses numbers merely as a means of separating the properties or elements into different classes or categories. The number assigned to the observation serves only as a "name" for the category to which the observation belongs, hence the title "nominal." We used the nominal scale of measurement when we defined a random variable which equaled 1 if a coin landed as a "head," and 0 if the coin landed "tail." We could, just as appropriately, have used the numbers 7.3 and 3.9 to represent head and tail respectively. Our choice of 1 and 0 was primarily for convenience when we later desired to count the total number of heads in several tosses of the coin. When twelve subjects are arbitrarily numbered 1 through 12, a nominal scale of measurement is being used, and the assignment of the numbers is a form of random variable. When classifying objects according to color, the categories may be labeled 1, 2, 3, or blue, yellow, red, or A, B, C. The numbers merely serve as category names. The numbers may be replaced by other unused numbers, as long as the categories remain intact.

The *ordinal scale* of measurement refers to measurements where only the comparisons "greater," "less," or "equal" between measurements are relevant. The numeric value of the measurement is used only as a means of arranging the elements being measured in order, from the smallest to the largest. It is this need to "order" the elements, on the basis of the relative size of their measurements, that gives the name to the "ordinal" scale. If some of the elements have equal measurements, we say "ties" exist, and the ordering is no longer unique. For many statistical analyses a unique ordering is desired, and so it is advisable to exercise sufficient care in measurement so that the number of ties is minimized wherever possible. When a person is asked to assign the number 1 to the most preferred of three brands, the number 3 to the least preferred, and the number 2 to the remaining brand, he is using an ordinal scale of measurement, and he is using the numbers merely as a convenient way of representing his order of preferences. Instead of the numbers 1, 2, 3, he could have used any three numbers, say 16, 20, 75, as long as the numbers are assigned to the brands in such a way that the relative order of the number represents the relative preference of the brand.

The third scale, the *interval scale* of measurement, considers as pertinent information not only the relative order of the measurements as in the ordinal scale, but also the size of the interval between measurements, that is, the size of the difference (in a subtraction sense) between two measurements. The interval scale involves the concept of a unit distance, and the distance between any two measurements may be expressed as some number of units. A good example is the scale by which we usually represent temperature. One unit (degree) increase in temperature is defined by a particular change in volume of mercury in a thermometer, and consequently the difference between any two temperatures may be measured in units, or degrees. The actual numerical value of the temperature is merely a comparison with an arbitrary point called "zero degrees." The interval scale of measurement requires a zero point as well as a unit distance (it is not possible to have the latter without the former) but it is not important which measurement is declared to be "zero" or which distance is defined to be the unit distance. Temperature has been measured quite adequately for some time by both the Fahrenheit and the Centigrade scales, which have different "zero" temperatures and different definitions of "one degree" or unit. The principle of interval measurement is not violated by a change in scale or location or both.

Finally, the *ratio scale* of measurement is used when not only the order and interval size are important, but also the ratio between two measurements is meaningful. If it is reasonable to speak of one quantity being "twice" another quantity, then the ratio scale is appropriate for the measurement, such as when measuring crop yields, distances, weights, heights, income, and so on. Actually the only distinction between the ratio scale and the interval scale is that the ratio scale has a natural measurement which is called zero, while the zero measurement is defined arbitrarily in the interval scale. As in the interval scale, the "unit distance" of the ratio scale is arbitrarily defined.

It is not possible to look at the measurements themselves in order to tell which scale of measurement is appropriate. Rather, one looks at the quantities being measured, and the method of measurement, and then determines the amount of meaning which may be attached to the numeric value of the measurement.

Most of the usual parametric statistical methods require an interval (or stronger) scale of measurement. Most nonparametric methods assume either the nominal scale or the ordinal scale to be appropriate. Of course, each scale of measurement has all of the properties of the weaker measurement scales, and therefore statistical methods requiring only a weaker scale may be used with the stronger scales also.

Thus far in this section we have been concerned with populations, samples from populations, and measurement scales for measuring sample properties of interest. Measurement scales relate to random variables, because a system for measuring elements of the sample is in reality a random variable. Therefore, measurement scales relate to statistics, because a statistic is a random variable. To a mathematical statistician the term "statistic" is interchangeable with the term "random variable." But popular usage of the word statistic indicates that it is more than just a random variable.

The word "statistic" originally referred to numbers published by the "state," where the numbers were the result of a summarization of data collected by the government. And so some people think of a statistic as a number that is based on several numbers, such as the average of several numbers in a sample, the proportion of a population that is in a particular category, and so on. In this sense a statistic is just a number. However, if we stop to consider that the numbers being averaged may vary from one sample to the next, or the population may change from one year to the next, then we can justify extending our idea of a statistic from being only a number to being a rule for finding the number. Then "the average of the numbers in the sample" is the statistic, and the actual average obtained in one sample is a value of the statistic. As a rule for obtaining a number, a statistic meets the requirements of being a random variable, a function which assigns numbers to the points in the sample space (for an appropriately defined sample space). A statistic also conveys the idea of a summarization of data, so usually a statistic is considered to be a random variable which is a function of several other random variables. Then a value assumed by the statistic is implicitly assumed to be the result of some arithmetic operations performed on other numbers (the data) which in turn are the values assumed by several random variables. Since a random variable is a function defined on a sample space, a statistic may be defined as a function defined on a special sample space, a sample space whose points are the possible values of an n-variate random variable. A formal definition and an example may clarify the concept.

Definition 3. A *statistic* is a function which assigns real numbers to the points of a sample space, where the points of the sample space are possible values of some multivariate random variable. In other words, a *statistic* is a function of several random variables.

Each sentence in Definition 3 would suffice as a definition of "statistic." Both sentences are included for clarity.

Example 2. Let $X_1, X_2, \ldots, X_n$ represent test scores of n students. Then each X_i is a random variable. Let W equal the average of the test scores.

$$(2) \qquad W = \frac{X_1 + X_2 + \ldots + X_n}{n} = \frac{1}{n} \sum_{i=1}^{n} X_i$$

Then W is a statistic. If $X_1 = 76$, $X_2 = 84$, and $X_3 = 85$ represent the scores of three students, then W equals $(1/3)(76 + 84 + 85) = 81\frac{2}{3}$. The statistic W satisfies the second sentence in Definition 3 by being a function of the random variables $X_1, X_2, \ldots,$ and X_n, and also the first sentence in Definition 3 is satisfied because W assigns real numbers to the values of the multivariate random variable $(X_1, X_2, \ldots, X_n)$. In this case, if (X_1, X_2, X_3) assumes the multivariate value (76, 84, 85), then W assumes the value $81\frac{2}{3}$, as shown above. This particular statistic is used often in statistics (the science). It is called the "sample mean," and will be discussed further in the next section.

We shall often have occasion to use a particular class of statistics called *order statistics*, particularly when we are dealing with ordinal type measurements. Suppose an observation $(x_1, x_2, \ldots, x_n)$ on a multivariate random variable $(X_1, X_2, \ldots, X_n)$ is "ordered"; that is, the elements are arranged from smallest to largest. We shall denote the *ordered observation* by $x^{(1)} \leq x^{(2)} \leq \ldots \leq x^{(n)}$.

Definition 4. The *order statistic of rank k, $X^{(k)}$* is the statistic that takes as its value the kth smallest element $x^{(k)}$ in each observation $(x_1, x_2, \ldots, x_n)$ of $(X_1, X_2, \ldots, X_n)$.

Therefore $X^{(1)}$, the order statistic of rank 1, always takes the smallest element in $(x_1, x_2, \ldots, x_n)$ as its value, and $X^{(n)}$ takes the largest. In Example 2, $X^{(1)}$ equals 76, $X^{(2)}$ equals 84, and $X^{(3)}$ equals 85. If another observation on (X_1, X_2, X_3) yields $(93, 73, 81)$, then the values of the order statistics are $X^{(1)} = 73$, $X^{(2)} = 81$, and $X^{(3)} = 93$. If $(X_1, X_2, \ldots, X_n)$ is a *random sample*, sometimes $(X^{(1)} \le X^{(2)} \le \ldots \le X^{(n)})$ is called the *ordered random sample*.

The next section introduces many other useful statistics. There we shall further discuss some uses of statistics in the analysis of experimental results.

PROBLEMS

1. A congressional committee wishes to examine the effect of proposed legislation on the nation's high schools. Therefore it randomly selects five high schools from the Washington D.C. area and conducts a study on those five schools.
   (a) What is the target population?
   (b) What is the sampled population?

2. A Topeka television station asks the question, "Should liquor by the drink be allowed in Kansas," and reported 372 phone calls, of which 164 said "no" and the remainder said "yes."
   (a) What was the target population?
   (b) What was the sampled population?
   (c) Three statistics were indicated above. What were they, and what numerical values did they assume?
   (d) Which measurement scale was used in the voting counts?

3. A certain track meet awards a trophy to the team that accumulates the most points. A team receives 5, 3, or 1 points each time a member of that team finishes first, second, or third, respectively, in competition.
   (a) What measurement scale is used in awarding points?
   (b) Which statistic is (implicitly) mentioned, and what is it used for?

4. Football players on a team wear numbers on their uniforms. What measurement scale do those numbers represent?

5. What measurement scale is used when time is measured using a
   (a) calendar?
   (b) clock?
   (c) stop watch?

6. If a random sample of size four is to be selected from among the integers one through seven, without replacement,
   (a) What is the total number of possible samples?
   (b) What is the probability of each sample?

(c) What is the probability that the sample has at least one odd number?

(d) What is the probability that the numbers in the sample sum to 12?

7. In order to select a law firm at random, a list of all lawyers of that city was obtained, and a lawyer was selected at random. The law firm to which the lawyer belonged was the selection. Was the law firm selected at random?

8. An experiment consists of n independent rolls of an unbalanced die. Let X_i be the number of spots showing on the ith roll. Does $X_1, X_2, \ldots, X_n$ constitute a random sample?

2.2.   ESTIMATION

One of the primary purposes of a statistic is to estimate unknown properties of the population. The unknown properties that may be estimated are necessarily numerical, and include such items as unknown proportions, means, probabilities, and so on. Actually the estimate is based on a sample, a random sample if probability statements are to be made, and the estimate is an educated guess concerning some unknown property of the probability distribution of a random variable, where that random variable represents some quantity of interest in the population. For example, we might use the proportion of defective items in a sample of radio tubes as a statistic to estimate the unknown proportion of defective radio tubes in some population of radio tubes. A statistic that is used to estimate is called, quite naturally, an *estimator*. In this section we shall discuss estimators such as the sample mean, the sample variance, and the sample quantiles. But first we shall introduce the *empirical distribution function*, an estimator of a somewhat different kind.

The true distribution function of a random variable is almost never known. Sometimes we make an educated guess at the form of the distribution function, and use our guess as an approximation to the true distribution function. One way of making a good guess is by observing several values of the random variable, and then constructing a graph $S(x)$ which may be used as an estimate of the entire unknown distribution function $F(x)$ of the random variable. The method of constructing the graph is best explained by an example, which follows this definition.

Definition 1.   Let $X_1, X_2, \ldots, X_n$ be a random sample. The *empirical distribution function* $S(x)$ is a function of x which equals the fraction of X_i's which are less than or equal to x for each x, $-\infty < x < \infty$.

Example 1.   In a physical fitness study five boys were selected at random from the boys in a certain high school. They were asked to run a mile, and the time it took each of them to run the mile was recorded. The times (converted to fractions of a minute) were 6.23, 5.58, 7.06, 6.42, 5.20, and are represented on the horizontal axis in Figure 1. The empirical distribution function $S(x)$ is the

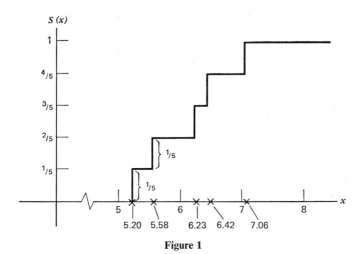

Figure 1

number of sample values less than or equal to x, and for this particular sample is represented graphically in Figure 1.

As in Example 1, the empirical distribution function is always a step function, where each step is of height $1/n$ and occurs only at the sample values. The vertical lines in Figure 1 are not part of the empirical distribution function, but are included partly for appearance and partly for later convenience in determining sample quantiles. As we look at the graph of the empirical distribution function from left to right we see that $S(x)$ equals zero until x equals the smallest value in the sample. Then $S(x)$ takes a step of $1/n$ in height. At each of the n sample values $S(x)$ rises in height another distance of $1/n$. At the largest of the sample values $S(x)$ reaches a height of 1.0 and remains 1.0 for all larger values of x. $S(x)$ resembles a distribution function in that it is a nondecreasing function that goes from zero to one in height. However, $S(x)$ is empirically (from a sample) determined and therefore its name.

Figure 1 represents merely one observation on $S(x)$. Another sample would have produced another and probably different graph of $S(x)$. This points out the random nature of $S(x)$. In a sense it is a random variable, but since it is a function, and its observed values are entire graphs rather than single numbers, $S(x)$ is more properly called a *random function*. It is used as an estimator, since it does a reasonably good job of estimating the distribution function of the random variable, which we shall call the population distribution function in order to distinguish it from the empirical (or "sample") distribution function.

In a sense, the observed value of an empirical distribution function may be considered a population distribution function. More precisely, an observed value of $S(x)$, based on the observations $x_1, x_2, \ldots, x_n$ in the sample, is identical to the distribution function of a random variable which may assume any of the numbers $x_1, x_2, \ldots, x_n$, each with probability $1/n$. The distribution function of such a random variable is a step function with jumps of height $1/n$ at each of the n numbers $x_1, x_2, \ldots, x_n$. We could find the mean, variance, and quantiles of the random variable simply by using the definitions of Chapter 1.

Example 2.   The random variable which has a distribution function identical to the function $S(x)$ of Example 1, is the random variable X with the following probability distribution.

$$P(X = 5.20) = .2$$
$$P(X = 5.58) = .2$$
$$P(X = 6.23) = .2$$
$$P(X = 6.42) = .2$$
$$P(X = 7.06) = .2$$

The graph of the distribution function of X is the same as the graph in Figure 1. The median of X is 6.23 by Definition 1.5.1. The mean of X, by Definition 1.5.3, is given by

$$E(X) = \sum_x x f(x)$$

$$= (5.20)(.2) + (5.58)(.2) + (6.23)(.2) + (6.42)(.2)$$
$$+ (7.06)(.2)$$

(1) $= 6.098$

Similarly the variance of X may be found from Definition 1.5.4,

$$\text{Var}(X) = \sum_x (x - E(X))^2 f(x)$$

(2) $= .424$

The mean, variance, and quantiles obtained from the sample, as illustrated in Example 2, will be called the *sample mean*, *sample variance*, and *sample quantiles* to distinguish them from the true "population" mean, variance, and quantiles. In the same way that the empirical distribution function serves as an estimator of the population distribution function, the sample mean, variance, and quantiles may be used as estimators of their population counterparts.

Definition 2.   Let $X_1, X_2, \ldots, X_n$ be a random sample. The *pth sample quantile* is that number Q_p which satisfies the two conditions:

(a) The fraction of the X_i's which are less than Q_p is $\leq p$.

(b) The fraction of the X_i's which exceed Q_p is $\leq 1 - p$.

The sample quantile may be found from the empirical distribution function in exactly the same way that the population quantile is obtained from the population distribution function. The pth sample quantile is that value of x where $S(x)$ equals p. If more than one number satisfies the condition that $S(x) = p$, we adopt the convention of using the average of the largest and the smallest numbers that satisfy $S(x) = p$, as we did with the population quantiles. The sample quantile Q_p depends on the random sample for its values; therefore it is a statistic. Note that for simplicity we have defined sample quantiles only for random samples.

Example 3. Six married women were selected at random from among the married women in a ladies civic club, and the number of children belonging to each was recorded. These numbers were 0, 2, 1, 2, 3, 4. The empirical distribution function is given in Figure 2. The sample median $Q_{.5}$ is 2. The sample quartiles $Q_{.25}$ and $Q_{.75}$ are 1 and 3 respectively. The 1/3 sample quantile $Q_{1/3}$ is the average of 1 and 2 by our convention, which equals 1.5. These numbers are our estimates of the unknown population quantiles.

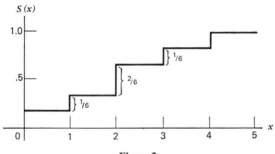

Figure 2

The sample mean and sample variance may be found in a simpler manner than in Example 2, by noting that $f(x)$ equals $1/n$ in Equations (1) and (2), and may be factored out of the summation. This leaves us with simpler computation methods, which are given in the following definitions.

Definition 3. Let $X_1, X_2, \ldots, X_n$ be a random sample. The *sample mean* $\bar{X}$ is defined by

$$(3) \qquad \bar{X} = \frac{1}{n} \sum_{i=1}^{n} X_i$$

The *sample variance* S^2 is defined by

$$(4) \qquad S^2 = \frac{1}{n} \sum_{i=1}^{n} (X_i - \bar{X})^2$$

which is equivalent to

(5)
$$S^2 = \frac{1}{n} \sum_{i=1}^{n} X_i^2 - \bar{X}^2$$

The *sample standard deviation* S is the square root of the sample variance.

Example 4.   In the random sample 0, 2, 1, 2, 3, 4 of Example 3 the sample mean is

$$\bar{X} = \tfrac{1}{6}(0 + 2 + 1 + 2 + 3 + 4)$$

(6) $= 2$

and the sample variance is

$$S^2 = \tfrac{1}{6}(2^2 + 0 + 1^2 + 0 + 1^2 + 2^2)$$

(7) $= 1\tfrac{2}{3}$

Therefore our estimate of the unknown mean is 2 and our estimate of the unknown variance is $1\tfrac{2}{3}$.

The estimators introduced thus far provide a "point estimate" of the unknown population quantity, with the possible exception of the empirical distribution function. That is, our estimate of the unknown mean in the preceding example was provided by the statement, "Our estimate of the mean is 2." The single point "2" is the estimate. It is usually preferred, but more difficult, to state the estimate as follows, "We are ninety-five percent confident that the unknown mean lies between 1.3 and 2.7." Such an estimate is called an *interval estimate*. An *interval estimator* consists of two statistics, one for each end of the interval, and the *confidence coefficient* is the probability that the unknown population quantity will lie within its interval estimate. The confidence coefficient in the above statement is .95. The interval estimator and the confidence coefficient together are usually called a *confidence interval* for the unknown quantity.

Point estimation is easy. To make a point estimate we need only to think of a number, any number. However, some point estimators are much better than others. Criteria for comparing point estimators, in order to determine which estimator we prefer, may be found in almost any introductory text in probability or statistics. We shall not discuss them here.

In a sense, point estimation is always a nonparametric statistical method, because no knowledge of the form of the unknown distribution function is required in order to make a point estimate. This was shown by the examples in this section, where point estimates were made without knowing anything about the unknown distribution function.

It is more difficult to tell whether the methods of forming confidence intervals are parametric or nonparametric. If no knowledge of the form of the distribution function is required in order to find a confidence interval, then that method is clearly nonparametric. If the method requires that the unknown distribution function be continuous, the method is still nonparametric. On the other hand, if the method requires that the unknown distribution function be a normal distribution function (see Definition 1.6.3), or some other specified form, then the method is parametric. Several nonparametric methods of forming confidence intervals will be presented later, in Sections 3.1, 3.2, 5.2, 5.4, 6.1, and 7.1.

PROBLEMS

1. Ten persons were selected at random from among all persons living in a particular community. Their taxable incomes for the previous calendar year were $4300, $7600, $8100, $8200, $14,800, and no income for the other five people.
   (a) Draw a graph of the empirical distribution function.
   (b) Estimate the population median income.
   (c) Estimate the population mean income.
   (d) Estimate the population variance.
   (e) Estimate the population standard deviation.

2. In five consecutive games a certain basketball team had scores of 73, 68, 86, 78, and 65.
   (a) Draw a graph of the empirical distribution function.
   (b) Estimate the upper quartile.
   (c) Estimate the interquartile range.
   (d) Estimate the population mean.

3. Using the same procedure for finding point estimators used in this section, find a point estimator for the probability $P(Y \leq c)$, for a given number c, based on a random sample $X_1, X_2, \ldots, X_n$ with the same distribution function as Y. In other words, if $X_1, X_2, \ldots, X_n$ is a random sample with the distribution function $F(x)$, estimate $F(c)$. Use your estimator to estimate the probability that the score in the next game will exceed 80, in Problem 2.

4. Using the same procedure for finding point estimators used in this section, find a point estimator for the range of a random variable. Will this sample range ever be larger than the population range? Will it ever be smaller? Is the expected value of the sample range smaller than the population range?

5. Since an estimator is a random variable, given enough information we should be able to find the probability distribution of an estimator. Suppose a finite population consists of four elements, with the respective measurements 4, 6, 7, and 10. A random sample of size two is drawn without replacement from the population.
   (a) How many possible random samples are there?
   (b) List the possible samples.
   (c) What is the probability of drawing each of the samples listed in (b)?
   (d) What is the sample median for each of the samples listed in (b)?
   (e) What is the probability of getting each of the sample medians listed in (d)?
   (f) Graph the distribution function of the sample median.

(g) Using the same procedure that was used for finding the distribution function of the sample median, find the distribution function of the sample mean.

2.3. HYPOTHESIS TESTING

Statistical inference has many forms. The form that has received the most attention by the developers and users of nonparametric methods is called hypothesis testing, and is treated in this section and the next.

Hypothesis testing is the process of inferring from a sample whether or not to accept a certain statement about the population. The statement itself is called the hypothesis. Examples of hypotheses include such statements as these.

1. Women are more likely than men to have automobile accidents.
2. Nursery school helps a child achieve better marks in elementary school.
3. The defendant is guilty.
4. Toothpaste A is more effective in preventing cavities than toothpaste B.

In each case the hypothesis is tested on the basis of the evidence contained in the sample. The hypothesis is either *rejected*, meaning the evidence from the sample casts enough doubt on the hypothesis for us to say with some degree of confidence that the hypothesis is false, or else the hypothesis is *accepted*, simply meaning that it is not rejected.

A test of a particular hypothesis may be very simple to perform. We may observe a set of data related to the hypothesis, or a set of data not related to the hypothesis, or perhaps no data at all, and arrive at a decision to accept or reject the hypothesis, although that decision may be of doubtful value. However, the type of hypothesis test we shall discuss is more properly called a statistical hypothesis test, and the test procedure is well defined. Here is a brief outline of the steps involved in such a test.

1. The hypotheses are stated in terms of the population.
2. A test statistic is selected.
3. A rule is made, in terms of possible values of the test statistic, for deciding whether to accept or reject the hypothesis.
4. On the basis of a random sample from the population, the test statistic is evaluated, and a decision is made to accept or reject the hypothesis.

A more precise description of the testing procedure follows this example.

Example 1.   A certain machine manufactures parts. The machine is considered to be operating properly if five percent or less of the manufactured parts are defective. If more than five percent of the parts are defective the machine needs remedial attention. The *null hypothesis*

$$H_0: \text{The machine is operating properly}$$

is the hypothesis to be tested. The *alternative hypothesis* is

$$H_1: \text{The machine needs attention}$$

H_0 will be tested on the basis of a random sample of ten parts, from the population of all parts being produced by the machine. The assumption is made that each part has the same probability p of being defective, independently of whether or not the other parts are defective. Therefore, in the assumed model, the original hypotheses H_0 and H_1 are equivalent to

$$H_0: p \leq .05$$

$$H_1: p > .05$$

We feel that if too many parts are defective, we should reject H_0. So let the test statistic T be the total number of defective items. Then, according to Example 1.4.5, T has the binomial distribution with parameters p, and 10 for n. From Table 3 we see that if H_0 is true ($p \leq .05$), then

(1) $$P(T \leq 2) \leq .9885$$

equaling .9885 if $p = .05$, and therefore

(2) $$P(T > 2) \leq .0115$$

equaling .0115 if $p = .05$. We decide to reject H_0 if T exceeds 2. The set of points in the sample space which correspond to values of T greater than 2 is called the *critical region*. Because the probability of getting a point in the critical region, when H_0 is true, is quite small (less than .0115) the decision rule is this: Reject H_0 if the observed outcome is in the critical region (when T exceeds 2); otherwise, accept H_0.

The procedure used in Example 1 will now be carefully examined. The hypothesis to be tested is called the *null hypothesis* and is denoted by H_0. The *alternative hypothesis*, denoted by H_1, is the negation of the null hypothesis, and usually consists of a statement equivalent to saying "H_0 is not true." As we mentioned earlier, the decision to reject H_0 is equivalent to the opinion "H_0 is false," and is equivalent to acceptance of H_1, or the opinion "H_1 is true." The decision to accept H_0 is *not* equivalent to the opinion "H_0 is true," but rather represents the opinion "H_0 has not been shown to be false," which could be the result of insufficient evidence. Therefore, if we wish to determine if a statement concerning the population is false, we make it the null hypothesis. If we wish to determine whether a statement is true, we make it the alternative hypothesis. In the example we wanted to determine whether the machine needs attention in the form of inspection and repair, so that statement became the alternative hypothesis.

Then assumptions are made concerning the conditions under which the data are collected, and concerning the type of data collected. These assumptions are tantamount to forming a model, or idealized experiment. "Under the model" means "under these assumptions."

Under the model, the original hypotheses may be restated in an equivalent form, usually using statistical terminology. These hypotheses may be classified as either *simple* or *composite*.

Definition 1.   The hypothesis is *simple* if the assumption that the hypothesis is true leads to only one probability function defined on the sample space.

Definition 2.   The hypothesis is *composite* if the assumption that the hypothesis is true leads to two or more probability functions defined on the sample space.

In the example, the model induces the probability $p^k(1-p)^{10-k}$ on each sample point with k defective items and $10-k$ nondefective items. This represents a whole class of probability functions defined on the sample space, depending on what value p has. (For each point, k is known.) Assume H_0 is true. Still, p may be any value from 0 to .05, so there are several possible probability functions, and H_0 is a composite hypothesis. The same is true for H_1. The hypothesis "$p=.05$" would be a simple hypothesis, because, assuming $p=.05$ is true, the probability function assigns the probability $(.05)^k(.95)^{10-k}$ to a point representing k defective parts, and that probability function is well defined (no unknown parameters) and the only one possible.

Definition 3.   A *test statistic* is a statistic used to help make the decision in a hypothesis test.

A desirable property of a test statistic is that it should assign real numbers to the points in the sample space in such a way that the points are arranged in some order corresponding to their ability to distinguish between a true H_0 and a false H_0. For example, the points that indicate most strongly that the experimenter should reject H_0 might be given large values by the test statistic, and the points that indicate that the experimenter should accept H_0 might be given small values by the test statistic. Then the larger the value assumed by the test statistic, the more the outcome of the experiment indicates that H_0 should be rejected. In this way, all values of the test statistic greater than a certain number might result in the decision to reject H_0. Furthermore, this enables the experimenter to determine objectively how much smaller, or larger, the rejection region might have been and still result in the same decision. Such a test, where the rejection region corresponds to the largest values of the test statistic, is called a *one-tailed test*. Similarly, if the ordering is reversed, so that the rejection region corresponds to the smallest values of the test statistic, the test is still called a one-tailed test. The test in the example was one tailed. If the test statistic is selected so that the largest values of the test statistic and the smallest values of the test statistic, combined, correspond to the rejection region, the test is called a *two-tailed test*, since the rejection region corresponds to both "tails" of the test statistic's possible values.

Definition 4.   The *critical region* is the set of all points in the sample space which result in the decision to reject the null hypothesis.

Sometimes the critical region is called the *rejection region*, and the set of all points in the sample space not in the critical region is called the *acceptance region*, for obvious reasons.

There are two ways of making an incorrect decision in hypothesis testing. If the null hypothesis is true we might make the mistake of rejecting it, thus committing an error known as an *error of the first kind*, or a *type I error*. That is, a type I error occurs when H_0 is true and yet the outcome of our experiment is in the critical region.

Definition 5.   A *type I error* is the error of rejecting a true null hypothesis.

The second way of committing an error in hypothesis testing is by accepting the null hypothesis when the null hypothesis is false. This error is known as an *error of the second kind*, or a *type II error*.

Definition 6.   A *type II error* is the error of accepting a false null hypothesis.

These two error types have associated with them certain probabilities of the errors being made. Consider first the probability of making a type I error.

Definition 7.   The *level of significance*, or α, is the maximum probability of rejecting a true null hypothesis.

The level of significance may be found by first assuming H_0 is true, and then ascertaining the probability of getting a point in the critical region. If H_0 is a simple hypothesis, then the assumption that H_0 is true leads to only one probability function defined on the sample space, and α may be found by adding the probabilities of all points in the critical region. Usually, however, it is easier to find α by computing the probability that the test statistic will assume one of the values that results in rejection of H_0, under the assumption that H_0 is true.

If H_0 is a composite hypothesis, then α is the *maximum* probability of rejecting H_0, where the maximum is obtained by considering all of the probability distributions possible when H_0 is true. In the example H_0 was composite, and the probability of rejecting a true null hypothesis was

$$P(\text{reject a true } H_0) = P(T > 2 \mid H_0 \text{ is true})$$

(3)
$$= \sum_{i=3}^{10} \binom{10}{i} p^i (1 - p)^{10-i}; \qquad p \leq .05$$

which differs for each value of p. However the probability in (3) is a maximum when p is a maximum. The maximum value of p, under H_0, is .05, so the level of significance is given by

$$\alpha = \text{maximum } P(T > 2 \mid H_0 \text{ is true})$$
$$= P(T > 2 \mid p = .05)$$
(4) $$= .0115$$

from Table 3, or from (2).

The level of significance is sometimes called the *size of the critical region*, for obvious reasons. If H_0 is true the maximum probability of rejecting H_0 is α, and therefore the minimum probability of accepting H_0, making the correct decision, is $1 - \alpha$.

The probability of committing an error of the second kind is denoted by β. Obviously it is desirable in hypothesis testing for α and β to be close to zero. In practice the sample size helps determine how small α and β may become. Only when the sample includes all of the information contained in the population may the possibility of error be completely eliminated.

If H_0 is false the decision may be to accept H_0, with a probability β, or to reject H_0, with a probability $1 - \beta$. This latter probability represents the power of the test to detect a false null hypothesis.

Definition 8.   The *power*, denoted by *1-β*, is the probability of rejecting a false null hypothesis.

Unlike α, the power is not always a unique number. If H_1 is simple, then the assumption that H_1 is true (equivalent to "H_0 is false") leads to one probability function, and hence one probability of rejecting H_0, or getting a point in the critical region. Thus when H_1 is simple, $1 - \beta$ is unique. If H_1 is composite, then each probability function, under H_1, has a possibly different value for $1 - \beta$, so the power depends on the various possible probability functions.

The Decision

		Accept H_0	Reject H_0
The true situation	H_0 is true	Correct decision probability $= 1 - \alpha$	Type I error probability $= \alpha$ (level of significance)
	H_0 is false	Type II error probability $= \beta$	Correct decision probability $= 1 - \beta$ (power)

Figure 1

Now that the error types have been discussed, we can return to the topic of the critical region. Although the critical region was discussed, no mention was made concerning how it is selected. If the test statistic has been chosen so as to result in a one- or two-tailed test, the selection of a critical region depends only on the experimenter's preference concerning the size of the critical region, the level of significance. Usually a desirable decrease in the level of significance α is accompanied by an undesirable increase in β. Our two objectives in hypothesis testing are to reject H_0 as seldom as possible if H_0 is true, and as often as possible if H_0 is false. As a result, the critical region is usually the set of points with the largest value of $1 - \beta$, from among those sets of points of some fixed size α. By convention more than any other reason α is usually chosen near .05 or .01, and the critical region is then selected in terms of possible values of the test statistic.

The results of a hypothesis test are much more meaningful if the value of the *critical level* is also stated.

Definition 9.   The *critical level* $\hat{\alpha}$ is the smallest significance level at which the null hypothesis would be rejected for the given observation.

In Example 1, T equaled 4, and $P(T \geq 4 \mid p = .05)$ equals .001, so the critical level is .001. The critical level is also known as the probability level (p-level) and significance level. (See, for instance, Dempster and Schatzoff, 1965.)

Example 2.   In order to see if children with nursery school experience perform differently academically than children without nursery school experience, twelve third-grade students are selected for study, four of which attended nursery school. The hypothesis to be tested is

H_0: The academic performance of third-grade children does not depend on whether or not they attended nursery school.

the alternative hypothesis is

H_1: There is a dependence between academic performance and attendance at nursery school.

The model assumes that the twelve children are a random sample of all third-grade children, and also that the children can be ranked from 1 to 12 (best to worst) academically. The "dependence" in the hypotheses is assumed to mean either the nursery school children tend to do better as a group or they tend to do worse than the nonnursery school children. Under the model the hypotheses may be restated as

H_0: The ranks of the four children with nursery school experience are a random sample of the ranks from 1 to 12.

H_1: The ranks of the children with nursery school experience tend to be higher or lower as a group than a random sample of four ranks out of twelve.

We choose as a test statistic T the sum of the ranks of the four children who attended nursery school. We decide to let the critical region correspond to values of T that are either very large or very small, so the test is two tailed.

Each possible outcome consists of four numbers from 1 to 12, corresponding to the ranks of the four children who attended nursery school. Therefore there are $\binom{12}{4} = 495$ points in the sample space. To decide which of these points to include in the critical region, we shall assume H_0 is true, and keep an eye on α as we decide on the critical region.

If H_0 is true, the ranks of the four children should behave as a *random* sample of four ranks out of the twelve possible. Therefore, each selection of four ranks is equally likely, and so each point in the sample space has equal probability, 1/495. Thus H_0 is a simple hypothesis. Since we decided on a two-tailed test, we examine the points that correspond to high and low values of T. The highest and lowest possible values of T are 42 and 10, corresponding to the points (12, 11, 10, 9) and (1, 2, 3, 4) respectively. Other high and low values of T and their corresponding experimental outcomes are given as follows.

T	Point	T	Point
10	(1, 2, 3, 4)	42	(9, 10, 11, 12)
11	(1, 2, 3, 5)	41	(8, 10, 11, 12)
12	(1, 2, 3, 6)	40	(7, 10, 11, 12)
12	(1, 2, 4, 5)	40	(8, 9, 11, 12)
13	(1, 2, 3, 7)	39	(6, 10, 11, 12)
13	(1, 2, 4, 6)	39	(7, 9, 11, 12)
13	(1, 3, 4, 5)	39	(8, 9, 10, 12)
14	(1, 2, 3, 8)	38	(5, 10, 11, 12)
14	(1, 2, 4, 7)	38	(6, 9, 11, 12)
14	(1, 2, 5, 6)	38	(7, 8, 11, 12)
14	(1, 3, 4, 6)	38	(7, 9, 10, 12)
14	(2, 3, 4, 5)	38	(8, 9, 10, 11)

Note that there are twelve points that correspond to values of $T \leq 14$, and twelve points that correspond to values of $T \geq 38$. If the critical region consists of all points that correspond to values of $T \leq 14$ or ≥ 38, then α is given by

$$\alpha = \frac{\text{number of points in critical region}}{\text{number of points in sample space}}$$

$$= \frac{24}{495}$$

(5) $\qquad\qquad = .0485$

since all points in the sample space have equal probability under H_0. Our decision rule is this: If the observed value of T is ≤ 14 or ≥ 38 we reject H_0; otherwise we accept H_0.

The sample is observed, and the academic ranks of the children who attended nursery school are 2, 5, 6 and 9, providing a value of

(6) $$T = 22$$

so we accept H_0. The critical level may be approximated using the normal distribution (see Example 1.6.7). To reject H_0 with $T = 22$ the symmetric two-tailed critical region would include values of $T \leq 22$ and ≥ 30, with a critical level

$$\hat{\alpha} = P(T \leq 22 \text{ or } T \geq 30)$$
$$= 1 - P(22 < T < 30)$$
$$\cong 1 - P\left(\frac{22 - 26}{5.9} < Z < \frac{30 - 26}{5.9}\right)$$

(7) $$= 1 - P(-.68 < Z < +.68)$$

where Z has the standard normal distribution. Table 1 then shows that

$$\hat{\alpha} \cong 1 - 2(1 - .75)$$
(8) $$= .50$$

showing that the experimental result is well within the acceptance region.

The test procedure explained in Example 2 is known as the Mann-Whitney test or the Wilcoxon test, and will be discussed extensively in Chapter 5, along with its many variations. The data in Example 2 have the ordinal scale of measurement. We did not need to know the numerical value of the academic achievement for each child. In fact such information usually has little value because each school, even each teacher, has its own interpretation of such numbers, while ranks have a universal interpretation.

Example 1 illustrated the analysis of nominal type data, "defective" or "not defective." The test of Example 1 was based on the binomial distribution. In Chapter 3 this test and other tests based on the binomial distribution will be presented formally.

PROBLEMS

1. A new teaching method is being tested to see if it is better than the existing teaching method. What are the appropriate H_0 and H_1?

2. A defendant is being tried by a judge, and it is assumed that the defendant is innocent until proven guilty.
(a) Who is doing the hypothesis testing?
(b) What are H_0 and H_1?
(c) What are the sample and the population?
(d) What do "level of significance" and "power" mean in this problem?

3. What is the appropriate H_1 for each of the following?
(a) H_0: Fertilizer B is at least as good as fertilizer A.
(b) H_0: My opponent is not cheating.
(c) H_0: The occurrence of sun spots does not affect the economic cycle.

4. What is the appropriate H_0 for each of the following?
   (a) H_1: The subject has extrasensory perception.
   (b) H_1: The dowsing rod is effective in finding water.
   (c) H_1: Our average yearly temperatures are rising.

5. The sample space contains 25 points. If H_0 is true all 25 points are equally likely to occur. The critical region has 2 points in it. What is α?

6. The sample space contains 10 points, only one of which is in the critical region. If H_0 is true, all of the points are equally likely. If H_1 is true, the point in the critical region has probability .91, and the other points each have probability .01. What is α? What is the power?

7. A coin is tossed five times, and the sequence of heads and tails observed is the outcome. The critical region is the event "at least four heads." If H_0 is true, all outcomes in the sample space are equally likely. What is α?

8. There are 12 plastic chips in a jar, and the chips are numbered consecutively from 1 to 12. An experiment consists of drawing three chips without replacement. The outcome of the experiment consists of the three numbers on the chips, without regard to the order in which they were drawn. Let the test statistic X be the sum of the numbers on the drawn chips and let the critical region correspond to values of X that are less than 11. Suppose that if H_0 is true the drawing of the chips is random.
   (a) How many points are in the sample space?
   (b) How many points are in the critical region?
   (c) If H_0 is true, what is the probability of each point in the sample space?
   (d) What is α?

9. In Problem 8 redefine the outcome of the experiment so that the numbers on the chips are recorded in the order in which they are drawn. Then the outcome $(1, 3, 4)$ is different than the outcome $(3, 1, 4)$.
   (a) How many points are in the sample space?
   (b) How many points are in the critical region?
   (c) If H_0 is true, what is the probability of each point in the sample space?
   (d) What is α?

10. Seven chips numbered consecutively from 1 to 7 are placed independently of each other into either of two boxes labeled A and B. The outcome of the experiment consists of the numbers on the chips in box A without regard to the order in which they were placed there. Let the test statistic X be the sum of the numbers on the chips in box A, and let the critical region correspond to values of X less than 6. Assume that if H_0 is true each chip has probability 1/2 of being placed into box A.
    (a) How many outcomes result in the event "exactly four chips are in box A"?
    (b) How many points are in the sample space?
    (c) How many points are in the critical region?
    (d) If H_0 is true, what is the probability of each point in the sample space?
    (e) What is α?

11. In Problem 10, assume that if H_0 is false then each chip has probability 1/3 of being placed into box A.

(a) If H_0 is false what is the probability of the outcome $(1, 3)$, which means only chips 1 and 3 are in box A.

(b) If H_0 is false what is the probability of no chips being placed into box A?

(c) What is the power?

12. Determine whether the following hypotheses are simple or composite:
    (a) H_0 in Problem 8
    (b) H_0 in Problem 9
    (c) H_0 in Problem 10
    (d) H_1 in Problem 11

13. Determine whether the following tests are one tailed or two tailed:
    (a) The test presented in Problem 8
    (b) The test presented in Problem 10
    (c) The test presented in Problem 10, if the critical region also corresponds to values of X greater than 23.

2.4. SOME PROPERTIES OF HYPOTHESIS TESTS

Once the hypotheses are formulated, there are usually several hypothesis tests available for testing the null hypothesis. In order to select one of these tests, we consider carefully several properties of the various tests. One of the most important questions is, "Are the assumptions of this test valid assumptions in my experiment?" If the answer is, "No," then that test probably should be discarded. However, before discarding the test, one should be sure that he understands the assumptions behind the test. For example, in most parametric tests one of the stated assumptions is that the random variable being examined has a normal distribution. Further investigation usually reveals that if the random variable has a distribution only slightly resembling a normal distribution, then the test is still approximately valid. So the implied assumption is "approximate normality," and the test should not be discarded if the assumptions are "approximately true." However, at least one black mark may be registered against that test. Another result of this criterion is that the test with the fewer assumptions in the model compares favorably with the test which has more assumptions.

The use of a test in a situation where the assumptions of the test are not valid is dangerous for two reasons. First, the data may result in rejection of the null hypothesis not because the data indicate that the null hypothesis is false, but because the data indicate that one of the assumptions of the test is invalid. Hypothesis tests in general are sensitive detectors not only of false hypotheses but also of false assumptions in the model. The second danger is that sometimes the data indicate strongly that the null hypothesis is false, and a false assumption in the model is also affecting the data, but these two effects neutralize each other in the test, so that the test reveals nothing and the null hypothesis is accepted.

From among the tests that are appropriate, based on the above criterion, the best test may be selected on the basis of other properties. These properties, which involve terms that will be defined later in this section, are as follows.

1. The test should be unbiased.
2. The test should be consistent.
3. The test should be more efficient in some sense than the other tests.

Sometimes we are content if one or two of the three criteria are met. Only rarely are all three met. The rest of this section discusses the terms unbiased, consistent, efficiency, and more about the power of a test.

If H_1 is composite, the power may vary as the probability function varies. If H_1 is stated in terms of some unknown parameter, the power usually may be given as a function of that parameter. Such a function is appropriately called a *power function*, and may be represented algebraically or graphically. Unlike the power, which is the probability of rejecting H_0 when H_1 is true, the power function is usually defined for all values of the parameter under both H_0 and H_1. In that sense the power function gives more than just the power; it gives the probability of rejecting H_0 whether or not H_0 is true.

Example 1. In Example 2.3.1 the critical region consisted of all points with more than two defectives in the ten items examined. Under the assumptions of the model, the probability of getting a point in the critical region, the same as the probability of rejecting H_0, is given by

$$(1) \qquad P(\text{reject } H_0) = \sum_{i=3}^{10} \binom{10}{i} p^i (1-p)^{10-i} = 1 - \sum_{i=0}^{2} \binom{10}{i} p^i (1-p)^{10-i}$$

where p is the probability of a defective item. The probability of rejecting H_0 is a function of p, and a rough graph of the power function may be drawn with the aid of Table 3.

$P(\text{reject } H_0)$	p	$P(\text{reject } H_0)$	p
.0000	0	.9453	.50
.0115	.05	.9726	.55
.0702	.10	.9877	.60
.1798	.15	.9952	.65
.3222	.20	.9984	.70
.4744	.25	.9996	.75
.6172	.30	.9999	.80
.7384	.35	1.0000	.85
.8327	.40	1.0000	.90
.9004	.45	1.0000	1.00

As indicated in Figure 1, the null hypothesis states that p is between zero and .05. The maximum value of the curve, when H_0 is true, is the level of significance, and is shown by Figure 1 and also by Equation (2.3.4) to equal .0115. The power is seen to range from .0115, for p close to .05, to 1.0000 for p equal to 1.0.

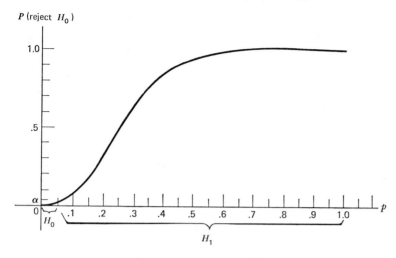

Figure 1.   A power function.

Two tests may be compared on the basis of their power functions. This basis of comparison is discussed again later in this section when relative efficiency is defined.

It is obviously desirable for a test to be more likely to reject H_0 when H_0 is false than when H_0 is true.

Definition 1.   An *unbiased test* is a test in which the probability of rejecting H_0 when H_0 is false is always greater than or equal to the probability of rejecting H_0 when H_0 is true.

Thus an unbiased test is one where the power is always at least as large as the level of significance. A test which is not unbiased is called a *biased* test. The test described in Example 2.3.1 and discussed further in Example 1 of this section is an unbiased test, a fact that is readily apparent from Figure 1.

Another desirable property of a test is that of being *consistent*. Although we refer to a test as being "consistent" or "not consistent," the term consistent actually applies to a sequence of tests, because the term applies when the sample size approaches the population size. For convenience we shall call the population size "infinity" even though it may be finite. Technically, for each different sample size we have a different test, because the sample space and the critical region depend on the sample size. Thus as the sample size increases we consider a sequence of tests, one for each sample size.

Definition 2.   A sequence of tests is *consistent against all alternatives in the class H_1* if the power of the tests approaches 1.0 as the sample size approaches infinity, for each fixed alternative possible under H_1. The level of significance of each test in the sequence is assumed to be as close as possible to but not exceeding some constant $\alpha > 0$.

Example 2.   We wish to determine whether human births tend to produce more babies of one sex, rather than both sexes being equally likely. We are testing

H_0: A human birth is equally likely to be male or female.

against the alternative hypothesis

H_1: Male births are either more likely, or less likely, to occur than female births.

The sampled population consists of births registered in a particular country. The sample consists of the last n births registered, for some selected value of n. It is assumed that this method of sampling is equivalent to random sampling, as far as the characteristics "male" and "female" are concerned. It is also assumed that the probability p (say) of a male birth remains constant from birth to birth, and that the births are mutually independent as far as the events "male" and "female" go. Then the hypotheses are equivalent to the following:

$$H_0 : p = 1/2$$
$$H_1 : p \neq 1/2$$

Let the test statistic T be the number of male births. The critical region is chosen to correspond symmetrically to the largest values and the smallest values of T, called the upper and lower tails of T, of the largest size not exceeding .05.

Thus we have described an entire sequence of tests, one for each value of n, the sample size. Each test is two tailed, has a level of significance of .05 or smaller, and T has a binomial distribution. For the various tests the critical regions are given by Dixon (1953) as follows.

		Values of T Corresponding to the Critical Region		
n				α
5		None		0
6	$T = 0$	and	$T = 6$	.03125
8	$T = 0$	and	$T = 8$	.00781
10	$T \leq 1$	and	$T \geq 9$	.02148
15	$T \leq 3$	and	$T \geq 12$	.03516
20	$T \leq 5$	and	$T \geq 15$	.04139
30	$T \leq 9$	and	$T \geq 21$	.04277
60	$T \leq 21$	and	$T \geq 39$	.02734
100	$T \leq 39$	and	$T \geq 61$	.03520

Note that for $n \leq 20$ these same values can be obtained from Table 3. For $n > 20$ the normal approximation (Example 1.6.6) could have been used, but the exact tables are preferred.

P (reject H_0)

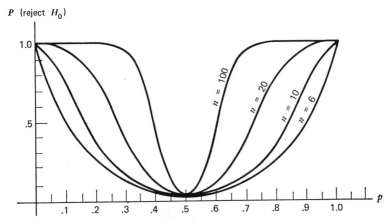

Figure 2.   A comparison of several power functions.

To see if this sequence of tests is consistent the power functions of the tests are compared. Several of these power functions are plotted on the same graph in Figure 2, from tables given by Dixon (1953). We can see that as the sample size increases, the power at each fixed value of p (except $p = .5$) increases toward 1.0.

This example merely demonstrates the idea behind the term consistent as it applies to a sequence of tests. This demonstration is not a proof that the sequence of tests is consistent. A rigorous proof of consistency usually requires more mathematics than we care to use in this introductory book, and so we shall merely state whether or not a sequence of tests (or "a test") is consistent.

Many other properties of statistical tests have been defined, and may be found in various books such as one by Lehmann (1959). We shall limit our discussion to one more property, that of *efficiency*. Efficiency is a relative term and is used to compare the sample size of one test with that of another test under similar conditions. Suppose two tests may be used to test a particular H_0 against a particular H_1. Also suppose that the two tests have the same α and the same β, and therefore are "comparable" with respect to level of significance and power. (Note that the condition that β be the same for both tests usually excludes consideration of composite alternative hypotheses, since those usually have more than one value of β.) Then the test requiring the smaller sample size is preferred over the other test, because a smaller sample size means less cost and effort is required in the experiment. The test with the smaller sample size is said to be "more efficient" than the other test, and its *relative efficiency* is greater than one.

Definition 3.   Let T_1 and T_2 represent two tests that test the same H_0 against the same H_1, with the critical regions of the same size α, and with the same values of β. *The relative efficiency of T_1 to T_2* (or "efficiency of T_1 relative to T_2") is the ratio n_2/n_1, where n_1 and n_2 are the sample sizes of the tests T_1 and T_2 respectively.

According to Definition 3 if n_1 is smaller than n_2, the efficiency of T_1 relative to T_2 is greater than unity, satisfying our preconceived notions.

If the alternative hypothesis is composite, the relative efficiency may be computed for *each* probability function defined by the alternative hypothesis, resulting in a multitude of values for relative efficiency which may then be represented in a table or, occasionally, graphically.

Example 3. Two tests are available for testing the same H_0 against the same H_1. Both tests have $\alpha = .01$ and $\beta = .14$. The first test requires a sample size of 75. The second test requires a sample size of 50. The first test is therefore less efficient than the second test. The relative efficiency of the first test to the second test is

$$\frac{50}{75} = .67$$

and the efficiency of the second test relative to the first is

$$\frac{75}{50} = 1.5$$

If we know that the efficiency of the first test relative to the second test at $\alpha = .05$, $\beta = .30$, $n_1 = 40$, is .75, then the sample size required by the second test may be obtained.

$$\text{relative efficiency} = \frac{n_2}{n_1}$$

$$.75 = \frac{n_2}{40}$$

$$n_2 = 30$$

A sample of size 30 will provide as good an analysis using the second testing method as a sample of size 40 would using the first.

The relative efficiency depends on the choice of α, the choice of β, and the particular alternative being considered if H_1 is composite. In order to provide an overall comparison of one test with another it is clear that relative efficiency leaves much to be desired. We would prefer a comparison which does not depend on our choice of α, or β, or a particular alternative possible under H_1 if H_1 is composite, which it usually is. One way this sometimes may be accomplished is described briefly as follows.

Consider a sequence of tests, all with the same fixed α. If the sequence of tests is consistent, β will become smaller as the sample size n_1 gets larger.

Instead of allowing β to become smaller, we could consider a different alternative each time (under the composite alternative hypothesis) for each different value of n_1, where, each time, the alternative considered is one that allows β to remain constant from test to test. Thus, as n_1 becomes larger, α and β remain fixed and the alternative being considered varies. This may be illustrated by considering Figure 2 again. An n_1 becomes larger, the graphs in Figure 2 show that β can remain constant by considering consecutive values of the parameter p that approach closer to $p = .5$. For each value of n_1, a value of n_2 is calculated so the second test has the same α and β under the alternative considered. Then there is a sequence of values of relative efficiency n_2/n_1, one for each test in the original sequence of tests. If n_2/n_1 approaches a constant as n_1 becomes large, and if that constant is the same no matter which values of α and β are being used, then that constant is called the *asymptotic relative efficiency* of the first test to the second test or, more correctly, the first sequence of tests to the second sequence of tests. Sometimes the name *Pitman's efficiency* is used for this definition of asymptotic relative efficiency, to distinguish it from other definitions of asymptotic relative efficiency.

Definition 4.   Let n_1 and n_2 be the sample sizes required for two tests T_1 and T_2 to have the same power under the same level of significance. If α and β remain fixed, then the limit of n_2/n_1, as n_1 approaches infinity, is called the *asymptotic relative efficiency* (A.R.E.) of the first test to the second test, if that limit is independent of α and β.

The A.R.E. of two tests is usually difficult to calculate. A comprehensive study of the A.R.E. of various pairs of tests could be the subject of a book by itself. A book by Noether (1967) contains many of the more important results of studies of A.R.E. See also Stuart (1954) and Ruist (1955) for further discussions.

So the A.R.E. may be given instead of tables of values of relative efficiency, but of what use is A.R.E. since it considers the infinite (and thus impossible) sample size? Studies of the exact relative efficiency for very small sample sizes show that the A.R.E. provides a good approximation to the relative efficiency in many situations of practical interest. Thus the A.R.E. often provides a compact summary of the relative efficiency between two tests.

The term *conservative* is another term we shall sometimes use when discussing a test.

Definition 5.   A test is *conservative* if the actual level of significance is smaller than the stated level of significance.

At times it is difficult to compute the exact level of significance of a test, and then some methods of approximating α are used. The approximate value is then reported as being the level of significance. If the approximate level of significance is larger than the true (but unknown) level of significance, the test is conservative, and we know the risk of making a type I error is not as great as it is stated to be.

PROBLEMS

1. In Problem 2.3.10, assume that if H_0 is false then each chip has a probability p of falling into box A, where $p \neq 1/2$.
   (a) What is the power function?
   (b) Is the test unbiased?

2. In Problem 2.3.10, assume that if H_0 is false then each chip has a probability p of falling into box A, where $p > 1/2$.
   (a) What is the power function?
   (b) Is the test unbiased?

3. Suppose the probability of rejecting H_0 is given by

$$Pr = p^3 + (1 - p)^3 \qquad 0 \leq p \leq 1$$

   and suppose $1/4 \leq p \leq 3/4$ only if H_0 is true, and $p < 1/4$ or $p > 3/4$ only if H_0 is false.
   (a) What is the power function?
   (b) What is α?
   (c) Is the test unbiased?

4. Let $T_1, T_2, \ldots$, represent a sequence of tests, and suppose the power function P_n for test T_n is given by $P_n = n/(n + 10)$. Is the sequence of tests consistent?

5. In Problem 4, suppose $P_n = n/(2n + 10)$. Is the sequence of tests consistent?

6. Let $T_1, T_2, \ldots$, represent a sequence of tests involving a parameter p, and suppose the power function P_n for T_n is given by

$$P_n = 1 - p^n \qquad \text{for} \qquad 0 < p < 1$$

   Is the sequence of tests consistent?

7. The hypothesis $H_0 : p = 1/2$ is being tested against $H_1 : p = 3/4$ using two tests T_1 and T_2 at the same level of significance. If T_1 uses a sample of size 20, then T_2 requires a sample of size 35 in order for the power of T_2 to equal the power of T_1.
   (a) What is the efficiency of T_2 relative to T_1?
   (b) What is the efficiency of T_1 relative to T_2?

8. The hypothesis $H_0 : p = 1/2$ is being tested against $H_1 : p \neq 1/2$ using two tests T_1 and T_2 at the same level of significance. T_2 needs a sample of size 30 when T_1 has a sample of size 15, in order for their power functions to be equal at the particular alternative $p = 1/3$.
   (a) What is the efficiency of T_2 relative to T_1?
   (b) Is the efficiency necessarily the same at the alternative $p = 2/3$?

9. Suppose that the asymptotic efficiency of T_2 relative to T_1 is .75, and suppose that the efficiency for finite sample sizes is always greater than the asymptotic efficiency. If an experimenter prefers to use test T_2, but wishes to have at least

as much power as if he had used test T_1 with a sample of size 24, what should be his minimum sample size?

2.5. NONPARAMETRIC STATISTICS

This book purports to be one on nonparametric statistical methods, and yet we have not yet defined the adjective nonparametric, except for an unsatisfactory hand waving distinction between parametric and nonparametric given in the Introduction. There is no agreement among statisticians as to the meaning of the word nonparametric. In fact, there is not even agreement among statisticians concerning whether certain tests should be classified as parametric or nonparametric. For instance, the test presented in Example 2.3.1 is claimed by both camps.

On one extreme, Walsh (1962) states the following.

"The viewpoint adopted in this handbook is that a statistical procedure is of a nonparametric type if it has certain properties which are satisfied to a reasonable approximation when some assumptions that are at least of a moderately general nature hold. That is, generality of application is the criterion for deciding on the nonparametric character of a statistical procedure. As an example, the binomial distribution is considered to be of nonparametric interest even though it is of an elementary parametric form."

At the other extreme consider the following statement by Kendall and Sundrum (1953).

"It is important to confine the adjectives 'parametric' and 'nonparametric' to statistical *hypotheses*. They should not be applied to statistics, tests, or types of inference. This may sound rather austere, but we have found a great deal of confusion arising from the use of phrases like 'nonparametric tests' and 'nonparametric inference.' We ourselves are far from guiltless in this respect in previous writings but hope to atone for past errors by correct usage in the future."

Furthermore, Kendall and Sundrum distinguish between the terms "nonparametric" and "distribution-free"; the latter term they refer to as "a convenient though not a perfect, term introduced by American writers." Then they go on to permit the adjective "distribution-free" to apply to test statistics, the distribution of the test statistic, the critical region determined by the test statistic for the null hypothesis, and so on, if the element "does not depend on the parent population association with the (null) hypothesis."

There are other definitions of the terms "nonparametric" and "distribution-free" by other well-qualified statisticians. See, for instance, Bell (1964).

We shall adopt the low-brow convention of using the two terms interchangeably. We shall also further confuse the already confused situation by offering our own definition of the term "nonparametric." Since this definition is appearing here for the first time, we can claim no acceptance, universal or otherwise, of this definition in the world of statisticians.

Definition 1. A statistical method is nonparametric if it satisfies at least one of the following criteria:

1. The method may be used on data with a nominal scale of measurement.

2. The method may be used on data with an ordinal scale of measurement.

3. The method may be used on data with an interval or ratio scale of measurement, where the distribution function of the random variable producing the data is either unspecified or specified except for an infinite number of unknown parameters.

The test in Example 2.3.1 analyzed data with a nominal scale of measurement (defective or nondefective) and therefore the test is nonparametric by the first criterion. The test in Example 2.3.2 analyzed data with an ordinal scale of measurement and therefore, by the second criterion, it too is nonparametric. Nearly all nonparametric hypothesis tests satisfy one of these two criteria. The point estimates of Section 2.2 satisfy the third criterion, and so do the procedures that assume symmetric distributions in Sections 5.1, 5.2 and 7.4. Therefore we consider them to be nonparametric. There are undoubtedly situations where Definition 1 is inadequate, but we shall advocate its use, at least temporarily, as a convenient yardstick.

This book is primarily concerned with hypothesis testing and the forming of confidence intervals, which is just hypothesis testing in disguise. Unfortunately, this emphasis often gives the experimenter the false impression that if he does not test some hypotheses or form some confidence interval he is not using a statistical analysis. Other forms of statistical inference are just as important, such as description of the population, interpretation of the data, prediction of unknown events, and point estimation. These other forms of inference depend to a great extent on the experimenter's maturity and good judgment, rather than on complicated probabilistic arguments, and therefore we consider them difficult to present in a book. We are attempting to assist the experimenter already possessing maturity and good judgment by spelling out the complicated probabilistic arguments associated with hypothesis testing and confidence intervals. The author can merely offer his apologies for not presenting an adequate discussion of the other forms of inference. An article by Blum and Fattu (1954) may provide more insight into these other areas.

CHAPTER 3

Some Tests Based on the Binomial Distribution

3.1. THE BINOMIAL TEST

One example of the binomial test has already been presented. In Example 2.3.1 the binomial test was applied to a quality control problem. This entire chapter (Chapter 3) is little more than an elaboration of Example 2.3.1, showing the many uses and amazing versatility of that simple little binomial test. With a little ingenuity the binomial test may be adapted to test almost any hypothesis, with almost any type of data amenable to statistical analysis. In some situations the binomial test is the most powerful test, and in those situations the test is claimed by both parametric and nonparametric statistics. In other situations more powerful tests are available, and the binomial test is claimed only by nonparametric statistics. However, even in those situations where more powerful tests are available, the binomial test is sometimes preferred because it is usually simple to perform, simple to explain, and sometimes powerful enough to reject the null hypothesis when it should be rejected.

We shall now formally present the binomial test, and at the same time introduce the format for presenting tests. While the following method of presenting tests is not superior to other methods, we feel that there is a need for some format in presenting tests, for the convenience of the reader.

The Binomial Test

DATA. The sample consists of the outcomes of n independent trials. Each outcome is in either "class 1" or "class 2," but not both. The number of observations in class 1 is 0_1 and the number of observations in class 2 is $0_2 = n - 0_1$.

ASSUMPTIONS
1. The n trials are mutually independent.
2. Each trial has probability p of resulting in the outcome "class 1," where p is the same for all n trials.

HYPOTHESES. Let p^* be some specified constant, $0 \leq p^* \leq 1$. The hypotheses may take one of the following three forms.

A. (two-tailed test)

$$H_0 : p = p^*$$
$$H_1 : p \neq p^*$$

B. (one-tailed test)

$$H_0 : p \leq p^*$$
$$H_1 : p > p^*$$

C. (one-tailed test)

$$H_0 : p \geq p^*$$
$$H_1 : p < p^*$$

TEST STATISTIC. Since we are concerned with the probability of the outcome "class 1," we shall let the test statistic T be the number of times the outcome is "class 1." That is

(1) $$T = 0_1$$

DECISION RULE. Depending on which hypothesis is being tested, A, B, or C, the different decision rules are as follows. (Because the test statistic T is discrete, α will seldom be a nice round number.)

A. (two-tailed test) The critical region of size α corresponds to the two tails of the binomial distribution with parameters p^* and n, where the size of the upper tail is α_1, the size of the lower tail is α_2, and $\alpha_1 + \alpha_2$ equals α. That is, from Table 3 for the particular values of p^* and n, find the number t_1 such that

(2) $$P(Y \leq t_1) = \alpha_1$$

and find the number t_2 such that

(3) $$P(Y > t_2) = \alpha_2$$

or, equivalently,

$$(4) \qquad\qquad P(Y \leq t_2) = 1 - \alpha_2$$

where Y is a binomial random variable with parameters p^* and n.

The values of α_1 and α_2 should be approximately equal to each other. Then reject H_0 if T exceeds t_2 or if T is less than or equal to t_1. Otherwise accept H_0.

B. (one-tailed test) Since large values of T indicate that H_0 is false, the critical region of size α consists of all values of T greater than t, where t is the number obtained from Table 3, using p^* and n, such that

$$(5) \qquad\qquad P(Y > t) = \alpha$$

or equivalently

$$(6) \qquad\qquad P(Y \leq t) = 1 - \alpha$$

where Y has the binomial distribution with parameters p^* and n. Reject H_0 if T is greater than t. Accept H_0 if T is less than or equal to t.

C. (one-tailed test) Since small values of T indicate that H_0 is false, the critical region of size α consists of all values of T less than or equal to t, where t is obtained from Table 3, using p^* and n, so that

$$(7) \qquad\qquad P(Y \leq t) = \alpha$$

where Y has the binomial distribution with parameters p^* and n. Reject H_0 if T is less than or equal to t. Otherwise accept H_0.

Example 1. Under simple Mendelian inheritance a cross between plants of two particular genotypes may be expected to produce progeny 1/4 of which are "dwarf" and 3/4 of which are "tall." In an experiment to determine if the assumption of simple Mendelian inheritance is reasonable in a certain situation, a cross results in progeny having 243 dwarf plants and 682 tall plants. If "class 1" denotes "tall," then $p^* = 3/4$ and T equals the number of tall plants. The null hypothesis of simple Mendelian inheritance is equivalent under the model to the hypothesis

$$H_0 : p = 3/4$$

The alternative of interest is

$$H_1 : p \neq 3/4$$

Since n equals 925 (243 + 682), the critical region of approximate size $\alpha = .05$ may be obtained using the large sample approximation given at the end of Table 3. Thus the critical region corresponds to all values of T less than or equal to t_1, where

$$t_1 = np^* + w_{.025}\sqrt{np^*(1 - p^*)}$$

$$= (925)(\tfrac{3}{4}) + (-1.960)\sqrt{(925)(\tfrac{3}{4})(\tfrac{1}{4})}$$

$$(8) \qquad\qquad = 667.95$$

and all values of T greater than t_2, where

$$t_2 = np^* + w_{.975} \sqrt{np^*(1 - p^*)}$$

$$= (925)(\tfrac{3}{4}) + (1.960) \sqrt{(925)(\tfrac{3}{4})(\tfrac{1}{4})}$$

(9) $$= 719.55$$

The value of T obtained is 682 in this experiment. Therefore the null hypothesis is accepted.

The critical level $\hat{\alpha}$ may be found by considering that the acceptance region in our test is some region on both sides of $np^* = 693.75$. Since the observed value 682 is smaller than np^*, half of $\hat{\alpha}$ is found by calculating $P(T \leq 682)$, assuming the null hypothesis is true. As in Example 1.6.6,

$$P(T \leq 682) = P\left(\frac{T - np^*}{\sqrt{np^*(1 - p^*)}} \leq \frac{682 - np^*}{\sqrt{np^*(1 - p^*)}} \right)$$

(10) $$\cong P\left(Z \leq \frac{682 - 693.75}{52.6} \right)$$

where Z has the standard normal distribution as given in Table 1.

$$\frac{\hat{\alpha}}{2} = P(T \leq 682) \cong P(Z \leq -.223)$$

(11) $$\cong .41$$

Therefore $\hat{\alpha} = 2(.41) = .82$. A level of significance of at least .82 would be required to reject H_0. Thus the data are in good agreement with the null hypothesis.

The previous example illustrates the two-tailed form of the binomial test. The one-tailed binomial test is illustrated in Example 2.3.1.

Theory. That the test statistic in the binomial test has a binomial distribution is easily seen by comparing the assumptions in the binomial test with the assumptions in Examples 1.4.5 and 1.3.9. That is, if T equals the number of trials which result in the outcome "class 1," where the trials are mutually independent, and where each trial has probability p of resulting in that outcome (as stated by the assumptions), then T has the binomial distribution with parameters p and n. The size of the critical region is a maximum when p equals p^*, under the null hypothesis, and so Table 3 is entered with n and p^* to determine the exact value of α.

As we mentioned earlier, hypothesis testing is only one branch of statistical inference. We shall now discuss another branch, known as interval estimation. If we are attempting to make some inferences regarding an unknown parameter associated with some population, it is reasonable to examine a random sample from that population, and on the basis of that sample to make some statement regarding the population parameter. Such a statement might be "the population parameter lies between *a* and *b*," where *a* and *b* are two real numbers obtained from the sample. The numbers *a* and *b* are computed from the sample, and therefore are realizations of two statistics. The two statistics that furnish us with the lower and upper boundary points for the interval will be denoted by *L* and *U* respectively, for "lower" and "upper." The interval from *L* to *U* is called the *interval estimator*. The probability that the unknown population parameter lies within its interval estimate is called the *confidence coefficient*. The interval estimator together with the confidence coefficient provide us with the *confidence interval*.

A method for finding a confidence interval for *p*, the unknown probability of any particular event occurring, is closely related to the binomial test.

Confidence Interval for a Probability

DATA. A sample consisting of observations on *n* independent trials is examined, and the number *Y* of times the specified event occurs is noted.

ASSUMPTIONS
1. The *n* trials are mutually independent.
2. The probability *p* of the specified event occurring remains constant from one trial to the next.

METHOD A. For confidence coefficients of .95 or .99 use the charts in Table 4. Read from the lower left corner if *Y/n* is less than .50, and the upper right corner if *Y/n* is greater than .50. Read horizontally across the chart to the value obtained for *Y/n*, and then read vertically from there to the two lines labeled with the correct sample size *n*, interpolating if necessary. The ordinates of these two intersections provide the values for *L* and *U*, obtained from the left side if *Y/n* is less than .50 and the right side otherwise. (Note that the values for *L* and *U* depend on the values of the random variable *Y*, which in turn is a function of the random sample, which shows that *L* and *U* are statistics.)

METHOD B. For confidence coefficients other than .95 or .99, and for *n* less than or equal to 20, Table 3 may be used. Let the desired confidence coefficient be denoted by $1 - \alpha$. Compute $P_1 = 1 - \alpha/2$. Enter Table 3 with the sample size *n*, and read across the row for $y = Y - 1$, until the entry in the table equals P_1 (approximately). The value of *p* found at the top of that column containing P_1 is the value of the lower limit *L*. Interpolate if necessary.

Read across the next row ($y = Y$) until the entry $P_2 = \alpha/2$ is reached (approximately). The value of *p* at the top of the column containing P_2 is the value for the upper limit *U*. Interpolate if necessary.

METHOD C. For n greater than 20 the normal approximation may be used. Use

(12) $$L = \frac{Y}{n} - x_{1-\alpha/2} \sqrt{Y(n-Y)/n^3}$$

and

(13) $$U = \frac{Y}{n} + x_{1-\alpha/2} \sqrt{Y(n-Y)/n^3}$$

where $x_{1-\alpha/2}$ is the quantile of a normally distributed random variable, obtained from Table 1.

For the sake of illustration, all three methods of computing confidence intervals are used in the following example.

Example 2. In a certain state twenty high schools were selected at random to see if they met the standards of excellence proposed by a national committee on education. It was found that seven schools did qualify and accordingly were designated "excellent." What is a 95% confidence interval for p, the proportion of all high schools in the state which would qualify for the designation "excellent"?

First we assume that the number of high schools in the state is large enough so the high schools are classified "excellent or "not excellent" independently of one another. (Actually, for any finite number of schools, the fact that one school is classified one way tends to increase the chances of the next school being classified the other way, since there would then be a slightly higher proportion of schools in the other category among those not yet selected.) Because we assumed the selection was random, p is the same for all schools, and represents the probability of a randomly selected school being designated "excellent."

Since n equals 20 and Y equals 7, Y/n equals .35. For Method A we consult Table 4, read across the bottom of the chart to .35, and up to the two curves with "20" (for n) written on them. At the lower curve we read $p = .15 = L$ on the left, and at the upper curve the point of intersection is $p = .59 = U$. Therefore the 95% confidence interval for p is $(L, U) = (.15, .59)$, or

(14) $$P(.15 < p < .59) = .95$$

Method B involves the use of Table 3. Reading across the row $y = Y - 1 = 6$ in the table for $n = 20$, we are looking for the probability $1 - (1/2)(.05) = .975$. Interpolating between $p = .15$ for the entry .9781, and $p = .20$ for the entry .9133, we obtain $p = .15$ as a value for the lower bound L. To find the upper bound U we read across the next line, $y = 7$, to the entry $.05/2 = .025$. This involves interpolating between .0580 ($p = .55$) and .0210 ($p = .60$). The result is $p = .59 = U$. The result is the same as before.

(15) $$P(.15 < p < .59) = .95$$

Method C, the use of the normal approximation based on the Central Limit Theorem, gives

$$L = \frac{Y}{n} - x_{.975} \sqrt{Y(n - Y)/n^3}$$

$$= .35 - (1.960) \sqrt{(7)(13)/(20)^3}$$

$$= .35 - .21$$

(16) $$= .14$$

and

$$U = .35 + .21$$

(17) $$= .56$$

The confidence interval furnished by method C is

(18) $$P(.14 < p < .56) = .95$$

Both Methods A and B are exact methods, insofar as graph readings in Table 4 or interpolation in Table 3 may be considered to be exact. However, Method C is an approximation that becomes better as n gets larger, and is better for values of p near .5 than it is for values of p near 0 or 1. For n as small as twenty, the example shows that the approximation is still pretty close.

We should remark at this point that objections are often made to statements in the form of (14), namely,

$$P(.15 < p < .59) = .95$$

The objection is that either $.15 < p < .59$ is true or else it is not true, and so the probability is either 1.0 or .0, according to the situation. It is fine to say

$$P(L < p < U) = .95$$

because L and U are random variables. We feel that the objection is too subtle to concern us, and we shall consider (incorrectly, perhaps) our definition of probability to be stretched sufficiently to include the above statements.

Theory. For the exact Methods A and B described above, the confidence interval consists of all values of p^* such that the data obtained in the sample would result in acceptance of

$$H_0 : p = p^*$$

if one were using the two-tailed binomial test. More precisely, if we want to form a $(1 - \alpha)$ confidence interval, we observe the sample and determine Y. Then we ask, "For the given value of Y, which values may we use for p^* in the hypothesis

$$H_0 : p = p^*$$

such that a two-tailed binomial test (at level α) would result in *acceptance* of H_0?" Those values of $p*$ would be in our confidence interval. The values of $p*$ that would result in *rejection* of H_0 would not be in the confidence interval. Since each tail of the binomial test has probability $\alpha/2$, the value of L is selected as that value of $p*$ which would barely result in rejection of H_0, for the given value of Y, say y, or a larger value. Thus p_1^* is selected so that

$$(19) \qquad\qquad P(Y \geq y \mid p = p_1^*) = \frac{\alpha}{2}$$

and then $L = p_1^*$. Next, another value of $p*$ is selected so the same value y is barely in the lower tail. That is, p_2^* is selected so

$$(20) \qquad\qquad P(Y \leq y \mid p_2^*) = \frac{\alpha}{2}$$

and we set $U = p_2^*$. Because Table 3 gives $P(Y \leq y)$, Equation (20) may be solved by finding the value of p that gives

$$P(Y \leq y) = \frac{\alpha}{2}$$

as described by Method B. Equation (19) needs a slight rearrangement into the form

$$(21) \qquad\qquad P(Y \leq y - 1 \mid p = p_1^*) = 1 - \frac{\alpha}{2}$$

and then Table 3 may be used.

More information on confidence intervals for the binomial parameter p may be found in Clopper and Pearson (1934).

The large sample approximations for L and U may be obtained by considering Example 1.6.6 which states that if Y is a binomially distributed random variable with parameters p and large n, then

$$(22) \qquad\qquad Z = \frac{Y - np}{\sqrt{npq}}$$

is a random variable whose distribution may be approximated by the standard normal distribution. Then, if $x_{1-\alpha/2}$ is the $(1 - \alpha/2)$ quantile from Table 1, and because $x_{\alpha/2} = -x_{1-\alpha/2}$, we have

$$1 - \alpha = P\left(-x_{1-\alpha/2} < \frac{Y - np}{\sqrt{npq}} < x_{1-\alpha/2} \right)$$

$$= P(-x_{1-\alpha/2}\sqrt{npq} < Y - np < x_{1-\alpha/2}\sqrt{npq})$$

Multiplication by (-1) reverses the inequalities

$$1 - \alpha = P(x_{1-\alpha/2}\sqrt{npq} > np - Y > -x_{1-\alpha/2}\sqrt{npq})$$

and reversal of the reading order gives

$$1 - \alpha = P(-x_{1-\alpha/2}\sqrt{npq} < np - Y < x_{1-\alpha/2}\sqrt{npq})$$
$$= P(Y - x_{1-\alpha/2}\sqrt{npq} < np < Y + x_{1-\alpha/2}\sqrt{npq})$$

Now we divide through by n

$$(23) \qquad 1 - \alpha = P\left(\frac{Y}{n} - x_{1-\alpha/2}\sqrt{\frac{pq}{n}} < p < \frac{Y}{n} + x_{1-\alpha/2}\sqrt{\frac{pq}{n}}\right)$$

Using a further approximation, the estimator Y/n for p under the radical in (23), gives

$$1 - \alpha \cong P\left(\frac{Y}{n} - x_{1-\alpha/2}\sqrt{\frac{Y}{n}\left(1 - \frac{Y}{n}\right)\Big/ n}\right.$$
$$\left. < p < \frac{Y}{n} + x_{1-\alpha/2}\sqrt{\frac{Y}{n}\left(1 - \frac{Y}{n}\right)\Big/ n}\right)$$

$$(24) \qquad\qquad \cong P(L < p < U)$$

where L and U are the same as in (12) and (13). This latter approximation of Y/n for p results in a slight difference between the confidence interval and the hypothesis test, when the large sample approximations are used for both.

Multiplication by the sample size n in the above procedure gives nL and nU as the lower and upper bounds of the confidence interval for np, the mean of a binomial random variable. Also the binomial test may be used to test hypotheses involving the mean of a binomial random variable, because

$$H_0: p = p^*$$

is equivalent to

$$H_0: np = np^*$$

Other methods of obtaining binomial confidence limits are given by Anderson and Burstein (1967 and 1968). Methods dealing with simultaneous confidence intervals for multinomial proportions are given by Quesenberry and Hurst (1964) and Goodman (1965).

PROBLEMS

1. It is known that twenty percent of a certain species of insect exhibit a particular characteristic A. Eighteen insects of that species are obtained from an unusual environment, and none of these have characteristic A. Is it reasonable to assume that insects from that environment have the same probability of .20 that the species in general has?

2. Of sixteen cars inspected during a safety campaign, six were found to be unsafe. Test the hypothesis that no more than ten percent of the cars in the population are unsafe. (Which assumption is most likely to be false in this application?)

3. In a dice game a pair of dice were thrown 180 times. The event "seven" occurred on 38 of these throws.
   (a) Is the probability of "seven" what it should be if the dice were fair?
   (b) Find a 95 percent confidence interval for P(seven) using Table 4.
   (c) Find a 95 percent confidence interval for P(seven) using the large sample approximation.

4. In Problem 2, what is a 90 percent confidence interval for the true proportion of unsafe cars in the population?

5. Twenty independent observations on a random variable X with the unknown distribution function $F(x)$ resulted in the numbers

142	134	98	119	131
103	154	122	93	137
86	119	161	144	158
165	81	117	128	103

Find a 95 percent confidence interval for $F(100)$.

6. A civic group reported to the town council that at least sixty percent of the town residents were in favor of a particular bond issue. The town council then asked a random sample of one hundred residents if they were in favor of the bond issue. Forty-eight said "yes." Is the report of the civic group reasonable?

3.2. THE QUANTILE TEST

The binomial test may be used to test hypotheses concerning the quantiles of a random variable, in which case we call it the quantile test. For example, we may examine a random sample of values of some random variable X to see if the median of X is greater than zero, or equal to 17 (say). The measurement scale is usually at least ordinal for the quantile test, although the binomial test only required the weaker nominal scale for its measurements. If the random variable being examined is a continuous random variable, then the hypothesis being tested

$$H_0 : \text{The } p^*\text{th quantile of } X \text{ is } x^* \text{ (specified)}$$

is the same as

$$H_0 : P(X \leq x^*) = p^*$$

from the definition of the word *quantile*. If we represent the unknown probability $P(X \leq x^*)$ by p, then H_0 becomes

$$H_0:p = p^*$$

which is the same null hypothesis tested with the binomial test. The test statistic equals the number of sample values which are less than or equal to x^*, and the two-tailed binomial test may be used.

The situation is not as simple if the random variable is not assumed to be continuous. Then the null hypothesis

$$H_0:\text{The } p^*\text{th quantile of } X \text{ is } x^*$$

is the same as

$$H_0:P(X \leq x^*) \geq p^* \quad \text{and} \quad P(X < x^*) \leq p^*$$

Now the binomial test may be used, but the adaptation of the test to this hypothesis is a little tricky, so we shall present the procedure as a separate test.

The Quantile Test

DATA. Let $X_1, X_2, \ldots, X_n$ be a random sample. The data consist of observations on the X_i.

ASSUMPTIONS
1. The X_i's are a random sample (i.e., they are independent and identically distributed random variables).
2. The measurement scale of the X_i's is at least ordinal.

HYPOTHESES. Let x^* and p^* represent some specified numbers, $0 < p^* < 1$. The hypotheses may take one of the following three forms.

A. (two-tailed test)

$$H_0:\text{The } p^*\text{th population quantile is } x^*$$

(This is equivalent to $H_0:P(X \leq x^*) \geq p^*$, and $P(X < x^*) \leq p^*$, where X has the same distribution as the X_i's in the random sample.)

$$H_1:x^* \text{ is not the } p^*\text{th population quantile}$$

(This is equivalent to $H_1:\text{Either } P(X \leq x^*) < p^* \text{ or } P(X < x^*) > p^*$.)

B. (one-tailed test)

$$H_0:\text{The } p^*\text{th population quantile is at least as great as } x^*$$

(This is equivalent to $H_0:P(X < x^*) \leq p^*$.)

$$H_1:\text{The } p^*\text{th population quantile is less than } x^*$$

(This is the same as $H_1 : P(X < x^*) > p^*$.)

C. (one-tailed test)

H_0: The p^*th population quantile is no greater than x^*

(Or $H_0 : P(X \leq x^*) \geq p^*$.)

H_1: The p^*th population quantile is greater than x^*

(Or $H_1 : P(X \leq x^*) < p^*$.)

TEST STATISTIC. We shall use two test statistics in this test. Let T_2 equal the number of observations less than x^*, and let T_1 equal the number of observations less than *or equal to* x^*. Then T_1 equals T_2 if none of the numbers in the data exactly equals x^*. Otherwise T_1 is greater than T_2.

DECISION RULE. As in the binomial test, the test statistics have a discrete distribution and so α will seldom be a nice round number. The different decision rules, corresponding to the hypotheses A, B, or C, are given below.

A. (two-tailed test) The critical region corresponds to values of T_2 which are too large (indicating that perhaps $P(X < x^*)$ is greater than p^*) and to values of T_1 which are too small (indicating that perhaps $P(X \leq x^*)$ is less than p^*). The critical region is found by entering Table 3 with the sample size n and the hypothesized probability p^*, as in the two-tailed binomial test. Find the number t_1 such that

(1) $$P(Y \leq t_1) = \alpha_1$$

where Y has the binomial distribution with parameters n and p^*, and where α_1 is about half of the desired level of significance. Then find the number t_2 such that

(2) $$P(Y > t_2) = \alpha_2$$

or, equivalently

(3) $$P(Y \leq t_2) = 1 - \alpha_2$$

where α_2 is chosen so that $\alpha_1 + \alpha_2$ is about equal to the desired level of significance. Reject H_0 if T_1 is less than or equal to t_1, or if T_2 is greater than t_2. Otherwise accept H_0. The level of significance equals $\alpha_1 + \alpha_2$.

B. (one-tailed test) Since large values of T_2 indicate that H_0 is false, enter Table 3 with the sample size n and the hypothesized p^* as p. Find the number t_2 such that

(4) $$P(Y > t_2) = \alpha$$

which is the same as

(5) $$P(Y \leq t_2) = 1 - \alpha$$

for some acceptable level of significance α. Then reject H_0 if T_2 exceeds t_2. Accept H_0 if T_2 is less than or equal to t_2. (This is the same as decision rule B in the binomial test.)

C. (one-tailed test) Small values of T_1 indicate H_0 is false, so enter Table 3 with the sample size n and the specified probability p^*, to find t_1 such that

$$(6) \qquad\qquad P(Y \le t_1) = \alpha$$

for an acceptable level α, where Y has a binomial distribution with parameters n and p^*. Reject H_0 if T_1 is less than or equal to t_1. Accept H_0 if T_1 exceeds t_1. (This is the same as decision rule C in the binomial test.)

Example 1. Entering college freshmen have taken a particular high school achievement examination for many years, and the upper quartile is well established at a score of 193. A particular high school sends fifteen of its graduates to college, where they take the exam and get scores of

189	233	195	160	212
176	231	185	199	213
202	193	174	166	248

It is assumed that these fifteen students represent a random sample of all students from that high school who go on to college. One way of comparing college students from that high school with other college students is by testing the hypothesis that the above scores come from a population whose upper quartile is 193. That is,

$$H_0: \text{The upper quartile is 193}$$

is tested against the alternative

$$H_1: \text{The upper quartile is not 193}$$

where we are referring to the upper quartile of the test scores of all college students from that high school, past, present, or future.

The two-tailed quantile test is applied. A critical region of approximate size .05 is obtained by entering Table 3 with $n = 15$ and $p = .75$. There it is seen that, for the binomial random variable Y,

$$(7) \qquad\qquad P(Y \le 7) = .0173$$

and

$$P(Y \le 14) = .9866$$
$$(8) \qquad\qquad\qquad = 1 - .0134$$

The critical region of size

$$\alpha = .0173 + .0134$$
$$(9) \qquad\qquad\quad = .0307$$

corresponds to values of T_1 less than or equal to $t_1 = 7$, and values of T_2 greater than $t_2 = 14$.

In this example T_1 equals seven, the number of observations less than or equal to 193, and T_2 equals six, since one observation exactly equals 193.

Therefore T_1 is too small, and H_0 is rejected. The upper quartile for students from that high school does not seem to be 193.

Because the observed value of the test statistic T_1 was barely in the rejection region, the level of significance is as small as it could be to still result in rejection of H_0. Therefore the *critical level* in this example equals .0307, the same as the level of significance.

The one-tailed quantile test, with the large sample approximation, is illustrated in the following example.

Example 2.   The time interval between eruptions of Old Faithful geyser is recorded 112 times to see whether the median interval is less than or equal to 60 minutes (null hypothesis) or whether the median interval is greater than 60 minutes (alternative hypothesis). If the median interval is 60, then 60 is $x_{.50}$, or the median. If the median interval is less than 60, then 60 is a p quantile for some $p \geq .50$. Thus H_0 is $P(X \leq 60) \geq .50$, and H_1 is $P(X \leq 60) < .50$, where X is the time interval between eruptions. Assuming that the various intervals are independent and identically distributed, the one-tailed quantile test may be used, with decision rule C. The test statistic T_1 equals the number of intervals that are less than or equal to 60 minutes, and the critical region of size .05 corresponds to values of T_1 less than

$$t_1 = np^* + w_{.05}\sqrt{np^*(1 - p^*)}$$
$$= (112)(.50) - (1.645)\sqrt{(112)(.50)(.50)}$$
(10) $$= 47.3.$$

Of the 112 time intervals, 8 are 60 minutes or less, so T_1 equals 8 and H_0 is soundly rejected in favor of the alternative "the median time interval between eruptions is greater than 60 minutes." The critical level is

$$\hat{\alpha} = P(T_1 \leq 8)$$
$$= P\left(\frac{T_1 - np}{\sqrt{npq}} \leq \frac{8 - np}{\sqrt{npq}}\right)$$
(11) $$\cong P\left(Z \leq \frac{8 - (112)(.50)}{\sqrt{(112)(.50)(.50)}}\right)$$

where Z is a standard normal random variable. Then from Table 1,

$$\hat{\alpha} \cong P\left(Z \leq \frac{-48}{5.3}\right)$$
$$= P(Z \leq -9.05)$$
(12) $$\ll .0001$$

which is read "much less than .0001."

Theory.   First we shall explain why the hypotheses within the parentheses in A, B, and C are equivalent to the hypothesis not in parentheses.

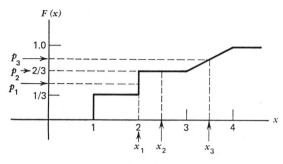

Figure 1

Perhaps this is most easily seen by referring to the graph of an arbitrary distribution function, as in Figure 1.

The distribution function at x^* may be in one of three phases: it may be rising vertically, as at x_1; it may be in a horizontal segment, as at x_2; or it may be rising gradually, as at x_3. H_0 in the third set of hypotheses (set C) states that the p^*th population quantile (x_{p*}) is no greater than x^*, or $x_{p*} \leq x^*$. Because every value of x^* is some sort of a quantile, we can say that x^* is the pth quantile for some p, say p_0. (We are temporarily ignoring our convention of choosing only the midpoint of the horizontal segments as the quantile, and adhering directly to the definition of quantile.) Because the graph of the distribution function never descends as x gets larger, $x_{p*} \leq x^*$ implies that $p^* \leq p_0$, which may be seen by imagining x^* as being in each of the three phases typified by x_1, x_2, and x_3 in Figure 1. Any value of x_{p*} to the left of x^* implies that the ordinate p^* at x_{p*} is no greater than the ordinate p_0 of x^*. From the definition of quantile, Definition 1.5.1,

$$(13) \qquad P(X > x^*) \leq 1 - p_0$$

which is the same as

$$(14) \qquad p_0 \leq 1 - P(X > x^*) = P(X \leq x^*)$$

Since $p^* \leq p_0$, this implies

$$(15) \qquad p^* \leq P(X \leq x^*)$$

which is the equivalent form of H_0 in set C of hypotheses. The negation of H_0 is H_1, and the negation of (15) is

$$(16) \qquad p^* > P(X \leq x^*)$$

as stated in the alternative hypothesis. The same reasoning is used to show the other hypotheses to be equivalent.

Briefly, Figure 1 is used to visualize that $x_{p_0} \leq x_{p*}$ (H_0 in B) implies that $p_0 \leq p^*$. If x^* equals x_{p_0}, then by Definition 1.5.1

$$\text{(17)} \qquad\qquad P(X < x^*) \leq p_0 \leq p^*$$

is true, which furnishes the equivalent form of H_0.

The binomial test is applied directly to test the hypotheses in parentheses. H_0 in B is tested by defining the "class 1" of the binomial test as those observations less than x^*. H_0 in C is tested by considering "class 1" to represent those observations less than or equal to x^*. The two tests in B and C are combined to give the two-tailed test in A. The assumptions of independence, and constant probability p in the binomial test are satisfied because the X_i are independent, and identically distributed (respectively).

In the previous section we showed how to find a confidence interval for a probability p. The same method is used to find a confidence interval for $F(x_0)$, the distribution function at some specified number x_0 (see Problem 3.1.5). That is, given a number x_0, we can find a "vertical" confidence interval (referring to a graph) for the unknown probability $F(x_0)$. Suppose, however, we are given a probability, say p^*, and asked to find a "horizontal" confidence interval for the unknown quantile x_{p*}. This second type of confidence interval, a confidence interval for a quantile, is found if we wish to make a statement concerning a specified quantile such as the median, the upper quartile, or any p^* quantile where p * is a specified constant, $0 \leq p^* \leq 1$. The statement then takes the form

$$\text{(18)} \qquad\qquad P(X^{(r)} \leq x_{p*} \leq X^{(s)}) = 1 - \alpha$$

where $1 - \alpha$ is a known *confidence coefficient*, and where $X^{(r)}$ and $X^{(s)}$ are order statistics (see Definition 2.1.4) with r and s specified. The values for r and s may be determined prior to drawing the sample in the manner described below, with knowledge of only the sample size n and the desired confidence coefficient. The sample $X_1, \ldots, X_n$ needs only to be random. No restrictions are made on the distribution function of the X_i. Thus this statistical method may be applied freely to any random sample from any population.

Confidence Interval for a Quantile

DATA. The data consist of observations on $X_1, X_2, \ldots, X_n$, which are independent and identically distributed random variables. Let $X^{(1)} \leq X^{(2)} \leq \ldots \leq X^{(r)} \leq \ldots \leq X^{(s)} \leq \ldots \leq X^{(n)}$ represent the ordered sample, where $1 \leq r < s \leq n$. We wish to find a confidence interval for the (unknown) p^*th quantile, where p^* is some specified number between zero and one.

ASSUMPTIONS
1. The sample $X_1, X_2, \ldots, X_n$ is a random sample.
2. The measurement scale of the X_i's is at least ordinal.

METHOD A (*small samples*). For $n \leq 20$ Table 3 may be used to find r and s. Enter Table 3 with the sample size n and the probability $p = p^*$. Read down the column for $p = p^*$ until reaching an entry approximately equal to $\alpha/2$, where $1 - \alpha$ is the approximate confidence coefficient desired. Call that entry α_1, and the corresponding value for y (to the far left of α_1) is $r - 1$. Add 1 to get r. Then continue down the column for $p = p^*$ until reaching an entry approximately equal to $1 - (\alpha/2)$, which we shall call $1 - \alpha_2$. The value of y corresponding to the entry $1 - \alpha_2$ is called $s - 1$, and 1 is added to obtain s. Thus we have determined α_1, α_2, r, and s. The exact confidence coefficient is $1 - \alpha_1 - \alpha_2$. The interval estimator is the interval between $X^{(r)}$ and $X^{(s)}$, whose values may be obtained from the data. Then

$$(19) \qquad P(X^{(r)} \leq x_{p*} \leq X^{(s)}) \geq 1 - \alpha_1 - \alpha_2$$

provides the confidence interval. If we assume that the unknown distribution function is continuous, then

$$(20) \qquad P(X^{(r)} \leq x_{p*} \leq X^{(s)}) = 1 - \alpha_1 - \alpha_2$$

as stated in (18) also.

METHOD B (*large sample approximation*). For n greater than 20 the approximation based on the Central Limit Theorem may be used. (See the end of Table 3, or Example 1.6.6.) Compute

$$(21) \qquad r^* = np^* + w_{\alpha/2}\sqrt{np^*(1 - p^*)}$$

and

$$(22) \qquad s^* = np^* + w_{1-\alpha/2}\sqrt{np^*(1 - p^*)}$$

where the quantiles w_p are obtained from Table 1, and where $1 - \alpha$ is the desired confidence coefficient. In general, r^* and s^* will not be integers. Let r and s be the integers obtained by rounding r^* and s^* upward to the next highest integers. Then the approximate confidence interval is given by (19), or (20) if the unknown distribution function is continuous.

A one-sided confidence interval may be formed by finding only r or s as described above. One-sided confidence intervals are of the form

$$(23) \qquad P(X^{(r)} \leq x_{p*}) = 1 - \alpha_1$$

and

$$(24) \qquad P(x_{p*} \leq X^{(s)}) = 1 - \alpha_2$$

if the distribution function is continuous, or

$$(25) \qquad P(X^{(r)} \leq x_{p*}) \geq 1 - \alpha_1$$

and

(26) $$P(x_{p*} \leq X^{(s)}) \geq 1 - \alpha_2$$

otherwise.

Example 3. Sixteen radio tubes are selected at random from a large batch of radio tubes, and are tested. The number of hours until failure is recorded for each one. We wish to find a confidence interval for the upper quartile, with a confidence coefficient close to 90%. Table 3 is entered with $n = 16$ and $p = .75$. Reading down the column for $p = .75$ the probability .0271 is selected as being close to .05. The value of y associated with $\alpha_1 = .0271$ is $y = 8$, therefore r equals 9. The probability closest to .95 is $.9365 = 1 - \alpha_2$, which has a corresponding y of 14. Therefore s equals 15. The confidence interval is

(27) $$P(X^{(9)} \leq x_{.75} \leq X^{(15)}) = .9094$$

(It is reasonable to assume the time to failure is a continuous random variable, so we can use Equation (20).)

The results of the testing, arranged in increasing order, are as follows.

$X^{(1)} = 46.9$	$X^{(5)} = 56.8$	$X^{(9)} = 63.3$	$X^{(13)} = 67.1$
$X^{(2)} = 47.2$	$X^{(6)} = 59.2$	$X^{(10)} = 63.4$	$X^{(14)} = 67.7$
$X^{(3)} = 49.1$	$X^{(7)} = 59.9$	$X^{(11)} = 63.7$	$X^{(15)} = 73.3$
$X^{(4)} = 56.5$	$X^{(8)} = 63.2$	$X^{(12)} = 64.1$	$X^{(16)} = 78.5$

Because $X^{(9)}$ equals 63.3 and $X^{(15)}$ equals 73.3, we may say "the interval from 63.3 hours to 73.3 hours, inclusive, is a 90.94% confidence interval for the upper quartile."

The large sample approximation, furnished by Equations (21) and (22), yields

$$r* = (16)(.75) + (-1.645)\sqrt{(16)(.75)(.25)}$$
$$= 12 - 2.86$$
(28) $$= 9.14$$

and

$$s* = 12 + 2.86$$
(29) $$= 14.86$$

Therefore r equals 10 and s equals 15, so the 90% confidence interval becomes (63.4, 73.3) slightly smaller than the more precise method used above.

Theory. Consider first the simpler case where the distribution function is continuous. If x_{p*} is the $p*$th quantile, we have the exact relationship

(30) $$P(X \geq x_{p*}) = P(X > x_{p*}) = 1 - p*$$

where the distribution function of X is the same as that of the random sample.

The order statistic of rank 1, $X^{(1)}$, will assume a value larger than some specified constant only if the smallest value in the sample is larger than the constant. Therefore $X^{(1)}$ is greater than the constant only if all n values in the sample are greater than the constant. Choosing x_{p*} as the constant, we may conclude

$$
\begin{aligned}
P(x_{p*} < X^{(1)}) &= P(\text{all sample values exceed } x_{p*}) \\
&= P(x_{p*} < X_1, x_{p*} < X_2, \ldots, x_{p*} < X_p) \\
&= P(x_{p*} < X_1) \cdot P(x_{p*} < X_2) \cdot \ldots \cdot P(x_{p*} < X_p) \\
&= (1 - p^*)^n
\end{aligned}
$$

(31)

because the X_i's are independent, and they all have the same $p*$th quantile x_{p*}.

If x_{p*} is less than $X^{(2)}$, then either exactly $n - 1$ observations are greater than x_{p*}, in which case $X^{(1)} \leq x_{p*} < X^{(2)}$, or else exactly n observations are greater than x_{p*}, in which case $x_{p*} < X^{(1)} < X^{(2)}$. Therefore

$$
\begin{aligned}
P(x_{p*} < X^{(2)}) &= P(x_{p*} < X^{(1)}) + P(X^{(1)} \leq x_{p*} < X^{(2)}) \\
&= P(\text{at least } n - 1 \text{ of the } X_i \text{ exceed } x_{p*}) \\
&= P(1 \text{ or fewer of the } X_i \text{ are } \leq x_{p*})
\end{aligned}
$$

(32)

Now the probability in (32) is given by the binomial distribution function, because each X_i has probability p^* of being less than or equal to x_{p*}, and the X_i are independent. Therefore (32) leads to

$$
(33) \qquad P(x_{p*} < X^{(2)}) = \sum_{i=0}^{1} \binom{n}{i} (p^*)^i (1 - p^*)^{n-i}
$$

With the aid of the binomial distribution function given by Equation (1.4.8), the above argument may be extended as follows.

$$
\begin{aligned}
P(x_{p*} < X^{(r)}) &= P(\text{at least } n - r + 1 \text{ of the } X_i \text{ exceed } x_{p*}) \\
&= P(r - 1 \text{ or fewer of the } X_i \text{ are } \leq x_{p*}) \\
&= \sum_{i=0}^{r-1} \binom{n}{i} (p^*)^i (1 - p^*)^{n-i}
\end{aligned}
$$

(34)

The confidence coefficient is given by

$$
\begin{aligned}
1 - \alpha &\cong P(X^{(r)} \leq x_{p*} \leq X^{(s)}) \\
&= P(x_{p*} \leq X^{(s)}) - P(x_{p*} < X^{(r)})
\end{aligned}
$$

(35)

Therefore r and s may be selected, with the aid of (34) and Table 3, so that

$$
(36) \qquad 1 - \alpha_2 = P(x_{p*} \leq X^{(s)}) = 1 - \frac{\alpha}{2}
$$

and

$$(37) \qquad \alpha_1 = P(x_{p^*} < X^{(r)}) = \frac{\alpha}{2}$$

Then the confidence coefficient will be $1 - \alpha_2 - \alpha_1 \cong 1 - \alpha$. Note that because the distribution function is assumed to be continuous, we have

$$(38) \qquad P(x_{p^*} \leq X^{(s)}) = P(x_{p^*} < X^{(s)})$$

so that Table 3 may be used to find s.

If the distribution function of X, and therefore of the X_i's, is not necessarily continuous, then Equation (30) is not necessarily true. Rather, by Definition 1.5.1 we have

$$(39) \qquad P(X > x_{p^*}) \leq 1 - p^*$$

and

$$(40) \qquad P(X \geq x_{p^*}) \geq 1 - p^*$$

First we shall consider how (39) affects (34), and therefore our method for finding r in (37). Because (39) is true, the probability of each observation exceeding x_{p^*} may be smaller than it was when X was continuous. Therefore there may be less of a tendency for each of the order statistics to exceed x_{p^*} than was formerly the case That is, the probability $P(x_{p^*} < X^{(r)})$ may be smaller than it was in the continuous case, which was then given by Equation (34). So in general the following holds true instead of (34).

$$(41) \qquad P(x_{p^*} < X^{(r)}) \leq \sum_{i=0}^{r-1} \binom{n}{i} (p^*)^i (1 - p^*)^{n-i}$$

If Table 3 is used to find r in the manner described above, then

$$(42) \qquad P(x_{p^*} < X^{(r)}) \leq \alpha_1$$

Now we shall consider how Equation (40) affects the probability $1 - \alpha_2$ resulting from our method of selecting s. Because (40) is true, there may be a larger probability for each observation to be greater than *or equal to* x_{p^*} than in the continuous case. Therefore the number of observations exceeding or equaling x_{p^*} may tend to be larger, and the probability of $X^{(s)} \geq X_{p^*}$ may be larger than in the continuous case. As a result, (34) is modified in the general case to read

$$(43) \qquad P(x_{p^*} \leq X^{(s)}) \geq \sum_{i=0}^{s-1} \binom{n}{i} (p^*)^i (1 - p^*)^{n-i}$$

Therefore if Table 3 is used to find s in the manner described above, we have

$$P(x_{p*} \leq X^{(s)}) \geq 1 - \alpha_2 \qquad (44)$$

Equations (42) and (44) which are true for all distributions may be used as follows.

$$P(X^{(r)} \leq x_{p*} \leq X^{(s)}) = P(x_{p*} \leq X^{(s)}) - P(x_{p*} < X^{(r)})$$
$$\geq P(x_{p*} \leq X^{(s)}) - \alpha_1$$
$$\geq 1 - \alpha_2 - \alpha_1 \qquad (45)$$

Thus the method of finding a confidence interval for a quantile has been justified for the case where exact tables of the binomial distribution function are available.

The large sample method of obtaining r and s is based on the use of the standard normal distribution to approximate the binomial distribution. Different arguments may be advanced for the different possible ways of converting r^* and s^* into the integers r and s, but the method given here of simply rounding upward to the next higher integer seems to provide a sufficiently close approximation.

The usage of one-sided confidence intervals for quantiles in life testing situations is discussed by Barlow and Gupta (1966).

PROBLEMS

1. Twenty independent observations on a random variable resulted in the numbers

142	134	98	119	131
103	154	122	93	137
86	119	161	144	158
165	81	117	128	103

Test the hypothesis that the median is 103.

2. In Problem 1 test the hypothesis that the upper quartile is at least 150.
3. In Problem 1 test the hypothesis that the third decile is no greater than 100.
4. In Problem 1 find an approximate 90% confidence interval for the median. What is the exact confidence coefficient? Also compare the results obtained using the exact method with the results obtained using the large sample approximation.
5. An automobile manufacturer wishes to allow enough headroom in his automobiles to comfortably accommodate all but the tallest 5% of the people who drive. Former studies indicate that the 95th percentile was 70.3 inches. In order to see if the results of the former studies are still valid, a random sample of size

100 is selected. It is found that the twelve tallest persons in the sample have the following heights.

72.6	70.0	71.3	70.5
70.8	76.0	70.1	72.5
71.1	70.6	71.9	72.8

Is it reasonable to use 70.3 as the 95th percentile?

6. In Problem 5, what is a 95% confidence interval for the 95th percentile of the population of drivers from which the sample was selected?

3.3. TOLERANCE LIMITS

The confidence intervals of the two preceding sections provide interval estimates for unknown population parameters, such as the unknown probability p or the unknown quantile x_p, and a certain probability $1 - \alpha$ (confidence coefficient) that the unknown parameter is within the interval. Tolerance limits differ from confidence intervals in that tolerance limits provide an interval within which at least a proportion q of the population lies, with probability $1 - \alpha$ or more than the stated interval does indeed "contain" the proportion q of the population. A typical application would be in a situation where we are about to draw a random sample of size n, X_1, X_2, ..., X_n, and we want to know how large n should be so that we can be ninety-five percent sure that at least ninety percent of the population lies between $X^{(1)}$ and $X^{(n)}$, the largest and smallest observations in our sample. We may generalize somewhat, and consider the question, "How large must the sample size n be so that at least a proportion q of the population is between $X^{(r)}$ and $X^{(n+1-m)}$ with probability $1 - \alpha$ or more " The numbers q, r, m, and $1 - \alpha$ are known (or selected) beforehand, and only n needs to be determined.

The above tolerance limit would be a two-sided tolerance limit. One-sided tolerance limits are of the form, "At least a proportion q of the population is greater than $X^{(r)}$, with probability $1 - \alpha$," or, "At least a proportion q of the population is less than $X^{(n+1-m)}$, with probability $1 - \alpha$." One-sided tolerance limits are identical with one-sided confidence intervals for quantiles as will be shown in this section.

The population referred to above is either infinite, or else if the population is finite the sample is drawn with replacement so the X_i are independent. For finite populations where the sampling is without replacement, and where the sample size n is small compared to the population size N, these methods are fairly accurate. More precise methods for finite populations may be found in Wilks (1962).

Tolerance Limits

DATA. Choose a confidence coefficient $1 - \alpha$, a pair of positive integers r and m, and a fraction q between zero and one. We wish to determine the size n of a random sample $X_1, X_2, \ldots, X_n$, for which we can make the statement, "The probability is $1 - \alpha$ that the random interval from $X^{(r)}$ to $X^{(n+1-m)}$ inclusive contains a proportion q or more of the population." Note that we are using the convention $X^{(0)} = -\infty$ and $X^{(n+1)} = +\infty$, so that one-sided tolerance limits may be obtained by setting either r or m equal to zero.

ASSUMPTIONS
1. The $X_1, X_2, \ldots, X_n$ constitute a random sample.
2. The measurement scale is at least ordinal.

METHOD. If $r + m$ equals 1, that is, if either r or m equals zero as in a one-sided tolerance limit, read n directly from Table 5 for the appropriate values of α and q. If $r + m$ equals 2, read n directly from Table 6 for the appropriate values of α and q. If Tables 5 and 6 are not appropriate, use the following approximation,

(1)
$$n \cong \tfrac{1}{4}x_{1-\alpha}\frac{1 + q}{1 - q} + \tfrac{1}{2}(r + m - 1)$$

where $x_{1-\alpha}$ is the $(1 - \alpha)$ quantile of a chi-square random variable with $2(r + m)$ degrees of freedom, obtained from Table 2.

Then with a sample of size n, there is probability $1 - \alpha$ that at least q [or $(100)(q)\%$] of the population is between $X^{(r)}$ and $X^{(n+1-m)}$ inclusive. That is

(2) $P(X^{(r)} \leq$ at least a fraction q of the population $\leq X^{(n+1-m)}) = 1 - \alpha$

For one-sided tolerance regions let either r equal zero or m equal zero, where $X^{(0)}$ and $X^{(n+1)}$ are considered to be $-\infty$ and $+\infty$ respectively, and proceed as described above.

Example 1. Probably the most widely used two-sided tolerance limits are those where $r = 1$ and $m = 1$. A certain manufactured product varies in length from one item to the next. In order to obtain the proper length box for shipping the item it is desirable that we obtain an upper and a lower limit, within which, we are ninety percent certain, lies eighty percent or more of all of the manufactured items. What must n be so that $X^{(n)}$ and $X^{(1)}$ furnish our upper and lower limits?

Table 6 is entered with $q = .80$ and $1 - \alpha = .90$. The obtained value for n is 18. The approximation furnished by Equation (1) is

$$n \cong \tfrac{1}{4}x_{1-\alpha}\frac{1 + q}{1 - q} + \tfrac{1}{2}(r + m - 1)$$

$$= \tfrac{1}{4}(7.779)\frac{1.80}{.20} + \tfrac{1}{2}$$

(3) $= 18.003$

A sample of size eighteen is drawn. The largest value in the sample is

$$X^{(18)} = 7.57 \text{ inches}$$

and the smallest value is

$$X^{(1)} = 7.21 \text{ inches}$$

Therefore there is probability .90 that at least eighty percent of the manufactured items are equal to or between 7.21 and 7.57 inches in length.

The following is an example of a one-sided tolerance limit.

Example 2.   Along with each lot of steel bars, a certain manufacturer guarantees that at least ninety percent of the bars will have a breaking point above a number specified for each lot. Because of variable manufacturing conditions the guaranteed breaking point is established separately for each lot by breaking a random sample of bars from each lot and setting the guaranteed breaking point equal to the minimum breaking point in the sample. How large should the sample be so that the manufacturer can be ninety-five percent sure his guarantee statement is correct?
Table 5 is entered with $q = .90$ and $1 - \alpha = .95$, with the result $n = 29$. In each lot a sample of size twenty-nine is selected at random, and the smallest breaking point of these bars in the sample is stated as the guaranteed breaking point, at which at least ninety percent of the bars in the lot will still be intact, with probability .95.

Theory.   A careful examination of the statement furnished by the one-sided tolerance limit reveals the similarity it has with the one-sided confidence interval for quantiles. That is, the one-sided tolerance limit says

(4) $P(\text{at least } q \text{ of the population is} \leq X^{(n+1-m)}) \geq 1 - \alpha$

However, "at least q of the population is $\leq X^{(n+1-m)}$" is the same as saying "the q quantile is $\leq X^{(n+1-m)}$"; the two statements are merely different ways of stating the same idea. So we have

$$P(\text{at least } q \text{ of the population is} \leq X^{(n+1-m)})$$

$$= P(\text{the } q \text{ quantile is} \leq X^{(n+1-m)})$$

(5) $$= P(x_q \leq X^{(n+1-m)})$$

The probability in (5) was given in Equation (3.2.43) as

(6) $$P(x_q \leq X^{(n+1-m)}) \geq \sum_{i=0}^{n-m} \binom{n}{i} q^i (1-q)^{n-i}$$

The right side of (6) is examined to find the smallest value for n such that the right side of (6) exceeds $1 - \alpha$. This may be accomplished by entering Table 3 with $y = n - m$, the parameter p equal to q, and then by searching for the lowest value of n for which the entry is greater than or equal to $1 - \alpha$. Because the value for y changes as n changes, it is more convenient to rewrite the right side of (6) as

$$(7) \qquad \sum_{i=0}^{n-m} \binom{n}{i} q^i (1-q)^{n-i} = 1 - \sum_{i=n-m+1}^{n} \binom{n}{i} q^i (1-q)^{n-i}$$

which is possible because the sum of all the binomial probabilities equals unity. A change of index, $j = m - i$, on the right side of (7) results in

$$(8) \qquad \sum_{i=0}^{n-m} \binom{n}{i} q^i (1-q)^{n-i} = 1 - \sum_{j=0}^{m-1} \binom{n}{j} (1-q)^j q^{n-j}$$

Equation (8) could have been obtained immediately by saying, "The probability of $n - m$ or fewer successes equals the probability of m or more failures, which equals 1 minus the probability of $m - 1$ or fewer failures." The combination of Equations (8) and (6) shows that we could find n by solving for the smallest value of n that satisfies

$$(9) \qquad \sum_{j=0}^{m-1} \binom{n}{j} (1-q)^j q^{n-j} \leq \alpha$$

which is obtained from the inequality

$$(10) \qquad 1 - \sum_{j=0}^{m-1} \binom{n}{j} (1-q)^j q^{n-j} \geq 1 - \alpha$$

Then Table 3 may be entered with $y = m - 1$, and $p = 1 - q$, and the pages turned until the entry in the table is less than or equal to α. That value of n is the sample size selected.

The other one-sided tolerance limit is

$$(11) \qquad P(X^{(r)} \leq \text{ at least } q \text{ of the population}) \geq 1 - \alpha$$

which is equivalent to the statement

$$(12) \qquad P(X^{(r)} \leq x_{1-q}) \geq 1 - \alpha$$

because at least $1 - q$ of the population is greater than or equal to x_{1-q}. Equation (12) becomes

$$(13) \qquad \alpha \geq 1 - P(X^{(r)} \leq x_{1-q}) = P(x_{1-q} < X^{(r)})$$

From Equation (3.2.41) we see that the solution to (13) is the smallest value of n such that

$$(14) \qquad \alpha \geq \sum_{i=0}^{r-1} \binom{n}{i} (1-q)^i q^{n-i}$$

just as was done above in (9).

In fact, it can be shown, with the aid of calculus (see Noether, 1967a) that for the two-sided tolerance limits, and for both types of one-sided tolerance limits, the sample size n depends on the solution to

$$(15) \qquad \alpha \geq \sum_{i=0}^{r+m-1} \binom{n}{i} (1 - q)^i q^{n-i}$$

which is somewhat surprising, in that (15) depends on the sum $r + m$, but does not depend on whether we wish to choose as our interval all values to the right of $X^{(r+m)}$, or all values to the left of $X^{(n+1-r-m)}$, or all values between $X^{(r)}$ and $X^{(n+1-m)}$, or any combination of two order statistics whose ranks have a difference of $n + 1 - m - r$.

The use of Table 3 to solve (15) is, at best, frustrating. Therefore Tables 5 and 6 are given for the most popular values $r + m = 1$ and $r + m = 2$. The approximation in Equation (1) is furnished without proof by Scheffé and Tukey (1944). Graphs to aid in finding n are given by Murphy (1948) and Birnbaum and Zuckerman (1949). More extensive tables are given by Owen (1962).

Tolerance limits may also be used with two samples (Danziger and Davis, 1964), with a single censored sample (Bohrer, 1968), or for deciding from which of two possibly multivariate populations a sample was obtained (Quesenberry and Gessaman, 1968). Usage of tolerance limits on discrete random variables is examined by Hanson and Owen (1963). An application of tolerance intervals to the regression problem is discussed by Bowden (1968). Other articles dealing with tolerance limits are given by Mack (1969) and Goodman and Madansky (1962).

PROBLEMS

1. What must the sample size be to be 90% sure that at least 95% of the population lies within the sample range?
   (a) Use the exact table.
   (b) Use the approximation.

2. What must the sample size be to be 95% certain that at least 90% of the population equals or exceeds $X^{(1)}$?
   (a) Use the exact table.
   (b) Use the approximation.

3. What must the sample size be in order for there to be a probability .90 that at least 85% of the population lies below $X^{(n)}$?
   (a) Use the exact table.
   (b) Use the approximation.
   (c) Use Table 3 as a check, and to find the exact value of $1 - \alpha$.

4. What must the sample size be in order for there to be 95% of a chance that 99% of the population lies above $X^{(2)}$?
   (a) Use the exact table.
   (b) Use the approximation.
5. What must the sample size be so there is probability .90 that at least 50% of the population is between $X^{(5)}$ and $X^{(n-4)}$?

3.4. THE SIGN TEST AND SOME VARIATIONS

After straying from hypothesis testing somewhat, at least in the previous section, we shall now return to discuss the oldest of all nonparametric tests, the sign test. Actually the sign test is just the binomial test, with $p^* = 1/2$. But the sign test deserves special consideration because of its versatility, its age (dating back to 1710), and because $p^* = 1/2 = 1 - p^*$ makes it even simpler than the binomial test. The sign test is useful for testing whether two populations have the same medians, where the observations come in pairs with one element of each pair being from each population. Also it may be used to test for trend in a series of ordinal measurements, or as a test for correlation. In many situations where the sign test may be used, more powerful nonparametric tests are available for the same model. However, the sign test is usually simpler and easier to use, and special tables to find the critical region are sometimes not needed.

The sign test as it is presented here is used with ordinal measurements which naturally occur in pairs (X, Y), to test whether the median of the X's is the same as the median of the Y's.

The Sign Test

DATA. The data consist of observations on a bivariate random sample (X_1, Y_1), $(X_2, Y_2), \ldots, (X_{n'}, Y_{n'})$, where there are n' pairs of observations. There should be some natural basis for pairing of the observations, otherwise the X's and Y's are independent, and the more powerful Mann-Whitney test of Chapter 5 is more appropriate.

Within each pair (X_i, Y_i) a comparison is made, and the pair is classified as "$+$" or "plus" if $X_i < Y_i$, as "$-$" or "minus" if $X_i > Y_i$, or as "0" or "tie" if $X_i = Y_i$. Thus the measurement scale needs only to be ordinal.

ASSUMPTIONS
1. The bivariate random variables (X_i, Y_i), $i = 1, 2, \ldots, n'$, are mutually independent.
2. The measurement scale is at least ordinal within each pair. That is, each pair (X_i, Y_i) may be determined to be a "plus," "minus," or "tie."

3. The pairs (X_i, Y_i) are internally consistent, in that if $P(+) > P(-)$ for one pair (X_i, Y_i), then $P(+) > P(-)$ for all pairs. The same is true for $P(+) < P(-)$, and $P(+) = P(-)$.

HYPOTHESES

A. (two-tailed test)

$$H_0 : P(X_i < Y_i) = P(X_i > Y_i) \qquad \text{for all } i$$

$H_1 :$ Either $P(X_i < Y_i) < P(X_i > Y_i)$ for all i, or $P(X_i < Y_i) > P(X_i > Y_i)$ for all i

B. (one-tailed test)

$$H_0 : P(X_i < Y_i) \leq P(X_i > Y_i) \qquad \text{for all } i$$
$$H_1 : P(X_i < Y_i) > P(X_i > Y_i) \qquad \text{for all } i$$

C. (one-tailed test)

$$H_0 : P(X_i < Y_i) \geq P(X_i > Y_i) \qquad \text{for all } i$$
$$H_1 : P(X_i < Y_i) < P(X_i > Y_i) \qquad \text{for all } i$$

It should be noted that the sign test is unbiased and consistent when testing the above hypotheses. The sign test is also used for testing the following counterparts of the above hypotheses, in which case it is neither unbiased nor consistent unless additional assumptions concerning the distributions of (X_i, Y_i) are made.

A. (two-tailed test). The null hypothesis is interpreted as "X_i and Y_i have the same location parameter," and therefore

$$H_0 : E(X_i) = E(Y_i) \qquad \text{for all } i$$

is tested against the alternative

$$H_1 : E(X_i) \neq E(Y_i) \qquad \text{for all } i$$

to see if X_i and Y_i have different means. Similarly, the test may be a test of medians;

$$H_0 : \text{The median of } X_i \text{ equals the median of } Y_i \text{ for all } i$$
$$H_1 : X_i \text{ and } Y_i \text{ have different medians for all } i$$

B. (one-tailed test). The null hypothesis may be considered to indicate that the values of X_i tend to be larger than the values of Y_i, because H_0 states that X_i may be more likely to exceed Y_i than to be less than Y_i. Therefore this one-tailed sign test is sometimes used to test

$$H_0 : E(X_i) \geq E(Y_i) \qquad \text{for all } i$$

against the alternative

$$H_1 : E(X_i) < E(Y_i) \qquad \text{for all } i$$

A similar set of hypotheses may be stated in terms of the median.

C. (one-tailed test). The null hypothesis in C above may be considered to indicate that X_i has a tendency to assume smaller values than does Y_i, and hence this one-tailed sign test may be used to test

$$H_0 : E(X_i) \leq E(Y_i) \qquad \text{for all } i$$

against the alternative

$$H_1:E(X_i) > E(Y_i) \qquad \text{for all } i$$

A similar statement may be made concerning the medians.

TEST STATISTIC. Let the test statistic T equal the number of "plus" pairs; that is, T equals the number of pairs (X_i, Y_i) in which X_i is less than Y_i.

$$T = \text{total number of } +\text{'s}$$

DECISION RULE. First disregard all tied pairs, and let n equal the number of pairs that are not ties.

$$n = \text{total number of } +\text{'s and } -\text{'s}$$

Let α represent the approximate level of significance desired.

Use decision rules A, B, or C depending on whether the hypothesis being tested is classified under categories A, B, or C above.

A. (two-tailed test) For $n \leq 20$, use Table 3 with the proper value of n, and with $p = 1/2$. Select a table value of about $\alpha/2$, and call it α_1. The value of y corresponding to α_1 is called t. The critical region of size $2\alpha_1$ corresponds to values of T less than or equal to t, or greater than or equal to $n - t$. Reject H_0 if $T \leq t$ or if $T \geq n - t$, at a level of significance of $2\alpha_1$. Otherwise accept H_0.

For n larger than 20 the approximation at the end of Table 3 is used to obtain

$$(1) \qquad\qquad t = \tfrac{1}{2}(n + w_{\alpha/2}\sqrt{n})$$

where $w_{\alpha/2}$ is obtained from Table 1. If α equals .05, then $w_{\alpha/2}$ equals (-1.96), and (1) becomes approximately

$$(2) \qquad\qquad t = \frac{n}{2} - \sqrt{n}$$

which may be easily remembered.

B. (one-tailed test) Large values of T indicate that a plus is more probable than a minus, as stated by H_1. Therefore the critical region corresponds to values of T greater than or equal to $n - t$, where t is found by entering Table 3 with $p = 1/2$ and n, and finding the table entry that approximately equals α, say α_1. The value of y corresponding to α_1 is t. For n greater than 20, t may be found from the approximation

$$(3) \qquad\qquad t = \tfrac{1}{2}(n + w_\alpha\sqrt{n})$$

where w_α is obtained from Table 1. H_0 is rejected at the level of significance α_1 (or α) if T is greater than or equal to $n - t$.

It is equivalent, and may be easier, to consider the test statistic T', equal to the number of "minus" pairs. Then H_0 is rejected if $T' \leq t$, where t is the same as above. As in (2) above, for $\alpha = .025$, t may be quickly computed using the equation

$$(4) \qquad\qquad t \cong n/2 - \sqrt{n}$$

C. (one-tailed test) Small values of T indicate that a minus is more probable than a plus, in agreement with H_1. Therefore t is found exactly as in B above. The critical region of size α_1 (or α) corresponds to values of T less than or equal to t. Reject H_0 if $T \leq t$, at a level of significance of α_1 (or α in the case of $n > 20$). Otherwise accept H_0.

Example 1.   An item A is manufactured using a certain process. Item B serves the same function as A, but is manufactured using a new process. The manufacturer wishes to determine whether B is preferred to A by the consumer, so he selects a random sample consisting of ten consumers, gives each of them one A and one B, and asks them to use the items for some period of time. The sign test (one tailed) will be used to test

$$H_0 : P(+) \leq P(-)$$

against

$$H_1 : P(+) > P(-)$$

where " $+$ " represents the event "item B is preferred over item A," and " $-$ " represents the event "item A is preferred over item B." In words, H_0 says, "Item B does not tend to be preferred to item A," while H_1 says, "Item B tends to be preferred to item A." The test statistic T is the number of $+$ signs, the number of consumers who prefer B over A. The critical region corresponds to values of T greater than or equal to $n - t$. However, we need to know how many ties there are before we can find n, and hence t.

At the end of the allotted period of time the consumers report their preferences to the manufacturer. Eight consumers preferred B to A, one preferred A to B, and one reported "no preference." Therefore

$$8 = \text{number of } +\text{'s}$$
$$1 = \text{number of } -\text{'s}$$
$$1 = \text{number of ties}$$
$$n = \text{number of } +\text{'s and } -\text{'s}$$
$$= 8 + 1 = 9$$
$$T = \text{number of } +\text{'s}$$
$$= 8$$

with $n = 9$ and $p = 1/2$, Table 3 is examined for an entry close to .05. The critical region of size $\alpha_1 = .0195$ corresponds to values of T greater than or equal to

$$n - t = 9 - 1 = 8$$

Since T equals 8, H_0 is rejected. The critical level $\hat{\alpha}$ equals .0195, because the observed value of T was barely in the rejection region.

The manufacturer decides that the consumer population prefers B to A.

A two-tailed sign test illustrating the use of the large sample approximation is presented in the next example.

Example 2.   In what was perhaps the first published report of a nonparametric
test, Arbuthnott (1710) examined the available London birth records of 82
years and for each year compared the number of males born with the number of
females born. If for each year we denote the event "more males than females
were born" by "$+$," and the opposite event by "$-$," (there were no ties), we
may consider the hypotheses to be

$$H_0 : P(+) = P(-)$$
$$H_1 : P(+) \neq P(-)$$

The test statistic T equals the number of $+$ signs, and the critical region of
size $\alpha = .05$ corresponds to values of T less than

$$t = .5(82 - (1.960)\sqrt{82})$$
$$= 32.1$$

and values of T greater than

$$n - t = 82 - 32.1$$
$$= 49.9$$

where t is calculated using Equation (3).

From the records, Arbuthnott obtained 82 plus signs, no minus signs and no
ties as we mentioned earlier. So T equals 82 and the null hypothesis is rejected.
In fact, H_0 could have been rejected at an α as small as

$$\hat{\alpha} = P(T = 0) + P(T = 82)$$
$$= (\tfrac{1}{2})^{82} + (\tfrac{1}{2})^{82}$$
$$= (\tfrac{1}{2})^{81}$$

Theory.   The event "$+$" represents the event "$Y_i > X_i$," or "$Y_i - X_i > 0$,"
which says that the difference $Y_i - X_i$ is positive. Similarly, "$-$" and
"0" represent the events $Y_i - X_i$ is negative, or zero respectively.
Therefore the sign test is a test for comparing the probability of a positive
difference with the probability of a negative difference. In the binomial
test these were called "class 1" and "class 2" probabilities, respectively.
By omitting ties, we have

(5) $$P(+) + P(-) = 1$$

and so the hypothesis

$$H_0 : P(+) = P(-)$$

is the same as saying

$$H_0 : P(+) = 1/2$$

which is in the same form as that of the binomial test with $p^* = 1/2$. So
the same binomial test procedure is used, although a slight simplification

results from the symmetry of

$$p^* = 1/2 = 1 - p^*$$

When the sign test is used with the original sets A, B, and C of hypotheses, the sign test is unbiased and consistent (Hemelrijk, 1952). Example 2.4.2 illustrated the binomial test with $p = 1/2$, which is the same as the sign test if there are no ties. Therefore the power functions graphed in Figure 2.4.2, in that example, are power functions for the sign test. It is evident from those graphs that the sign test is unbiased and consistent, although such evidence is not conclusive proof.

If, in addition to the assumptions in the sign test, we can also assume legitimately that the differences $Y_i - X_i$ are continuous random variables with a symmetric distribution function (the distribution function of a random variable Z is symmetric about some point c if $P(Z \leq c - x) = P(Z \geq c + x)$ for all x) then the Wilcoxon test for matched pairs is more appropriate (see Chapter 5). Furthermore, if the differences $Y_i - X_i$ are independent and identically distributed normal random variables, the appropriate parametric test is called the paired t-test. The A.R.E. compared to the paired t-test under these conditions is only $2/\pi = .637$. Also under these conditions the A.R.E. compared to the Wilcoxon test is $2/3$. Both small and large sample relative efficiencies have been examined by Walsh (1951), Dixon (1953), Hodges and Lehmann (1956), and Gibbons (1964), among others. Special tables for sample sizes to 1000 are given by MacKinnon (1964).

Data that occur naturally in pairs, as in the sign test, are usually analyzed by reducing the sequence of pairs to a sequence of single values, and then the data are analyzed as if only one sample were involved. That is, bivariate samples are usually analyzed using univariate techniques. In the sign test, the differences $Y_i - X_i$ were analyzed in the same manner that one would analyze a series of values to see if positive values are more likely than negative values. This principle of reducing bivariate (or even multivariate) data to a simple univariate sample is a useful one to remember.

Suppose now that the data are not ordinal as in the sign test, but nominal with two categories which we shall call "0" and "1." That is, each X_i is either 0 or 1 and similarly for each Y_i. Then a question sometimes asked is, "Can we detect a difference between the probability of (0, 1) and the probability of (1, 0)"? Such a question arises when the X_i in the pair (X_i, Y_i) represents the condition (or state) of the subject before the experiment, and Y_i represents the condition of the same subject after the experiment. The same procedure as used in the sign test may be used here also, but the test is well known by a different name.

The McNemar Test for Significance of Changes

DATA. The data consist of observations on n' independent bivariate random variables (X_i, Y_i), $i = 1, 2, \ldots, n'$. The measurement scale for the X_i and the Y_i is nominal with two categories, which we may call "0" and "1"; that is, the possible values of (X_i, Y_i) are $(0, 0)$, $(0, 1)$, $(1, 0)$, and $(1, 1)$. In the McNemar test the data are usually summarized in a 2 *by* 2 *contingency table*, as follows.

Classification of the Y_i

		$Y_i = 0$	$Y_i = 1$
Classification of the X_i	$X_i = 0$	a (the number of pairs where $X_i = 0$ and $Y_i = 0$)	b (the number of pairs where $X_i = 0$ and $Y_i = 1$)
	$X_i = 1$	c (the number of pairs where $X_i = 1$ and $Y_i = 0$)	d (the number of pairs where $X_i = 1$ and $Y_i = 1$)

ASSUMPTIONS
1. The pairs (X_i, Y_i) are mutually independent.
2. The measurement scale is nominal with two categories, for all X_i and Y_i.
3. The difference $P(X_i = 0, Y_i = 1) - P(X_i = 1, Y_i = 0)$ is negative for all i, or zero for all i, or positive for all i.

HYPOTHESES

$$H_0 : P(X_i = 0, Y_i = 1) = P(X_i = 1, Y_i = 0) \quad \text{for all } i$$
$$H_1 : P(X_i = 0, Y_i = 1) \neq P(X_i = 1, Y_i = 0) \quad \text{for all } i$$

These hypotheses may take a slightly different form, if we add $P(X_i = 0, Y_i = 0)$ to both sides of the equation in H_0 to get

$$H_0 : P(X_i = 0, Y_i = 1) + P(X_i = 0, Y_i = 0)$$
$$= P(X_i = 1, Y_i = 0) + P(X_i = 0, Y_i = 0)$$

The left side of H_0 includes all possibilities for Y_i, and hence equals $P(X_i = 0)$. Similarly, the right side includes all possibilities for X_i, and so equals $P(Y_i = 0)$. Therefore we have a new set of hypotheses in the form

$$H_0 : P(X_i = 0) = P(Y_i = 0) \quad \text{for all } i$$
$$H_1 : P(X_i = 0) \neq P(Y_i = 0) \quad \text{for all } i$$

Of course, these are also equivalent to

$$H_0 : P(X_i = 1) = P(Y_i = 1) \quad \text{for all } i$$
$$H_1 : P(X_i = 1) \neq P(Y_i = 1) \quad \text{for all } i$$

These latter sets of hypotheses are usually easier to interpret in terms of the experiment.

TEST STATISTIC. The test statistic for the McNemar test is usually written as

(6)
$$T_1 = \frac{(b - c)^2}{b + c}$$

However, for $b + c \leq 20$ the following test statistic is preferred

(7)
$$T_2 = b$$

Note that neither T_1 nor T_2 depends on a or d. This is because a and d represent the number of "ties," and ties are discarded in this analysis.

DECISION RULE. Let n equal $b + c$. If $n \leq 20$, use Table 3. If α is the desired level of significance, enter Table 3 with $n = b + c$, and $p = 1/2$, to find the table entry approximately equal to $\alpha/2$. Call this entry α_1, and the corresponding value of y is called t. Reject H_0 if $T_2 \leq t$, or if $T_2 \geq n - t$, at a level of significance of $2\alpha_1$. Otherwise accept H_0.

If n exceeds 20, use T_1 and Table 2. Reject H_0 at a level of significance α if T_1 exceeds the $(1 - \alpha)$ quantile of a chi-square random variable with one degree of freedom. Otherwise accept H_0.

Example 3.   Prior to a nationally televised debate between the two presidential candidates, a random sample of 100 persons stated their choice of candidates as follows. Eighty-four persons favored the Democratic candidate, and the remaining sixteen favored the Republican. After the debate the same hundred people expressed their preference again. Of the persons who formerly favored the Democrat, exactly one-fourth of them changed their minds, and also one-fourth of the people formerly favoring the Republican switched to the Democratic side. The results are summarized in the following 2 × 2 contingency table.

		After Dem.	After Rep.	Total Before
Before	Dem.	63	21	84
	Rep.	4	12	16
				100

The McNemar test may be used to test H_0: The population voting alignment was not altered by the debate, against H_1: There has been a change in the proportion of all voters who favor the Democrat. Consider the X_i in (X_i, Y_i) to be 0 if the ith person favored the Democrat before, or 1 if the Republican was favored before. Similarly Y_i represents the choice of the ith person after the debate. (Our choice of whether to represent the Democrat by 0 or 1 does not

affect the results, as long as the X_i and the Y_i use the same representation.)
The test statistic T_1 in the McNemar test becomes

$$T_1 = \frac{(b - c)^2}{b + c}$$

$$= \frac{(21 - 4)^2}{21 + 4}$$

$$= \frac{289}{25}$$

(8) $$= 11.56$$

The critical region of size $\alpha = .05$ corresponds to all values of T_1 greater than 3.841, the .95 quantile of a chi-square random variable with one degree of freedom, obtained from Table 2. Because 11.56 exceeds 3.841, the null hypothesis is rejected, and the conclusion is that the voter alignment has been altered. The critical level is less than .001.

Theory.   This test is a variation of the sign test, where the event $(0, 1)$ was called "$+$," the event $(1, 0)$ was called "$-$," and the events $(1, 1)$ and $(0, 0)$ were called ties. The hypothesis of the McNemar test then takes the form

$$H_0: P(+) = P(-)$$

which is the same as H_0 in the two-tailed sign test. The critical region for T_2 is found just as in the sign test for $n \leq 20$.

For n greater than twenty the sign test suggests using the normal approximation, based on the idea that

(9) $$Z = \frac{T_2 - n(\frac{1}{2})}{\sqrt{n(\frac{1}{2})(\frac{1}{2})}} = \frac{b - n(\frac{1}{2})}{(\frac{1}{2})\sqrt{n}}$$

has approximately the standard normal distribution when H_0 is true (see Example 1.6.6). Because n equals $b + c$, (9) reduces to

$$Z = \frac{b - \dfrac{b + c}{2}}{(\frac{1}{2})\sqrt{b + c}}$$

(10) $$= \frac{b - c}{\sqrt{b + c}}$$

Therefore

$$T_1 = Z^2$$

has approximately a chi-square distribution with one degree of freedom (see Theorem 1.6.3). A two-tailed test involving T_2 or Z is comparable to using the upper tail of $T_1 = Z^2$ for a critical region.

As the sign test was presented in both the two-tailed and the one-tailed forms, so could the McNemar test take both forms. The easiest way of performing a one-tailed McNemar test is just to use the one-tailed sign test.

Another modification of the sign test is one introduced by Cox and Stuart (1955), and is used to test for the presence of *trend*. A sequence of numbers is said to have trend if the later numbers in the sequence tend to be greater than the earlier numbers (upward trend) or less than the earlier numbers (downward trend). This test involves pairing the later numbers with the earlier numbers, and then performing a sign test on the pairs thus formed. If there is a trend, one member of each pair will have a tendency to be higher or lower than the other member. On the other hand, if there is no trend and the sequence of numbers actually represents observations on independent and identically distributed random variables, then there will be no tendency for one particular member of each pair to exceed the other one.

Cox and Stuart Test for Trend

DATA. The data consist of observations on a sequence of random variables $X_1, X_2, \ldots, X_{n'}$, arranged in a particular order, such as the order in which the random variables are observed. It is desired to see if a trend exists in the sequence. Group the random variables into pairs (X_1, X_{1+c}), (X_2, X_{2+c}), ... $(X_{n'-c}, X_{n'})$ where c equals $n'/2$ if n' is even, and c equals $(n' + 1)/2$ if n' is odd. (Note that the middle random variable is eliminated using this scheme if n' is odd.) Replace each pair (X_i, X_{i+c}) with a " $+$ " if $X_i < X_{i+c}$, or a " $-$ " if $X_i > X_{i+c}$, eliminating ties. The number of untied pairs is called n.

It should be mentioned that this test may be used to detect any specified type of nonrandom pattern, such as a sine wave or other periodic pattern. The sequence of random variables is merely rearranged so that the smallest numbers, as predicted, will be near the beginning of the sequence, and the larger numbers near the end. Then presence of an upward trend in the rearranged sequence is evidence that the predicted pattern is present in the original arrangement of the sequence.

ASSUMPTIONS
1. The random variables $X_1, X_2, \ldots, X_{n'}$ are mutually independent.
2. The measurement scale of the X_i's is at least ordinal.
3. Either the X_i's are identically distributed, or else there is a trend, that is, the later random variables are more likely to be greater than, rather than be less than, the earlier random variables (or vice versa).

HYPOTHESES. The following hypotheses are comparable to their counterparts in the sign test.

A. (two-tailed test)

$$H_0: P(X_i < X_{i+c}) = P(X_i > X_{i+c}) \qquad \text{for all } i$$
$$H_1: P(X_i < X_{i+c}) \neq P(X_i > X_{i+c}) \qquad \text{for all } i$$

B. (one-tailed test)

$$H_0: P(X_i < X_{i+c}) \leq P(X_i > X_{i+c}) \qquad \text{for all } i$$
$$H_1: P(X_i < X_{i+c}) > P(X_i > X_{i+c}) \qquad \text{for all } i$$

C. (one-tailed test)

$$H_0: P(X_i < X_{i+c}) \geq P(X_i > X_{i+c}) \qquad \text{for all } i$$
$$H_1: P(X_i < X_{i+c}) < P(X_i > X_{i+c}) \qquad \text{for all } i$$

The usual interpretation given to the above hypotheses is the following.

A. H_0: No trend exists.
   H_1: There is either an upward trend or a downward trend.

B. H_0: There is no upward trend.
   H_1: There is an upward trend.

C. H_0: There is no downward trend.
   H_1: There is a downward trend.

TEST STATISTIC. The test statistic T, as in the sign test, equals the number of + pairs (the pairs where X_{i+c} exceeds X_i).

DECISION RULE. The decision rule for the Cox and Stuart test for trend is exactly the same as the decision rule for the sign test, and therefore is not repeated here.

The following is an example in which the two-tailed Cox and Stuart test for trend is applied.

Example 4.   The total annual precipitation is recorded each year for nineteen years, and this record is examined to see if the amount of precipitation is tending to increase or decrease. The precipitation in inches was 45.25, 45.83, 41.77, 36.26, 45.37, 52.25, 35.37, 57.16, 35.37, 58.32, 41.05, 33.72, 45.73, 37.90, 41.72, 36.07, 49.83, 36.24, 39.90. Because $n' = 19$ is odd, the middle number 58.32 is omitted. The remaining numbers are paired;

$$(45.25, 41.05) \qquad (45.27, 41.72)$$
$$(45.83, 33.72) \qquad (52.25, 36.07)$$
$$(41.77, 45.73) \qquad (35.37, 49.83)$$
$$(36.26, 37.90) \qquad (57.16, 36.24)$$
$$(35.37, 39.90)$$

There are no ties, so n equals 9. The test statistic T equals the number of pairs in which the second number exceeds the first number. The critical region of size .0390 corresponds to values of T less than or equal to 1, and values of T greater than or equal to $9 - 1 = 8$.

For the data obtained, T equals 4, well within the region of acceptance. The critical level is 1.0. Therefore the null hypothesis "no trend exists" is accepted.

In Example 4 the assumptions of the model on which the test is valid are reasonable assumptions. Thus the test is reasonably valid. However, the assumptions listed are not all necessary. We need only to assume enough to satisfy the model for the sign test. That is, we need only assume:

1. The bivariate random variables (X_i, X_{i+c}) are mutually independent.
2. The probabilities $P(X_i < X_{i+c})$ and $P(X_i > X_{i+c})$ have the same relative size for all pairs.
3. Each pair (X_i, X_{i+c}) may be judged to be a $+$, a $-$, or a tie.

These assumptions are not as readily understood as the set of assumptions given in the test, but they may prove more useful in some applications such as the following.

Example 5.   On a certain stream the average rate of water discharge is recorded each month (in cubic feet per second) for a period of 24 months. The hypothesis to be tested is

$$H_0: \text{The rate of discharge is not decreasing}$$

against the alternative

$$H_1: \text{The rate of discharge is decreasing}$$

The rate of discharge is known to follow a yearly cycle, so that nothing is learned by pairing stream discharges for two different months. However, by pairing the same months in two successive years the existence of a trend can be investigated. The following data were collected.

Month	First Year	Second Year	Month	First Year	Second Year
Jan	14.6	14.2	Jul	92.8	88.1
Feb	12.2	10.5	Aug	74.4	80.0
Mar	104	123	Sep	75.4	75.6
Apr	220	190	Oct	51.7	48.8
May	110	138	Nov	29.3	27.1
Jun	86.0	98.1	Dec	16.0	15.7

The test statistic T equals the number of pairs where the second year had a higher discharge than the first year, which is 5 in this example. Because the test is to detect a downward trend, the critical region of size .0730 corresponds to all values of T less than or equal to 3 (from Table 3, $n = 12$, $p = 1/2$). Therefore H_0 is accepted. The critical level $\hat{\alpha}$ is given by

$$\hat{\alpha} = P(T \le 5 \mid H_0 \text{ is true})$$
$$= .3872$$

which is too large to be an acceptable α.

The examples presented in this section represent only a few of the many ways the sign test may be applied to test different types of hypotheses. Two more applications conclude this section. In the first the sign test is used as a simple method of detecting correlation, that is, detecting whether high values of one random variable tend to be paired with high values of a second random variable, and low values with low values (positive correlation) or whether high values of one random variable tend to be paired with low values of the second random variable, and low values with high values (negative correlation). The test involves arranging the pairs (the pairs remain intact) so that one member of the pair (either the first member or the second) is arranged in increasing order. If there is correlation the other member of the pair will exhibit a trend, upward if the correlation is positive, and downward if the correlation is negative. The Cox and Stuart test for trend may be used on the sequence formed by the other member of the pair.

Example 6.   Cochran (1937) compares the reactions of several patients with each of two drugs, to see if there is a positive correlation between the two reactions for each patient.

Patient	Drug 1	Drug 2	Patient	Drug 1	Drug 2
1	+ .7	+1.9	6	+3.4	+4.4
2	−1.6	+ .8	7	+3.7	+5.5
3	− .2	+1.1	8	+ .8	+1.6
4	−1.2	+ .1	9	.0	+4.6
5	− .1	− .1	10	+2.0	+3.4

Ordering the pairs according to the reaction from drug 1 gives

Patient	Drug 1	Drug 2	Patient	Drug 1	Drug 2
2	−1.6	+ .8	1	+ .7	+1.9
4	−1.2	+ .1	8	+ .8	+1.6
3	− .2	+1.1	10	+2.0	+3.4
5	− .1	− .1	6	+3.4	+4.4
9	.0	+4.6	7	+3.7	+5.5

The one-tailed Cox and Stuart test for trend is applied to the newly arranged sequence of observations on drug 2. The five resulting pairs are $(+.8, +1.9)$, $(+.1, +1.6)$, $(+1.1, +3.4)$, $(−.1, +4.4)$, $(+4.6, +5.5)$. Because we are testing

$$H_0: \text{There is no positive correlation}$$

against the alternative

$$H_1: \text{There is positive correlation}$$

we are, in essence, testing for the presence of an upward trend (H_1). The test statistic T equals 5, because in all five pairs the second observation on drug 2 exceeds the first observation on drug 2. The critical region of size .0312 (obtained from Table 3 for $n = 5$, $p = 1/2$, and hence $t = 0$) corresponds to the single value $T = 5$. Therefore the null hypothesis is rejected, and we may conclude that there is a positive correlation between reactions to the two drugs. The critical level in this example is also .0312.

The final example illustrates how the sign test, or rather the Cox and Stuart test for trend, may be used to test for the presence of a predicted pattern.

Example 7.   The number of eggs laid by a group of insects in a laboratory is counted on an hourly basis during a 24-hour experiment, to test

H_0: The 24 egg counts constitute observations on 24 identically distributed random variables

against the alternative

H_1: The number of eggs laid tends to be a minimum at 2:15 P.M., increasing to a maximum at 2:15 A.M., and decreasing again until 2:15 P.M.

The hourly counts are as follows.

Time	No. of Eggs	Time	No. of Eggs	Time	No. of Eggs
9 A.M.	151	5 P.M.	83	1 A.M.	286
10 A.M.	119	6 P.M.	166	2 A.M.	235
11 A.M.	146	7 P.M.	143	3 A.M.	223
Noon	111	8 P.M.	116	4 A.M.	176
1 P.M.	63	9 P.M.	163	5 A.M.	176
2 P.M.	84	10 P.M.	208	6 A.M.	174
3 P.M.	60	11 P.M.	283	7 A.M.	139
4 P.M.	109	Midnight	296	8 A.M.	137

If the alternative hypothesis is true, the egg counts nearest 2:15 P.M. should tend to be the smallest, and the egg counts nearest 2:15 A.M. should tend to be the largest. Therefore the numbers of eggs are rearranged according to the times, from the times nearest 2:15 P.M. to the times nearest 2:15 A.M.

Time	No. of Eggs	Time	No. of Eggs
2 P.M.	84	8 A.M.	137
3 P.M.	60	9 P.M.	163
1 P.M.	63	7 A.M.	139
4 P.M.	109	10 P.M.	208
Noon	111	6 A.M.	174
5 P.M.	83	11 P.M.	283
11 A.M.	146	5 A.M.	176
6 P.M.	166	Midnight	296
10 A.M.	119	4 A.M.	176
7 P.M.	143	1 A.M.	286
9 A.M.	151	3 A.M.	223
8 P.M.	116	2 A.M.	235

If H_1 is true these numbers should exhibit an upward trend. The Cox and Stuart one-tailed test for trend is used. The first half of the sequence (first column) is paired with the last half of the sequence (second column) with the result that the two egg counts on each line above form a pair. In all twelve pairs the number in the second column exceeds the number in the first column, so T equals 12. For $n = 12$, $p = 1/2$, Table 3 shows the critical region of size $\alpha = .0193$ corresponds to values of T greater than or equal to $12 - 2 = 10$. Therefore H_0 is rejected, and we conclude that the predicted pattern does seem to be present. The critical level is given as

$$\hat{\alpha} = P(T \geq 12) = .0002$$

Therefore H_0 would have been rejected at any reasonable level of significance.

Theory.   The Cox and Stuart test for trend is an obvious modification of the sign test, and therefore the distribution of the test statistic when H_0 is true is obviously binomial. Also the test is unbiased and consistent when the first sets A, B, and C of hypotheses are being used, but not necessarily so when the later sets are used. Stuart (1956) shows that the test, when applied to random variables known to be normally distributed, has an A.R.E. of .78 with respect to the best parametric test, a test based on the regression coefficient. Under the same conditions it has an A.R.E. of .79 compared to Spearman's or Kendall's rank correlation tests used as tests of randomness which will be presented in Chapter 5.

If the test is altered so that the middle one-third of the observations are eliminated, and only the first one-third of the observations are paired with the last one-third of the observations, the A.R.E. increases to .83 when compared to the best parametric test, under ideal conditions for the parametric test. Apparently the loss of data is small as compared with the gain in larger differences. This suggests another variation, that of pairing from the ends of the sequence. That is, by forming the pairs (X_1, X_n), (X_2, X_{n-1}), etc., using all the data, perhaps the larger differences may be preserved, along with no loss in data. The test may still be performed as described above, because the distribution of the test statistic under the null hypothesis remains unchanged.

The test for correlation, described in Example 6, has not been investigated to see what its properties are. One of the difficulties in applying the test for correlation is that if many observations equal each other, there is more than one way of arranging the observations so that the test for trend can be applied. Therefore it is recommended that the original data pairs be arranged using the pair member which has the smallest number of ties. Of the arrangements still possible due to ties, the conservative approach is to choose that arrangement which will be least likely to result in rejection of H_0.

A bivariate sign test for location is discussed by Chatterjee (1966). Other modifications of the sign test may be used to test for trends in dispersion (Ury, 1966), or to compare several treatments with a control (Rhyne and Steel, 1965). Rao (1968) uses the Cox and Stuart test for testing trend in dispersion. The power of the test for trend is discussed further by Mansfield (1962). Olshen (1967) presents tests for testing quadratic trend versus linear trend.

PROBLEMS

In each of the following problems clearly state H_0, H_1, α, $\hat{\alpha}$, the decision, and the name of the test used, where such information is appropriate.

1. Six students went on a diet in an attempt to lose weight, with the following results:

Name	Abdul	Ed	Jim	Max	Phil	Ray
Wt. before	174	191	188	182	201	188
Wt. after	165	186	183	178	203	181

Is the diet an effective means of losing weight?

2. The reaction time before lunch was compared with the reaction time after lunch for a group of twenty-eight office workers. Twenty-two workers found their reaction time before lunch was shorter; four found their reaction time after lunch shorter; and two could detect no difference. Is the reaction time after lunch significantly longer than the reaction time before lunch?

3. Two different additives were compared to see which one is better for improving the durability of concrete. One hundred small batches of concrete were mixed under various conditions, and during the mixing each batch was divided into two parts. One part received additive A and the other part received additive B. After the concrete hardened, the two parts in each batch were crushed against each other and an observer determined which part appeared to be the more durable. In seventy-seven cases the concrete with additive A was rated more durable, and in twenty-three cases the concrete with additive B was rated more durable. Is there a significant difference between the effects of the two additives?

4. Twenty-two customers in a grocery store were asked to taste each of two types of cheese and to declare their preference. Seven customers preferred one kind, 12 preferred the other kind, and 3 had no preference. Does this indicate a significant difference in preference?

5. In a certain city the mortality rate per 100,000 citizens, due to automobile accidents, for each of the last 15 years was 17.3, 17.9, 18.4, 18.1, 18.3, 19.6, 18.6, 19.2, 17.7, 20.0, 19.0, 18.8, 19.3, 20.2, 19.9. Is there any basis for the statement that the mortality rate is increasing?

6. For each of the last 34 years a small midwestern college recorded the average heights of male freshmen. The average were 68.3, 68.6, 68.4, 68.1, 68.4, 68.2, 68.7, 68.9, 69.0, 68.8, 69.0, 68.6, 69.2, 69.2, 68.9, 68.6, 68.6, 68.8, 69.2, 68.8,

68.7, 69.5, 68.7, 68.8, 69.4, 69.3, 69.3, 69.5, 69.5, 69.0, 69.2, 69.2, 69.1, 69.9.
Do these averages indicate an upward trend in heights?

7. A manufacturer computes the average cost in dollars of producing a certain item for each of 44 months with the resulting averages 13.65, 13.41, 13.53, 13.23, 13.58, 13.43, 13.73, 13.40, 13.70, 13.58, 13.80, 13.40, 13.63, 13.69, 13.92, 13.68, 13.72, 13.42, 13.66, 13.98, 13.81, 13.60, 13.32, 13.45, 13.27, 13.26, 13.28, 13.29, 13.10, 13.09, 13.36, 13.40, 13.35, 13.53, 13.66, 13.10, 13.28, 13.33, 13.02, 13.09, 13.12, 13.16, 12.96, 12.95. Is there a statistically significant trend in these averages?

8. In an experiment to determine the influence of suggestion, twenty straight lines of varying lengths were shown one at a time to subjects A and B, and the subjects estimated aloud the length of each line. Subject A stated his preference first and, unknown to subject B, was under instructions to overestimate the first ten lines and underestimate the last ten lines. After hearing subject A's estimate, subject B stated his estimate. The errors of the estimates, measured by subtracting the true lengths of the lines from the estimated lengths of the lines, were as follows:

Line	1	2	3	4	5	6	7	8	9	10
Error by A	+ .3	+1.1	+ .9	+ .6	+1.0	+1.3	+ .8	+1.6	+1.2	+ .8
Error by B	− .1	+ .6	+1.0	+ .7	+ .2	+ .9	− .1	+ .2	.0	+ .5
Line	11	12	13	14	15	16	17	18	19	20
Error by A	−1.3	−1.1	−1.3	− .7	−1.4	−1.1	− .8	− .5	−1.2	−1.0
Error by B	− .6	−1.2	−1.0	− .7	−1.0	− .1	− .5	.0	− .4	− .3

Is there a significant positive correlation between subject A's errors and subject B's errors?

9. A certain major league baseball player had compiled the following record over twelve years.

	1953	1954	1955	1956	1957	1958	1959	1960
No. of home runs	7	14	17	15	9	19	16	17
Batting averages	.212	.232	.234	.210	.201	.256	.261	.247

	1961	1962	1963	1964
No. of home runs	22	17	13	10
Batting average	.255	.241	.238	.235

Is there a significant correlation between the number of home runs he hit and his batting average for that year?

10. Test the following data to see if there is a significant correlation between the yearly income of a family and the number of children in that family.

Income	No. of Children	Income	No. of Children	Income	No. of Children
$3720	3	$6660	3	$9470	3
3832	2	6787	4	9650	1
3861	4	6975	2	9687	3
3941	3	7012	3	9776	1
4000	4	7118	5	9831	1
4166	2	7272	2	9902	2
4328	0	7465	5	10,067	2
4392	3	7786	4	10,317	3
4568	6	7812	1	10,617	1
4747	5	7937	2	10,899	3
4916	2	8005	1	10,945	4
5050	1	8070	3	11,182	1
5111	6	8267	2	11,474	3
5218	3	8330	4	11,799	2
5486	5	8564	5	12,055	2
5691	2	8734	0	12,221	3
5979	0	8851	1	12,853	1
6096	8	8957	4	14,672	1
6106	1	9122	2	15,270	0
6317	4	9349	4	22,843	2

11. One hundred thirty five citizens were selected at random, and were asked to state their opinion regarding U.S. foreign policy. Forty-three were opposed to the U.S. foreign policy. After several weeks, during which they received an informative newsletter, they were again asked their opinion, and 37 were opposed, 30 of the 37 being persons who originally were not opposed to the U.S. foreign policy. Is the change in numbers of people opposed to the U.S. foreign policy significant?

12. In the previous problem, suppose all 37 of the persons opposed to the U.S. foreign policy after the experiment were also among those opposed to the U.S. foreign policy before the experiment. Is the change in the number of people opposed to the U.S. foreign policy significant?

13. A barber shop is considering raising the price of haircuts ten cents and then giving the customers a coupon worth one free refreshing drink at a nearby establishment. A survey was conducted and 200 people, selected at random

from the population of real and potential customers, were given an explanation of this proposal. There were 40 people who were customers, and 3 of them stated they would cease being customers with such a plan. The rest would remain customers. Of the 160 noncustomers 11 said they would become customers. Test the null hypothesis that the proposed change will not increase the total number of customers that receive haircuts in that barbershop.

CHAPTER 4

Contingency Tables

4.1. THE 2 × 2 CONTINGENCY TABLE

A contingency table is an array of natural numbers in matrix form, where those natural numbers represent counts, or frequencies. For example, an entomologist observing insects may say he observed 37 insects, or he may say he observed:

Moths	Grasshoppers	Others	Total
12	22	3	37

using a 1 × 3 (one by three) contingency table. He may wish to be more specific and use a 2 × 3 contingency table:

	Moths	Grasshoppers	Others	Total
Alive	3	21	3	27
Dead	9	1	0	10
Total	12	22	3	37

The totals, consisting of two *row totals*, three *column totals*, and one *grand total* are optional, and are usually included only for the reader's convenience.

In general an $r \times c$ contingency table is an array of natural numbers arranged into r rows and c columns, and thus with rc *cells* or places for the numbers. This section is concerned only with the case where r equals 2 and c equals 2, the 2 × 2 contingency table. Because there are four cells, the 2 × 2 contingency table is also called the *fourfold* contingency table.

One application of the 2 × 2 contingency table arises when N objects (or persons), possibly selected at random from some population, are classified into one of two categories before a treatment is applied, or an event takes place. After the treatment is applied the N objects are again examined and classified into the two categories. The question to be answered is, "Does the treatment significantly alter the proportion of objects in each of the two categories." This use of the contingency table was introduced in Section 3.4, and the appropriate statistical procedure was seen to be a variation of the sign test known as the McNemar test. The McNemar test is often able to detect subtle differences, primarily because the same sample is used in the two situations (such as "before" and "after"). Another way of testing the same hypothesis tested with the McNemar test, is by drawing a random sample from the population before the treatment, and then comparing it with another random sample drawn from the population after the treatment. The additional variability introduced by using two different random samples is undesirable because it tends to obscure the changes in the population caused by the treatment. However, there are times when it is not practical, nor even possible, to use the same sample twice. Then the procedure to be described in this section may be used.

In addition to the situation described above, the procedure may be, and usually is, used to analyze two samples drawn from two different populations, to see if both populations have the same or different proportions of elements in a certain category. More specifically, two random samples are drawn, one from each population, to test the null hypothesis that the probability of event A (some specified event) is the same for both populations.

The Chi-Square Test for Differences in Probabilities, 2 × 2

DATA. A random sample of n_1 observations is drawn from one population (or before a treatment is applied) and each observation classified into either class 1 or class 2, the total numbers in the two classes being 0_{11} and 0_{12}, respectively, where $0_{11} + 0_{12} = n_1$. A second random sample of n_2 observations is drawn from a second population (or the first population after some treatment is applied) and the number of observations in class 1 or class 2 is 0_{21} or 0_{22}, respectively, where $0_{21} + 0_{22} = n_2$. The data are arranged in the following 2×2 contingency table.

	Class 1	Class 2	Total
Population 1	0_{11}	0_{12}	n_1
Population 2	0_{21}	0_{22}	n_2
			$N = n_1 + n_2$

The total number of observations is denoted by N.

ASSUMPTIONS
1. Each sample is a random sample.
2. The two samples are mutually independent.
3. Each observation may be categorized either into class 1 or class 2.

HYPOTHESES. Let the probability that a randomly selected element will be in class 1 be denoted by p_1 in population 1 and p_2 in population 2.

A. (two-tailed test)
$$H_0: p_1 = p_2$$
$$H_1: p_1 \neq p_2$$

B. (one-tailed test)
$$H_0: p_1 \leq p_2$$
$$H_1: p_1 > p_2$$

Note that it is not necessary for p_1 and p_2 to be known. The hypotheses merely specify a relationship between them. Also note that in the one-tailed test, either population may be designated as population 1, just so the hypotheses are appropriate for the situation.

TEST STATISTIC. The test statistic T is given by

(1)
$$T = \frac{N(0_{11}0_{22} - 0_{12}0_{21})^2}{n_1 n_2 (0_{11} + 0_{21})(0_{12} + 0_{22})}$$

DECISION RULE. The exact distribution of T is difficult to tabulate because of all the different combinations of values possible for $0_{11}, 0_{12}, 0_{21}$, and 0_{22}. Therefore the large sample approximation is used for the distribution of T, which is the chi-square distribution with one degree of freedom.

A. (two-tailed test) Reject H_0 at the approximate level α if T exceeds $x_{1-\alpha}$, the $(1 - \alpha)$ quantile of the chi-square random variable with one degree of freedom.

B. (one-tailed test) First compute the proportions of the samples in class 1, $0_{11}/n_1$ and $0_{21}/n_2$. If $0_{11}/n_1$ is less than or equal to $0_{21}/n_2$, in agreement with H_0, accept H_0 immediately. If $0_{11}/n_1$ exceeds $0_{21}/n_2$, then compute T, and reject H_0 at the approximate level $\alpha/2$ if T exceeds $x_{1-\alpha}$, the $(1 - \alpha)$ quantile of a chi-square random variable with one degree of freedom.

For convenience the $(1 - \alpha)$ quantiles of a chi-square random variable with one degree of freedom are given below, as they appear on the first line of Table 2.

$$x_{.750} = 1.323 \qquad x_{.990} = 6.635$$
$$x_{.900} = 2.706 \qquad x_{.995} = 7.879$$
$$x_{.950} = 3.841 \qquad x_{.999} = 10.83$$
$$x_{.975} = 5.024$$

The above approximation for the distribution of T is a large sample approximation. Since all samples are of finite size, the approximation is always open to criticism. Few will dispute the approximation if there are at least five observations in each of the four cells. A less convenient, but more widely accepted rule states that each of the four quantities

$(0_{11} + 0_{21})n_1/N$	$(0_{12} + 0_{22})n_1/N$
$(0_{11} + 0_{21})n_2/N$	$(0_{12} + 0_{22})n_2/N$

should exceed 5. However, both of these rules are merely arbitrary rules of thumb, given only because accurate guidelines are lacking.

Example 1.   Two carloads of manufactured items are sampled randomly to determine if the proportion of defective items is different for the two carloads. From the first carload 13 of the 86 items were defective. From the second carload 17 of the 74 items were considered defective.

	Defective	Nondefective	Totals
Carload 1	13	73	86
Carload 2	17	57	74
Totals	30	130	160

The assumptions are met, and so the two-tailed test is used to test

$$H_0: \text{the proportion of defectives are equal}$$

using the test statistic

$$T = \frac{N(0_{11}0_{12} - 0_{12}0_{21})^2}{n_1 n_2(0_{11} + 0_{21})(0_{12} + 0_{22})}$$
$$= \frac{160((13)(57) - (73)(17))^2}{(86)(74)(30)(130)}$$
$$= 1.61$$

The .95 quantile of a chi-square random variable with one degree of freedom is 3.841. Therefore the critical region of approximate size .05 corresponds to values of T greater than 3.841. In this example T is less than 3.841, so H_0 is accepted. By interpolation the critical level is found to be about $1 - .78$, or

$$\hat{\alpha} \cong 1 - .78 = .22$$

Therefore the decision to accept H_0 seems to be a fairly safe one.

The following example illustrates the use of the one-tailed test.

Example 2. At the U.S. Naval Academy a new lighting system was installed throughout the midshipmen's living quarters. It was claimed that the new lighting system resulted in poor eyesight, due to a continual strain on the eyes of the midshipmen. Consider a (fictitious) study to test the null hypothesis,

H_0: The probability of a graduating midshipman having 20–20 (good) vision is the same or greater under the new lighting system than it was under the old lighting system.

against the one-sided alternative

H_1: The probability of good vision is less now than it was.

To match the above H_0 with the H_0

$$H_0: p_1 \leq p_2$$

population 1 is defined as the population of midshipmen with the old lighting system, population 2 as the population of midshipmen with the new lighting system, and "class 1" represents 20-20 vision. The random samples are taken to be the entire graduation class just prior to the installation of the new lights, for population 1, and the first graduation class to spend four years using the new lighting system for population 2. It is hoped that these samples will behave the same as would random samples from the entire population of graduating seniors, real and potential.

Suppose the results were as follows.

	Good Vision	Poor Vision	
Old lights	$0_{11} = 714$	$0_{12} = 111$	$n_1 = 825$
New lights	$0_{21} = 662$	$0_{22} = 154$	$n_2 = 816$
Totals	1376	265	$N = 1641$

No calculations are needed to see that $0_{11}/n_1$ exceeds $0_{21}/n_2$. Therefore T is computed.

$$T = \frac{N(0_{11}0_{22} - 0_{12}0_{21})^2}{n_1 n_2 (0_{11} + 0_{21})(0_{12} + 0_{22})}$$

$$= \frac{(1641)[(714)(154) - (111)(662)]^2}{(825)(816)(1376)(265)}$$

$$= 8.9$$

At a level of significance of .05, H_0 will be rejected if T exceeds $x_{.90} = 2.706$. Therefore H_0 is rejected. The critical level $\hat{\alpha}$ may be found by interpolation.

The obtained value of T is 8.9, which is approximately equal to the .996 quantile of a chi-square random variable with one degree of freedom (obtained by interpolation). In a one-tailed test this corresponds to an α of $(1/2)(1 - .996) = .002$. Therefore $\hat{\alpha}$ equals .002.

We may therefore conclude that the populations represented by the two graduation classes do differ with respect to the proportions having poor eyesight, and in the direction predicted. That is, population 2 (with the new lights) has poorer eyesight than population 1 (with the old lights). Whether the poorer eyesight is a *result* of the new lights has not been shown. However, an association of poor eyes with the new lights has been shown, in this hypothetical example.

Theory. The 2 × 2 contingency table just presented is actually a special case of the $r \times c$ contingency table presented in the next section, and so the theory involved is a special case of the theory behind the $r \times c$ case. However, the exact distribution of the test statistic is difficult to find unless r and c are very small, so the exact distribution of T is presented now. It should be mentioned, before we proceed, that the form of T given in (1) appears to be different from the general form of T given in the next section. However, with a little algebra the two expressions, for $r = 2$ and $c = 2$, may be seen to be equivalent.

The exact probability distribution of T, when $H_0: p_1 = p_2 = p$ (say) is true, may be calculated as illustrated in the following. For the sample from population 1, the probability of exactly x_1 items in class 1 and $n_1 - x_1$ items in class 2 is given by the binomial probability distribution.

$$(2) \quad P\left(\text{population 1} \begin{array}{|c|c|} \hline \text{class 1} & \text{class 2} \\ \hline x_1 & n_1 - x_1 \\ \hline \end{array}\right) = \binom{n_1}{x_1} p^{x_1}(1 - p)^{n_1 - x_1}$$

Similarly the probability of the sample from population 2 having exactly x_2 items in class 1 and $n_2 - x_2$ items in class 2 is given by

$$(3) \quad P\left(\text{population 2} \begin{array}{|c|c|} \hline \text{class 1} & \text{class 2} \\ \hline x_1 & n_2 - x_2 \\ \hline \end{array}\right) = \binom{n_2}{x_2} p^{x_2}(1 - p)^{n_2 - x_2}$$

Because the two samples are independent the probability of the joint event may be obtained by multiplying the right sides of Equations (2) and (3). Thus

$$(4) \quad P\left(\begin{array}{l|c|c} & \text{class 1} & \text{class 2} \\ \hline \text{pop. 1} & x_1 & n_1 - x_1 \\ \hline \text{pop. 2} & x_2 & n_2 - x_2 \\ \end{array}\right) = \binom{n_1}{x_1}\binom{n_2}{x_2} p^{x_1+x_2}(1 - p)^{N-x_1-x_2}$$

In the simple case where n_1 equals 2 and n_2 equals 2 there are nine different points in the sample space, corresponding to the nine possible tables on the facing page.

The undefined values for T arise from the indeterminate form $0/0$. However, since the two outcomes that result in undefined values for T are strongly indicative that H_0 is true, just as the fifth outcome is strongly indicative that H_0 is true, we may arbitrarily define T to be 0 for the first and last outcomes in agreement with the fifth outcome. Then T has the following probability distribution.

$$p = 1/2 \qquad\qquad p = 1$$
$$P(T = 0) = 3/8 \qquad\qquad P(T = 0) = 1$$
$$P(T = 4/3) = 1/2$$
$$P(T = 4) = 1/8$$

In a similar manner for any sample sizes n_1 and n_2 the exact probability distributions may be found, after the appropriate defining of the undefined values of T. However, the probability function is not unique even when H_0 is assumed to be true, as is seen in the above example, but rather depends on p. Hence the null hypothesis in the above test is a composite hypothesis. It is not easy to show, but the size of the critical region is a maximum when p is equal to $1/2$. Therefore α may be found in the above small sample case by setting p equal to $1/2$. If the critical region corresponds to the largest value of T, namely $T = 4$, then α equals .125.

It is not easy to show that the asymptotic distribution of T is the chi-square distribution with one degree of freedom, and so we shall not attempt it here. The interested and well-qualified reader may find the asymptotic distribution derived in Cramér (1946).

A "correction for continuity" was introduced by Yates (1934) to partially compensate for the inaccuracy introduced by the use of a continuous distribution function (the chi-square) to approximate the discrete distribution function of T. Yates' modification involves using

$$(5) \qquad T = \frac{N\left(|0_{11}0_{22} - 0_{12}0_{21}| - \dfrac{N}{2}\right)^2}{n_1 n_2 (0_{11} + 0_{21})(0_{12} + 0_{22})}$$

instead of (1). The correction consists of reducing the absolute value of $0_{11}0_{22} - 0_{12}0_{21}$ by an amount $N/2$, before squaring. However, this author tends to agree with Pearson (1947), Plackett (1964), and Grizzle (1967) in recommending against the use of Yates' correction. The use of (5) tends to be overly conservative, and (1) appears to be more in agreement with a chi-square random variable with one degree of freedom.

Tables	Probabilities if H_0 is true	$(p = 1/2)$	$(p = 1)$	T
<table><tr><td>2</td><td>0</td></tr><tr><td>2</td><td>0</td></tr></table>	p^4	1/16	1	undefined
<table><tr><td>2</td><td>0</td></tr><tr><td>1</td><td>1</td></tr></table>	$2p^3(1 - p)$	1/8	0	4/3
<table><tr><td>2</td><td>0</td></tr><tr><td>0</td><td>2</td></tr></table>	$p^2(1 - p)^2$	1/16	0	4
<table><tr><td>1</td><td>1</td></tr><tr><td>2</td><td>0</td></tr></table>	$2p^3(1 - p)$	1/8	0	4/3
<table><tr><td>1</td><td>1</td></tr><tr><td>1</td><td>1</td></tr></table>	$4p^2(1 - p)^2$	1/4	0	0
<table><tr><td>1</td><td>1</td></tr><tr><td>0</td><td>2</td></tr></table>	$2p(1 - p)^3$	1/8	0	4/3
<table><tr><td>0</td><td>2</td></tr><tr><td>2</td><td>0</td></tr></table>	$p^2(1 - p)^2$	1/16	0	4
<table><tr><td>0</td><td>2</td></tr><tr><td>1</td><td>1</td></tr></table>	$2p(1 - p)^3$	1/8	0	4/3
<table><tr><td>0</td><td>2</td></tr><tr><td>0</td><td>2</td></tr></table>	$(1 - p)^4$	1/16	0	undefined

Another use for the 2×2 contingency table appears when each observation in a single sample of size N is classified according to two properties, where each property may take one of two forms. Then there are $(2)(2) = 4$ different combinations of the two properties, and the 2×2 contingency table is a convenient means of tabulating the number of observations in each category. However, this use of the 2×2 contingency table is a special case of the $r \times c$ contingency table, and does not have any special variation (such as the one-sided test of this section) which would warrant a separate presentation. Therefore it is presented in the next section.

Confidence intervals may be formed for any unknown probabilities associated with the 2×2 contingency table, or any contingency table for that matter, by applying the procedure described in Section 3.1. Similarly, the test given in Section 3.1 may be used on contingency tables, whenever the hypotheses are pertinent and the assumptions of the test are met.

A short-cut rule for the one-sided test is given by Ott and Free (1969). Further discussion of the continuity correction may be found in Mantel and Greenhouse (1968). The power of the test is discussed by Harkness and Katz (1964). For methods of combining the test statistics in several 2×2 contingency tables see Radhakrishna (1965) and Nelson (1966).

PROBLEMS

1. A random sample of 135 people was drawn from each of two populations. In the first sample there were 43 "opposed." In the second sample there were 37 "opposed." Is there a difference in the proportion of people "opposed" in the two populations? Does a comparison of this problem with Problems 11 and 12 in Section 3.4 suggest an advantage in using the same persons in both samples whenever possible, such as in a "before" and "after" situation?

2. Sixty students were divided into two classes of thirty each, and taught how to write a program for a computer. One class used the conventional method of learning, and the other class used a new, experimental, method of learning. At the end of the courses, each student was given a test, which consisted of writing a computer program. The program was either correct or incorrect, and the results were tabulated as follows.

	Correct program	Incorrect program
Conventional class	23	7
Experimental class	27	3

Is there reason to believe the experimental method is superior? Or could the above differences be due to chance fluctuations?

3. One hundred men and one hundred women were asked to try a new toothpaste, and to state whether they liked or did not like the new taste. Thirty-two men and twenty-six women said they didn't like the new taste. Does this indicate a difference in preferences between men and women in general?

4. In the test for differences in probabilities, find the exact probability distribution of the test statistic when $n_1 = 2, n_2 = 3$. Also let the largest value of T correspond to the critical region, and find α.

5. Contingency tables may be used to present data representing scales of measurement higher than the nominal scale. For example, a random sample of size twenty was selected from the graduate students who are U.S. citizens, and their grade point averages were recorded.

3.42	3.54	3.21	3.63	3.22
3.80	3.70	3.20	3.75	3.31
3.86	4.00	2.86	2.92	3.59
2.91	3.77	2.70	3.06	3.30

Also, a random sample of twenty students was selected from the non-U.S. citizen group of graduate students at the same university. Their grade point averages were as follows.

3.50	4.00	3.43	3.85	3.84
3.21	3.58	3.94	3.48	3.76
3.87	2.93	4.00	3.37	3.72
4.00	3.06	3.92	3.72	3.91

Test the null hypothesis that the proportion of graduate students with averages of 3.50 or higher is the same for both the U.S. citizens and the non-U.S. citizens.

4.2. THE $r \times c$ CONTINGENCY TABLE

As an immediate generalization of the 2×2 contingency table of the previous section, we have the contingency table with r rows and c columns, called the $r \times c$ (r by c) contingency table. This contingency table may be used, as in the previous section, to present a tabulation of the data contained in several samples, where the data represents at least a nominal scale of measurement, and to test the hypothesis that the probabilities do not differ from sample to sample. Another use for the $r \times c$ contingency table is with the single sample, where each element in the sample may be classified into one of r different categories according to one criterion, and at the same time into one of c different categories according to a second criterion. Both of these applications are treated the same in the statistical analysis, but basic differences between the two applications justify separate discussions of the two situations. A third application, similar to the other two, will also be discussed.

First we shall consider the extension of the application presented in the previous section. Now, instead of only two samples, we have r samples, where each sample is tabulated in one of the r rows. Instead of each sample furnishing two categories (formerly called class 1 and class 2) we now consider c categories, corresponding to the c columns. Thus the entry in the (i, j) cell (ith row and jth column) is the number of observations from the ith sample which belong to the jth category.

The Chi-Square Test for Differences in Probabilities, $r \times c$

DATA. There are r populations in all, and one random sample is drawn from each population. Let n_i represent the number of observations in the ith sample (from the ith population) for $1 \leq i \leq r$. Each observation in each sample may be classified into one of c different categories. Let 0_{ij} be the number of observations from the ith sample which fall into category j, so

$$(1) \qquad n_i = 0_{i1} + 0_{i2} + \ldots + 0_{ic} \qquad \text{for all } i$$

The data are arranged in the following $r \times c$ contingency table.

	Class 1	Class 2	$\ldots$	Class c	Totals
Population 1	0_{11}	0_{12}	$\ldots$	0_{1c}	n_1
Population 2	0_{21}	0_{22}	$\ldots$	0_{2c}	n_2
$\ldots$	$\ldots$	$\ldots$	$\ldots$	$\ldots$	$\ldots$
Population r	0_{r1}	0_{r2}	$\ldots$	0_{rc}	n_r
Totals	C_1	C_2	$\ldots$	C_c	N

The total number of observations from all samples is denoted by N.

$$(2) \qquad N = n_1 + n_2 + \ldots + n_r$$

The number of observations in the jth column is denoted by C_j. That is, C_j is the total number of observations in the jth category or class, from all samples combined.

$$(3) \qquad C_j = 0_{1j} + 0_{2j} + \ldots + 0_{rj}, \qquad \text{for } j = 1, 2, \ldots c$$

ASSUMPTIONS

1. Each sample is a random sample.

2. The outcomes of the various samples are all mutually independent (particularly among samples, because independence within samples is part of the first assumption).

3. Each observation may be categorized into exactly one of the c categories or classes.

HYPOTHESES. Let the probability of a randomly selected value from the ith population being classified in the jth class be denoted by p_{ij}, for $i = 1, 2, \ldots, r$, and $j = 1, 2, \ldots, c$.

H_0: All of the probabilities in the same column are equal to each other, (i.e., $p_{1j} = p_{2j} = \ldots = p_{rj}$, for all j).

H_1: At least two of the probabilities in the same column are not equal to each other, (i.e., $p_{ij} \neq p_{kj}$ for some j, and for some pair i and k).

Note that it is not necessary to stipulate the various probabilities. The null hypothesis merely states that the probability of being in class j is the same for all populations, no matter what that probability might be (and no matter which category we are considering).

TEST STATISTIC. The test statistic T is given by

$$(4) \qquad T = \sum_{i=1}^{r} \sum_{j=1}^{c} \frac{(0_{ij} - E_{ij})^2}{E_{ij}}, \qquad \text{where} \qquad E_{ij} = \frac{n_i C_j}{N}$$

While the term 0_{ij} represents the observed number in cell (i, j), the term E_{ij} represents the *expected* number of observations in cell (i, j), if H_0 is really true. That is, if H_0 is true the number of observations in cell (i, j) should be close to the ith sample size n_i multiplied by the proportion C_j/N of all observations in category j.

An equivalent expression for T, more suited for machine computations, is given by

$$(5) \qquad T = \sum_{i=1}^{r} \sum_{j=1}^{c} \frac{0_{ij}^2}{E_{ij}} - N$$

If there are only two rows ($r = 2$) then the test statistic may be computed using one of the following simpler forms,

$$(6) \qquad T = \frac{1}{\dfrac{n_1}{N}\left(1 - \dfrac{n_1}{N}\right)} \sum_{j=1}^{c} \frac{\left(0_{1j} - \dfrac{n_1 C_j}{N}\right)^2}{C_j}$$

or, for machine computations,

$$(7) \qquad T = \frac{1}{\dfrac{n_1}{N}\left(1 - \dfrac{n_1}{N}\right)} \left(\sum_{j=1}^{c} \frac{0_{1j}^2}{C_j} - \frac{n_1^2}{N}\right)$$

Similarly, if $c = 2$, the corresponding equations are

$$(8) \qquad T = \frac{1}{\dfrac{C_1}{N}\left(1 - \dfrac{C_1}{N}\right)} \sum_{j=1}^{r} \frac{\left(0_{i1} - \dfrac{C_1 n_i}{N}\right)^2}{n_i}$$

and

$$(9) \qquad T = \frac{1}{\frac{C_1}{N}\left(1 - \frac{C_1}{N}\right)}\left(\sum_{i=1}^{r} \frac{0_{i1}^2}{n_i} - \frac{C_1^2}{N}\right)$$

If both r and c equal 2, then T reduces to

$$(10) \qquad T = \frac{N(0_{11}0_{22} - 0_{12}0_{21})^2}{n_1 n_2 C_1 C_2}$$

which is the same T as presented in the previous section.

DECISION RULE. Because of the difficulties involved in tabulating the exact distribution of T, the approximation based on the large sample distribution (where the E_{ij} are large) is used to find the critical region. The critical region of approximate size α corresponds to values of T larger than $x_{1-\alpha}$, the $1 - \alpha$ quantile of a chi-square random variable with $(r-1)(c-1)$ degrees of freedom, obtained from Table 2. Reject H_0 if T exceeds $x_{1-\alpha}$. Otherwise accept H_0.

Because the asymptotic distribution is used, the approximate value for α, found as above, is a good approximation to the true value of α if the E_{ij} are fairly large. However, if some of the E_{ij} are small, the approximation may be very poor. Cochran (1954) states that if any E_{ij} is less than 1, or if more than 20% of the E_{ij} are less than 5 the approximation may be poor. This seems to be a good rule of thumb to follow, unless all of the E_{ij} are about the same size. If all (or most) of the E_{ij} are nearly the same size, and if r and c are not too small, then this author feels that the E_{ij} may be as small as 1.0 without endangering the validity of the test. If some of the E_{ij} are too small, several categories may be combined to eliminate the E_{ij} which are too small. Just which categories should be combined is a matter of judgment. Generally, categories are combined only if they are similar in some respects, so that the hypotheses retain their meaning.

Example 1.   A sample of students randomly selected from private high schools and a sample of students randomly selected from public high schools were given standardized achievement tests with the following results.

	0–275	276–350	351–425	426–500	Totals
Private school	6	14	17	9	46
Public school	30	32	17	3	82
Totals	36	46	34	12	128

To test the null hypothesis that the distribution of test scores is the same for private and public high school students, the test for differences in probabilities is used. A critical region of approximate size $\alpha = .05$ corresponds to values of

T greater than 7.815, obtained from the chi-square distribution in Table 2 with $(r - 1)(c - 1) = (2 - 1)(4 - 1) = 3$ degrees of freedom.

The values of E_{ij} are computed using (4), and are given as follows.

	Column	1	2	3	4
	Row 1	12.9	16.5	12.2	4.3
E_{ij}:					
	Row 2	23.1	29.5	21.8	7.7

Note that the E_{ij} satisfy Cochran's criteria. Also note that the row and column sums for the E_{ij} are always the same as those for the 0_{ij}. This may be used as a check on the calculations.

For the cell in row 1, column 1, we have

$$\frac{(0_{ij} - E_{ij})^2}{E_{ij}} = \frac{(0_{11} - E_{11})^2}{E_{11}} = \frac{(6 - 12.9)^2}{12.9} = \frac{47.61}{12.9} = 3.69$$

A similar calculation is made for each cell and the result, using (4) for purposes of illustration, is

$$T = 3.69 + .38 + 1.89 + 5.14$$
$$+ 2.06 + .21 + 1.06 + 2.87$$
$$= 17.3$$

Since 17.3 is greater than 7.815, the null hypothesis is rejected. In fact, the null hypothesis could have been rejected using a level of significance as small as .001, so

$$\hat{\alpha} \cong .001$$

The conclusion is that test scores are distributed differently among public and private high school students.

In Example 1 the data (the test scores before grouping) possessed at least an ordinal scale of measurement, a stronger scale than the nominal scale of measurement considered to be more appropriate for the test used. Actually a more powerful nonparametric test based on ranks could have been used on the test scores before they were grouped, namely the Mann-Whitney test presented in the next chapter. However, the data were sufficient to result in a clear cut decision using this test, and so the more powerful (and more tedious) test was not needed in this case.

It should be mentioned that not all of the calculations in Example 1 were required to determine that H_0 could be rejected. By inspection a shrewd observer might have determined that cells (1, 1) and (1, 4) might yield the largest contributions to the test statistic, and computations for those two cells alone indicate that the test statistic is at least $3.69 + 5.14 = 8.83$, already in the region of rejection.

Theory.   The exact distribution of T in the $r \times c$ case may be found in exactly the same way as it was found in the previous section for the 2×2 case. That is, the row totals (sample sizes) are held constant, and then all possible contingency tables having those same row totals are listed, and their probabilities are calculated. The column totals may vary freely from one table to the next, but the row totals may not change. This is the essential difference between this application of the contingency table and the next to be described. In the next application the row totals are not fixed, and therefore a greater number of different contingency tables are possible. The only requirement is that the total number of observations N remains the same for all tables. Also, a third variation will be presented, sometimes known as Fisher's exact test. In that application the row totals and the column totals are all fixed, and do not vary from table to table. The number of possible tables is greatly reduced, and the exact distribution is then much easier to find.

In all three applications of contingency tables in this section, the asymptotic distribution of T is the same, namely chi-square with $(r - 1)(c - 1)$ degrees of freedom. Therefore this distribution is used to provide an approximate value for α, so that exact tables are not needed. The asymptotic distribution is derived in Cramér (1946).

The second application of the $r \times c$ contingency table involves a single random sample of size N, where each observation may be classified according to two criteria. There are r categories (rows) resulting from the first criterion, and c categories (columns) resulting from the second criterion. Each observation is classified according to both criteria, and thus ends up being assigned to a particular cell in the $r \times c$ contingency table. The cell entries represent the number of observations belonging to that cell. A nominal scale of measurement is all that is required, although higher scales may be used. The hypothesis tested is one of independence; loosely stated, the null hypothesis says that the rows and columns represent two independent classification schemes. A more precise description is now given.

The Chi-Square Test for Independence

DATA.   A random sample of size N is obtained. The observations in the random sample may be classified according to two criteria. Using the first criterion each observation is associated with one of the r rows, and using the second criterion each observation is associated with one of the c columns. Let 0_{ij} be the number of observations associated with row i and column j simultaneously. The cell counts 0_{ij} may be arranged in an $r \times c$ contingency table.

Column	1	2	3	...	c	Totals
Row 1	0_{11}	0_{12}	0_{13}	...	0_{1c}	R_1
2	0_{21}	0_{22}	0_{23}	...	0_{2c}	R_2
...	...	...	...	...	...	...
r	0_{r1}	0_{r2}	0_{r3}	...	0_{rc}	R_r
Totals	C_1	C_2	C_3	...	C_c	N

The total number of observations in row i is designated by R_i, (instead of n_i as the previous test, to emphasize that the row totals are now random rather than fixed), and in column j by C_j. The sum of the numbers in all of the cells is N.

ASSUMPTIONS

1. The sample of N observations is a random sample. (Each observation has the same probability as every other observation of being classified in row i and column j, independently of the other observations.)

2. Each observation may be classified into exactly one of r different categories according to one criterion, and into exactly one of c different categories according to a second criterion.

HYPOTHESES

H_0: The event "an observation is in row i" is independent of the event "that same observation is in column j," for all i and j.

By the definition of independence of events, H_0 may be stated as follows.

H_0: P(row i, column j) = P(row i) · P(column j),   for all i, j

The negation of H_0 is conveniently stated as

H_1: P(row i, column j) $\neq$ P(row i) · P(column j)   for some i, j

TEST STATISTIC. Let E_{ij} equal $R_i C_j / N$. Then the test statistic is given by

$$(11) \qquad T = \sum_{i=1}^{r} \sum_{j=1}^{c} \frac{(0_{ij} - E_{ij})^2}{E_{ij}}$$

or, for machine calculation,

$$(12) \qquad T = \sum_{i=1}^{r} \sum_{j=1}^{c} \frac{0_{ij}^2}{E_{ij}} - N$$

where the summation is taken over all cells in the contingency table. More convenient equations for the cases where $r = 2$ and/or $c = 2$ may be obtained by using (6) through (10) and replacing n_i by R_i, the new notation for row total.

DECISION RULE. Reject H_0 if T exceeds the $1 - \alpha$ quantile of a chi-square random variable with $(r - 1)(c - 1)$ degrees of freedom, obtained from Table 2. The approximate level of significance is then α. For more discussion of the decision rule see the previous test of this section. The same discussion of the decision rule applies to this test also.

Example 2.   A random sample of students at a certain university were classified according to the college in which they were enrolled, and also according to whether they graduated from a high school in the state or out of the state. The results were put into a 2 × 4 contingency table.

	Engineering	Arts and Sciences	Home Economics	Other	Totals
In state	16	14	13	13	56
Out of state	14	6	10	8	38
Totals	30	20	23	21	94

In order to test the null hypothesis that the college in which each student is enrolled is independent of whether his high school training was in state or out of state, the chi-square test for independence is selected. The rejection region corresponds to values of T greater than 7.815, the .95 quantile of a chi-square random variable with $(r - 1)(c - 1) = 3$ degrees of freedom, obtained from Table 2. Therefore α is approximately .05.

Using either (11) or the more convenient (6), T is computed for these data, resulting in

$$T = 1.55$$

Therefore H_0 is accepted. From Table 2 we may say that $\hat{\alpha}$ exceeds .25.

Theory.   The exact distribution of T may be found in the manner described earlier in this section, and illustrated below for the relatively simple case where $N = 4$. Let p_{ij} be the probability of an observation being classified in row i and column j (cell i, j). (Note that this p_{ij} is not the same as the p_{ij} of the previous test. Here the sum of the p_{ij} in all cells is one. In the previous test the p_{ij} in each row added to unity.) Then the probability of the particular outcome

Column
1   2

	1	2
Row 1	a	b
Row 2	c	d

$$\overline{N}$$

is given by

$$(13) \qquad \frac{N!}{a!\,b!\,c!\,d!}\,(p_{11})^a(p_{12})^b(p_{21})^c(p_{22})^d$$

because the number of ways N objects can result in the above cell counts is given by the multinomial coefficient $N!/a!\,b!\,c!\,d!$, and each result has probability

$$(14) \qquad (p_{11})^a(p_{12})^b(p_{21})^c(p_{22})^d$$

The maximum size of the upper tail of T, when H_0 is true, is found by setting all of the p_{ij} equal to each other, $1/4$ in this case (we shall not prove this). Therefore α is found by computing

$$(15) \qquad P\left(\begin{array}{|c|c|} \hline a & b \\ \hline c & d \\ \hline \end{array}\right) = \frac{N!}{a!\,b!\,c!\,d!}\left(\frac{1}{4}\right)^N$$

for each possible arrangement. For $N = 4$ there are thirty-five different contingency tables. These are listed in Figure 1, along with their probabilities, and the corresponding values of T. As before, we define zero divided by zero to be zero.

Figure 1 shows the exact distribution of T, when all p_{ij} equal $1/4$, to be

$$P(T = 0) = 84/256 = .33$$
$$P(T = 4/9) = 48/256 = .19$$
$$P(T = 4/3) = 96/256 = .37$$
$$P(T = 4) = 28/256 = .11$$

If the critical region corresponds to the largest value of T, $T = 4$, then α equals $.11$. This compares with $\alpha = .125$ for a similar situation discussed in the previous section. A comparison of the above distribution, where only N is fixed, with the distribution derived in the previous section where the row totals are also fixed, shows that in this case the distribution of T is more complicated to obtain because of the many more possible tables. Also additional values of T are now possible, and the probability distribution is altered somewhat.

Even though the exact distributions of T under the two applications differ somewhat, the asymptotic distributions are both chi-square with $(r - 1)(c - 1)$ degrees of freedom.

In the third application of the contingency table, not only are the row totals fixed, as in the first application, but the column totals are also fixed. Thus the exact distribution of T is easier to find than in both applications previously introduced However, easier or not, the exact distribution is still too complicated for practical purposes, unless extensive tables or a computer are

T = 0 Outcome	Probability	T = 4/9 Outcome	Probability	T = 4/3 Outcome	Probability	T = 4 Outcome	Probability
4 0 / 0 0	$(1/4)^4$	2 1 / 1 0	$12(1/4)^4$	2 1 / 0 1	$12(1/4)^4$	3 0 / 0 1	$4(1/4)^4$
0 4 / 0 0	$(1/4)^4$	1 2 / 0 1	$12(1/4)^4$	0 2 / 1 1	$12(1/4)^4$	0 3 / 1 0	$4(1/4)^4$
0 0 / 0 4	$(1/4)^4$	0 1 / 1 2	$12(1/4)^4$	1 0 / 1 2	$12(1/4)^4$	1 0 / 0 3	$4(1/4)^4$
0 0 / 4 0	$(1/4)^4$	1 0 / 2 1	$12(1/4)^4$	1 1 / 2 0	$12(1/4)^4$	0 1 / 3 0	$4(1/4)^4$
3 1 / 0 0	$4(1/4)^4$	Total = 48/256		2 0 / 1 1	$12(1/4)^4$	2 0 / 0 2	$6(1/4)^4$
0 3 / 0 1	$4(1/4)^4$			1 2 / 1 0	$12(1/4)^4$	0 2 / 2 0	$6(1/4)^4$
0 0 / 1 3	$4(1/4)^4$			1 1 / 0 2	$12(1/4)^4$	Total = 28/256	
1 0 / 3 0	$4(1/4)^4$			0 1 / 2 1	$12(1/4)^4$		
1 3 / 0 0	$4(1/4)^4$			Total = 96/256			
0 1 / 0 3	$4(1/4)^4$						
0 0 / 3 1	$4(1/4)^4$						
3 0 / 1 0	$4(1/4)^4$						
2 2 / 0 0	$6(1/4)^4$						
0 2 / 0 2	$6(1/4)^4$						
0 0 / 2 2	$6(1/4)^4$						
2 0 / 2 0	$6(1/4)^4$						
1 1 / 1 1	$24(1/4)^4$						
Total = 84/256							

Figure 1

available The chi-square approximation is recommended for finding the critical region and α.

The Chi-Square Test with Fixed Marginal Totals

DATA. The data are summarized in an $r \times c$ contingency table, as in the two previous applications, except that the row and column totals are determined beforehand, and are therefore fixed rather than random.

Column	1	2	...	c	Totals
Row 1	0_{11}	0_{12}	...	0_{1c}	$n_{1.}$
2	0_{21}	0_{22}	...	0_{2c}	$n_{2.}$
...	...	...	...	...	...
r	0_{r1}	0_{r2}	...	0_{rc}	$n_{r.}$
Totals	$n_{.1}$	$n_{.2}$	...	$n_{.c}$	N

The row and column totals are denoted by $n_{i.}$ and $n_{.j}$ respectively, to emphasize the fact that they are given rather than random. The total number of observations is N.

ASSUMPTIONS
1. Each observation is classified into exactly one cell.
2. The observations are observations on a random sample. Each observation has the same probability of being classified into cell (i, j) as any other observation.
3. The row and column totals are given, rather than random.

HYPOTHESES. The hypotheses may be either of the two sets of hypotheses introduced in the two previous applications in this section, under the condition that the row and column totals are fixed. Or the hypotheses may be tailored to fit the particular experimental situation. Usually the hypotheses are variations of the independence hypotheses of the previous test. See the two examples 3 and 4 for particular modifications which are dictated by the experiment.

TEST STATISTIC. Let $E_{ij} = n_{i.}n_{.j}/N$ be the expected number of observations in cell (i, j). Then the test statistic, as before, is given by

(16) $$T = \sum_{i=1}^{r} \sum_{j=1}^{c} (0_{ij} - E_{ij})^2/E_{ij}$$

where the summation is over all rc cells. If r equals 2 or c equals 2 the special equations (6) through (10) may be used, with the appropriate change in notation.

DECISION RULE. Reject H_0 if T exceeds the $1 - \alpha$ quantile of a chi-square random variable with $(r - 1)(c - 1)$ degrees of freedom, obtained from Table 2. The approximate level of significance is then α. For more details, see the decision rule given in the first test of this section. All of the comments made there are equally valid here also.

Example 3.   The chi-square test with fixed marginal totals may be used to test the hypothesis that two random variables X and Y are independent. Starting with a scatter diagram of twenty-four points, which represent independent observations on the bivariate random variable (X, Y), a contingency table may be constructed. The x coordinate of each point is the observed value of X and the y coordinate is the observed value of Y in each observation on (X, Y).

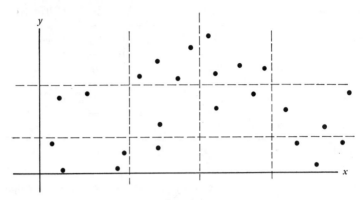

Figure 2

Assume the observed pairs (X, Y) are mutually independent. We wish to test

$$H_0: X \text{ and } Y \text{ are independent of each other}$$

against the alternative hypothesis of dependence.

To form the contingency table so that all E_{ij} are equal, we note that 3 and 4 both are factors of the sample size 24. Therefore we divide the points into 3 rows of eight points each, and 4 columns of 6 points each, using dotted lines as in Figure 2. (It is recommended that if the E_{ij} are small, they should be very nearly equal to each other. One way of accomplishing this is by having equal row totals, and equal column totals.) The resulting contingency table of counts is given as follows.

Column	1	2	3	4	Totals
Row 1	0	4	4	0	8
2	2	1	2	3	8
3	4	1	0	3	8
Totals	6	6	6	6	24

The critical region of approximate size .05 corresponds to values of T greater than 12.59, the .95 quantile of a chi-square random variable with $(r - 1)(c - 1) = (2)(3) = 6$ degrees of freedom, obtained from Table 2.

The test statistic is evaluated using (16), and $E_{ij} = (6)(8)/24 = 2$.

$$T = \sum_{i=1}^{r} \sum_{j=1}^{c} \frac{(0_{ij} - E_{ij})^2}{E_{ij}} = \sum_{i=1}^{3} \sum_{j=1}^{4} \frac{(0_{ij} - 2)^2}{2}$$

(17)

$$= 14$$

Because T exceeds 12.59, H_0 is rejected, and we conclude that X and Y are not independent. H_0 could have been rejected at a level of significance as small as .03, so $\hat{\alpha}$ equals .03.

Example 4. A psychologist asks a subject to learn twenty-five words. The subject is given twenty-five blue cards, each with one word on it. Five of the words are nouns, five are adjectives, five are adverbs, five are verbs, and five are prepositions. He must pair these blue cards with twenty-five white cards, each with one word on it and also containing the different parts of speech, five words each. The subject is allowed five minutes to pair the cards (one white card with each blue card) and five minutes to study the pairs thus formed. Then he is asked to close his eyes, and the words on the white cards are read to him one by one. When each word is read to him, he tries to furnish the word on the blue card associated with the word read.

The psychologist is not interested in the number of correct words, but rather in examining the pairing structure to see if it represents an ordering of some sort. The hypotheses are as follows.

H_0: There is no organization of pairs according to parts of speech against the alternative

H_1: The subject tends to pair particular parts of speech on the blue cards with particular parts of speech (not necessarily the same) on the white cards.

The pairings are summarized in a 5×5 contingency table.

		Noun	Adj.	Adv.	Verb	Prep.	Totals
				Blue Card			
White Card	Noun		3			2	5
	Adj.	4	1				5
	Adv.				5		5
	Verb			5			5
	Prep.	1	1			3	5
	Totals	5	5	5	5	5	25

The chi-square test with fixed marginal totals is selected because the experimenter feels that large values of T indicate H_1 is true. The marginal totals represent the number of words in each category, which was fixed in advance of the actual experiment. The critical region of approximate size .05 corresponds to values of T greater than 26.30, the .95 quantile of a chi-square random variable with $(r - 1)(c - 1) = (4)(4) = 16$ degrees of freedom, obtained from Table 2.

The observed value of T is obtained using (16).

$$E_{ij} = \frac{(5)(5)}{25} = 1 \qquad \text{for all } i \text{ and } j$$

$$T = \sum_{i=1}^{5} \sum_{j=1}^{5} \frac{(0_{ij} - 1)^2}{1}$$

(18)
$$= 66$$

Because T equals 66, H_0 is soundly rejected in favor of H_1. The critical level $\overset{\circ}{a}$ is less than .001.

Theory.   The exact distribution of T is found in a manner similar to the method described in the two previous applications of this section, and once in the previous section. We shall describe the method in the 2×2 case.

     Let A represent the event the following contingency table is obtained

	Class 1	Class 2	Total
Row 1	x_1	$n_1 - x_1$	n_1
Row 2	x_2	$n_2 - x_2$	n_2

where the row totals are considered to be fixed Then the probability of A, when H_0 is true, was computed in the previous section and was found to be given by Equation (4.1.4),

(19) $$P(A) = \binom{n_1}{x_1}\binom{n_2}{x_2} p^{x_1+x_2}(1-p)^{N-x_1-x_2}$$

Now let B be the event the column totals are $(x_1 + x_2)$ and $(N - x_1 - x_2)$ for columns 1 and 2 respectively. Since the probability of being in class 1 is denoted by p we have a situation where the binomial distribution applies, and so

(20) $$P(B) = \frac{N!}{(x_1 + x_2)!\,(N - x_1 - x_2)!} p^{x_1+x_2}(1-p)^{N-x_1-x_2}$$

We are interested in the probability of the event A given that the column totals are fixed, that is, given B is true. Because of the definition of conditional probability,

(21) $$P(A \mid B) = \frac{P(AB)}{P(B)}$$

we can find $P(A \mid B)$ by finding $P(AB)$, because $P(B)$ is given by (20).

Consider the fact that if A occurs (we obtain the 2 × 2 table specified by the event A) then B automatically occurs (the column totals are $x_1 + x_2$ and $N - x_1 - x_2$). The event AB occurs if and only if the event A occurs. So we have

$$(22) \qquad\qquad P(AB) = P(A)$$

where $P(A)$ is given by (19). Thus by dividing (19) by (20) we get

$$(23) \qquad\qquad P(A \mid B) = \frac{\binom{n_1}{x_1}\binom{n_2}{x_2}}{\binom{N}{x_1 + x_2}}$$

Therefore the probability for each 2 × 2 table, given that the row totals and the column totals are fixed, is given by (23), which may be recognized as the probability of the hypergeometric distribution. Note that we would have obtained the same result if we had fixed first the column totals and then the row totals.

If the row totals and column totals all equal 2, there are three possible contingency tables:

Table	Probability	T
$\begin{array}{c\|c} 2 & 0 \\ \hline 0 & 2 \end{array}$	$\dfrac{\binom{2}{2}\binom{2}{0}}{\binom{4}{2}} = 1/6$	4
$\begin{array}{c\|c} 1 & 1 \\ \hline 1 & 1 \end{array}$	$\dfrac{\binom{2}{1}\binom{2}{1}}{\binom{4}{2}} = 2/3$	0
$\begin{array}{c\|c} 0 & 2 \\ \hline 2 & 0 \end{array}$	$\dfrac{\binom{2}{0}\binom{2}{2}}{\binom{4}{2}} = 1/6$	4

Because the probability distribution of T is unique, H_0 is simple in this application.

Fixed row totals and fixed column totals greatly reduce the number of contingency tables possible, and so the exact distribution of T is more feasible in this case than in the previous two applications. When $r = 2$ and $c = 2$ the test is known as "Fisher's exact test," and extensive exact tables of probabilities are available (Finney, 1948). Programming Fisher's exact test is discussed by Robertson (1960).

For r and c in general, the exact probability of the table

Column	1	2	$\ldots$	c	Totals
Row 1	0_{11}	0_{12}	$\ldots$	0_{1c}	$n_{1.}$
2	0_{21}	0_{22}	$\ldots$	0_{2c}	$n_{2.}$
$\ldots$	$\ldots$	$\ldots$	$\ldots$	$\ldots$	$\ldots$
r	0_{r1}	0_{r2}	$\ldots$	0_{rc}	$n_{r.}$
Totals	$n_{.1}$	$n_{.2}$	$\ldots$	$n_{.c}$	N

with fixed marginal totals is given by

$$(24) \qquad \text{prob.} = \frac{\begin{bmatrix} n_{1.} \\ 0_{1i} \end{bmatrix}\begin{bmatrix} n_{2.} \\ 0_{2i} \end{bmatrix} \cdots \begin{bmatrix} n_{r.} \\ 0_{ri} \end{bmatrix}}{\begin{bmatrix} N \\ n_{.i} \end{bmatrix}}$$

where the multinomial coefficients are as defined by Rule 3 in Section 1.2.

Presumably, because there are fewer values for T in this third case. the chi-square approximation is not as good as in the first and second cases discussed in this section, and the approximation is probably the best in the second case.

The contingency tables of this section and Section 4.1 could be called *two-way contingency tables*, because the observations are classified two ways, by rows and by columns. An immediate extension may be made to include the situation where observations are classified according to three, or more, criteria, and thus the data are presented in the form of a three, or more, way contingency table. Of course, such a presentation of data becomes awkward on ordinary paper, and therefore is not often used. The next few remarks are made just in case such a presentation becomes necessary and the corresponding test is desired.

For convenience in extending the chi-square contingency table test, the two-way test statistic is rewritten as

$$(25) \qquad T = \sum_{i,j} \frac{\left[0_{ij} - N \frac{R_i}{N} \frac{C_j}{N} \right]^2}{N \frac{R_i}{N} \frac{C_j}{N}}$$

which has $(r - 1)(s - 1)$ degrees of freedom. In a three-way contingency table with r rows, s columns, and t blocks, denote the block totals by B_k, $k = 1, 2, \ldots, t$, to correspond to the row totals R_i and the column totals C_j. Let N still denote the total number of observations. Then

$$(26) \qquad R_i = \sum_{j,k} 0_{ijk}$$

$$(27) \qquad C_j = \sum_{i,k} 0_{ijk}$$

$$(28) \qquad B_j = \sum_{i,j} 0_{ijk}$$

where 0_{ijk} represents the total number of observations classified in row i, column j, and block k. Then E_{ijk}, the expected number of observations in row i, column j, and block k assuming the null hypothesis of row-column-block independence is true, may be estimated from

$$(29) \qquad E_{ijk} = N \frac{R_i}{N} \frac{C_j}{N} \frac{B_k}{N}$$

and the test statistic may be computed using

$$(30) \qquad T = \sum_{i,j,k} \frac{[0_{ijk} - E_{ijk}]^2}{E_{ijk}}$$

where the summation is over all $r \cdot s \cdot t$ cells. The test statistic is then tested for significance using the chi-square distribution with $(r - 1)(s - 1)(t - 1)$ degrees of freedom. The extension of the test to contingency tables of any dimension should be apparent. More detailed discussion of the multi-dimensional test is found in Goodman (1970) and Ireland, Ku, and Kullback (1969). A short book on analyzing contingency tables is by Maxwell (1961).

The exact distribution of the 2×3 contingency table test statistic is given for equal column totals and fixed row totals by Bennett and Nakamura (1963, 1964) and discussed by Healy (1969). Ireland and Kullback (1968) give a different test for contingency tables with given row and column totals. The power of chi-square tests for contingency tables is examined by Chapman and Meng (1966).

An excellent and readable survey article on contingency tables is one by Mosteller (1968). Haynam and Leone (1965) give an approximation to the exact distribution of T. Misclassification of data is the subject of an article by Mote and Anderson (1965). Tables with small or zero cell frequencies are discussed by Ku (1963) and Sugiura and Otake (1968). See Goodman (1964,

1968) and Bhapkar and Koch (1968) for information concerning tests for interaction. A class of bivariate contingency-type distributions is discussed by Plackett (1965), Mardia (1967b) and Steck (1968). Other methods for examining contingency tables are given by Ishii (1960), Gregory (1961), Claringbold (1961), Kullback, Kupperman and Ku (1962), Diamond (1963), Mielke and Siddiqui (1965), Hoeffding (1965), Gart (1966), and Chacko (1966). The large number of recent papers on contingency tables illustrates the usefulness and versatility of this type of analysis. Other applications of contingency tables are given in the following sections of this chapter.

PROBLEMS

1. Test whether the following observations indicate a dependence between the two variables observed: (3.6, 13), (4.7, 19), (1.4, 9), (5.5, 15), (4.8, 27), (4.3, 14), (3.0, 6), (4.2, 11), (6.0, 24), (6.8, 26), (4.1, 18), (3.2, 9), (4.0, 8), (1.9, 6), (.4, 7), (4.9, 14), (5.6, 18), (5.6, 20). Which test of this section is being used?

2. One horse was selected at random from each of 80 races and categorized according to post position (the position assigned to the horse for the start of the race) and the position in which the horse crossed the finish line (first, second, etc.).

		Finish			
		1	2	3	other
Post position	1–4	8	6	8	16
	5–9	3	6	5	28

Is the horse's position at the end of the race dependent on post position? Which test of this section is being used?

3. In another study, all of the horses in all of the races for three days were classified by post position and by the order in which they finished.

		Finish			
		1	2	3	other
Post position	1–4	15	14	15	52
	5–9	9	10	9	72

Is the horse's position at the end of the race dependent on post position? Which test of this section is being used?

4. Show that the two forms of T given by (4) and (5) are equivalent.

5. Show that if r equals 2 and c equals 2, then T may be given by (10).

4.3. THE MEDIAN TEST

The median test is designed to examine whether several samples came from populations having the same median. Actually the median test is not new to this chapter; it is merely a special application of the chi-square test with fixed marginal totals introduced in the previous section. It is a very useful application, however, and we consider it worth special treatment.

To test whether several (c) populations have the same median, a random sample is drawn from each population. (The scale of measurement is at least ordinal, or else the term "median" is without meaning.) A $2 \times c$ contingency table is constructed, and the two entries in the ith column are the numbers of observations in the ith sample which are above and below the grand median (the median of all observations combined). The usual chi-square test is then applied to the contingency table.

The Median Test

DATA. From each of c populations a random sample of size n_i is obtained, $i = 1, 2, \ldots, c$. The combined sample median is determined; that is, the number which is exceeded by about half of the observations in the entire array of N $(= n_1 + n_2 + \ldots + n_c)$ sample values is determined. This is called the *grand median*. Let 0_{1i} be the number of observations in the ith sample that exceed the grand median, and let 0_{2i} be the number in the ith sample that are less than or equal to the grand median. Arrange the frequency counts into a $2 \times c$ contingency table as follows.

Sample	1	2	$\ldots$	c	Totals
> median	0_{11}	0_{12}	$\ldots$	0_{1c}	a
≤ median	0_{21}	0_{22}	$\ldots$	0_{2c}	b
Totals	n_1	n_2	$\ldots$	n_c	$\overline{N}$

Let a equal the total number of observations above the grand median in all samples, and let b equal the total number of values less than or equal to the grand median. Then $a + b$ equals N, the total number of observations.

ASSUMPTIONS
1. Each sample is a random sample.
2. The samples are independent of each other.
3. The measurement scale is at least ordinal.

4. If all populations have the same median, then all populations have the same probability p of an observation exceeding the grand median.

HYPOTHESES

H_0: All c populations have the same median

H_1: At least two of the populations have different medians

TEST STATISTIC. The test statistic is obtained by a slight rearrangement of the test statistic given in the previous section, Equation (4.2.6), for the special case of two rows.

(1)
$$T = \frac{N^2}{ab} \sum_{i=1}^{c} \frac{\left(0_{1i} - \frac{n_i a}{N}\right)^2}{n_i}$$

If a calculator is being used, the following form is more convenient.

(2)
$$T = \frac{N^2}{ab} \sum_{i=1}^{c} \frac{0_{1i}^2}{n_i} - \frac{Na}{b}$$

If a is approximately equal to b, as it should be unless there are many values equal to the grand median, then the following simplification of the test statistic may be used.

(3)
$$T = \sum_{i=1}^{c} \frac{(0_{1i} - 0_{2i})^2}{n_i}$$

The simplified form for T given in (3) is exact if a exactly equals b. It is only approximate otherwise.

DECISION RULE. The exact distribution of T is difficult to tabulate, and so the large sample approximation is used to approximate the distribution of T. (See the discussion of the theory, later in this section, for the exact distribution of T.) The critical region of approximate size α corresponds to values of T greater than $x_{1-\alpha}$, the $(1 - \alpha)$ quantile of a chi-square random variable with $c - 1$ degrees of freedom, obtained from Table 2. If T exceeds $x_{1-\alpha}$, reject H_0. Otherwise accept H_0.

If some of the sample sizes n_i are too small, the above approximation may not be very accurate (the true α may vary considerably from the approximate α determined above). The same rule given in the previous section may be used as a rule of thumb. That is, the approximation may not be satisfactory if more than 20% of the n_i's are less than 10, or if any of the n_i's are less than 2. (This is almost equivalent to the rule stated in terms of the expected cell frequencies E_{ij} in the previous section). If most of the n_i's are about equal to each other, an exception to the above rule may be made, and n_i's as small as 2 may be allowed; however, the number of samples, c, should not be small in this case.

Example 1.   Four different methods of growing corn were randomly assigned
to a large number of different plots of land, and the yield per acre was computed
for each plot.

Method	1	2	3	4
	83	91	101	78
	91	90	100	82
	94	81	91	81
	89	83	93	77
	89	84	96	79
	96	83	95	81
	91	88	94	80
	92	91		81
	90	89		
		84		

In order to determine whether there is a difference in yields as a result of the
method used the median test was employed, because it was felt that a difference
in population medians could be interpreted as a difference in the value of the
method used. The hypotheses may be stated as follows.

H_0 : All methods have the same median yield per acre

H_1 : At least two of the methods differ with respect to the median yield per acre

A quick count reveals there are 34 observations in all, so the average of the 17th
and 18th smallest observations is the grand median. It is not necessary to order
all 34 observations to find the grand median. Rather, a count of the numbers
in the 70's and 80's reveals that 18 values are less than 90. Thus the average of
the two highest yields in the 80's is the grand median, and by inspection is seen
to be 89. Then for each method (sample) the number of values that exceed 89
and the number that are less than or equal to 89 are recorded in the following
form.

Method	1	2	3	4	Totals
> 89	6	3	7	0	16
≤ 89	3	7	0	8	18
Totals	9	10	7	8	34

The sample sizes are fairly small, but they are approximately equal, so we may
ignore the fact that three of them are less than 10 and use the chi-square approximation. The critical region corresponds to values of T greater than 7.815, the

.95 quantile of a chi-square random variable with $c - 1 = 3$ degrees of freedom, obtained from Table 2. T is computed using (1).

$$T = \frac{(34)^2}{(16)(18)} \left(\frac{\left(6 - \frac{(9)(16)}{34}\right)^2}{9} + \dots + \frac{\left(0 - \frac{(8)(16)}{34}\right)^2}{8} \right)$$

$$= 4.01(.34 + .29 + 1.97 + 1.78)$$

(4) $= 17.6$

Use of the more convenient (3) gives

$$T = \tfrac{9}{9} + \tfrac{16}{10} + \tfrac{49}{7} + \tfrac{64}{8}$$

(5) $= 17.6$

which is identical to the rounded-off value previously obtained.

Because the T of 17.6 exceeds the critical value 7.815, H_0 is rejected. Inspection of Table 2 shows the critical level $\hat{\alpha}$ to be slightly less than .001.

If the median test leads to rejection of the null hypothesis and it is desired to further inspect the samples to determine which population medians are different from each other, any subgroup of two or more populations may be analyzed using the median test, until the differences have been isolated. However, such "sorting out" of the populations by repeatedly using the same test on subgroups of the original data always distorts the true level of significance of all tests but the first. Such repeated testing procedures are for one's personal satisfaction, or for use as an objective "yardstick" for separating the various populations, but cannot receive the same interpretation as may legitimately be given to the first, overall, test. For a further discussion of repeated testing procedures see Gabriel (1966).

In the above example the experiment has been arranged in a so-called "completely randomized design," which assumes that the different methods are assigned to the different plots in some random manner (or a manner equivalent to a random manner). The usual parametric method of analyzing the data is called a "one-way analysis of variance."

The median test may be extended to become a "quantile test" for testing the null hypothesis that several populations have the same quantile, for any particular quantile chosen, merely by altering the data section of the test so that the observations are classified as being above or not above the grand quantile for the entire array of values. The remainder of this test remains the same, except that the approximation (3) will seldom be applicable. The exact distribution of T, as given in the theory to follow, remains the same.

Theory.   The row totals a and b are fixed, as in the third test of the previous section, because of the objective set of rules that is used to determine which observations are to be counted in the upper or lower cells of the contingency table. For example, if the test is an "upper quartile" test, then a is about $N/4$ and b is about $3N/4$, with allowances for tied, or equal, sample values. Thus the exact distribution of T is a conditional distribution which depends on the row and column totals. The probability of obtaining the table

(6)

0_{11}	0_{12}	$\cdots$	0_{1c}	a
0_{21}	0_{22}	$\cdots$	0_{2c}	b
n_1	n_2	$\cdots$	n_c	

with the row and column totals fixed, is the product of the binomial probabilities

(7)
$$P\left(\left|\begin{matrix} 0_{1i} \\ 0_{2i} \end{matrix}\right|\right) = \binom{n_i}{0_{1i}} p^{0_{1i}}(1-p)^{0_{2i}}; \quad i = 1, 2, \ldots, c$$

where p is the probability of an observation exceeding the grand median. Now H_0 merely states that all populations have the same median, and this does not necessarily imply that all populations have the same probability p of exceeding the grand (sample) median. On the other hand, H_0 does not preclude this latter situation, and the two statements (H_0, and the above situation) are very similar in intent. To find the distribution of T when H_0 is true, we need to require that the probabilities of exceeding the grand median be the same for all populations. This is why the fourth assumption was placed in the model.

(8)
$$P\left(\left|\begin{matrix} 0_{11} & 0_{12} & \cdots & 0_{1c} \\ 0_{21} & 0_{22} & \cdots & 0_{2c} \end{matrix}\right|\right) = \binom{n_1}{0_{11}}\binom{n_2}{0_{12}} \cdots \binom{n_c}{0_{1c}} p^a(1-p)^b$$

Because the samples are independent of each other we can obtain the joint probability by multiplication, using (7); where

$$a = 0_{11} + 0_{12} + \ldots + 0_{1c}$$

and

$$b = 0_{21} + 0_{22} + \ldots + 0_{2c}$$

The probability of the event in (8), given the row totals a and b, is found by dividing the probability in (8) by the probability of getting the row totals a and b, as was done in the latter part of the previous section.

The result is

$$
(9) \qquad P\left(\begin{array}{|c|c|c|c|}
\hline
0_{11} & 0_{12} & \cdots & 0_{1c} \\
\hline
0_{21} & 0_{22} & \cdots & 0_{2c} \\
\hline
n_1 & n_2 & \cdots & n_c
\end{array}
\begin{array}{c}
a \\[4pt]
b \\[4pt]
\overline{N}
\end{array}\right)
= \frac{\binom{n_1}{0_{11}}\binom{n_2}{0_{12}} \cdots \binom{n_c}{0_{1c}}}{\binom{N}{a}}
$$

which can be written in terms of multinomial coefficients as

$$
(10) \qquad \text{probability} = \frac{\left[\begin{array}{c} a \\ 0_{1i} \end{array}\right]\left[\begin{array}{c} b \\ 0_{2i} \end{array}\right]}{\left[\begin{array}{c} N \\ n_i \end{array}\right]}
$$

in agreement with Equation (24) of Section 4.2.

Thus the exact distribution of T can be, but almost never is, found using (9) or (10). Rather, the chi-square distribution with $(c$-1) degrees of freedom is used (because the number of rows is two) as in the previous section.

The above median test may be extended, so that more complex experiments may be analyzed. Because of the cumbersome notation involved in the extension of the median test, we shall introduce the test by presenting an example of its application.

An Extension of the Median Test

Example 2. Four different fertilizers are used on each of six different fields, and the entire experiment is replicated using three different types of seed. The yield per acre is calculated at the conclusion of the experiment, under each of the $(4)(6)(3) = 72$ different conditions with the following results.

	Seed 1				Seed 2				Seed 3			
Fertilizer:	1	2	3	4	1	2	3	4	1	2	3	4
Field: 1	80.5	90.1	87.0	88.0	79.1	87.0	82.6	81.5	85.4	92.3	92.0	89.3
2	87.0	83.4	89.1	90.3	77.6	82.0	81.4	87.9	89.2	90.1	90.2	93.6
3	86.1	82.4	91.0	86.1	84.1	80.6	89.0	80.4	90.0	88.1	87.2	90.8
4	82.1	84.9	84.4	83.1	83.3	79.5	86.3	83.1	83.4	85.3	94.3	87.6
5	79.3	87.1	92.2	90.8	76.6	86.2	84.0	87.4	87.1	86.3	88.4	93.7
6	84.2	89.3	85.3	84.7	81.0	84.1	88.1	85.0	82.3	92.9	95.1	82.9

To test the null hypothesis

H_0: There is no difference in median yields due to the different fertilizers let $x_{i_1 i_2 i_3}$ denote the observed yield using fertilizer i_1 in field i_2 with seed i_3. For example, x_{213} is the yield using fertilizer 2 in field 1 with seed 3, which is 92.3. Then x_{213} is compared with the median of $x_{113}, x_{213}, x_{313}$, and x_{413}, the four yields obtained under identical circumstances except for fertilizers (which H_0 claims to have no effect). Thus x_{213} is compared with the median of 85.4, 92.3, 92.0, and 89.3, which is

$$(\tfrac{1}{2})(89.3 + 92.0) = 90.65$$

If x_{213} exceeds 90.65, it is replaced in the table by a "1"; otherwise it is replaced by a zero.

Similarly, each $x_{i_1 i_2 i_3}$ is compared with the median of $x_{1 i_2 i_3}, x_{2 i_2 i_3}, \ldots, x_{c i_2 i_3}$, the observations obtained under similar conditions, except for the c different fertilizers. In our example each yield is compared with the median of the yields in the same row (field) and same block (seed), and replaced by one or zero according to whether it exceeds or does not exceed its respective median. The results are as follows.

	Seed 1				Seed 2				Seed 3			
Fertilizer:	1	2	3	4	1	2	3	4	1	2	3	4
Field: 1	0	1	0	1	0	1	1	0	0	1	1	0
2	0	0	1	1	0	1	0	1	0	0	1	1
3	0	0	1	0	1	0	1	0	1	0	0	1
4	0	1	1	0	1	0	1	0	0	0	1	1
5	0	0	1	1	0	1	0	1	0	0	1	1
6	0	1	1	0	0	0	1	1	0	1	1	0

Let 0_j be the number of yields in which fertilizer j was used and where the yield exceeded its respective median. Then 0_j is the total number of "ones" under fertilizer j in the above tables. The 0_j are given in the following table, for $j = 1, 2, \ldots, c$.

Fertilizer	1	2	3	4	Total
0_j = number of "ones"	3	8	14	10	$a = 35$
number of "zeros"	15	10	4	8	$b = 37$
	$n_1 = 18$	$n_2 = 18$	$n_3 = 18$	$n_4 = 18$	$N = 72$

The usual median test is then applied to the above table. Using Equation (3) we obtain

$$T = \frac{(144 + 4 + 100 + 4)}{18}$$

(11) $= 14.0$

A comparison of this value of T with the .95 quantile of a chi-square random variable with $c - 1 = 3$ degrees of freedom, obtained from Table 2 as $x_{.95} = 7.815$, results in rejection of H_0. The critical level $\hat{\alpha}$ in this experiment is about .004.

PROBLEMS

1. Test the hypothesis that the following samples were obtained from populations having the same medians.
   Sample 1: 35, 42, 42, 30, 15, 31, 29, 29, 17, 21
   Sample 2: 34, 38, 26, 17, 42, 28, 35, 33, 16, 40
   Sample 3: 17, 29, 30, 36, 41, 30, 31, 23, 38, 30
   Sample 4: 39, 34, 22, 27, 42, 33, 24, 36, 29, 25
2. Do the experimental results of Example 2 indicate a difference among seeds?
3. Do the experimental results of Example 2 indicate a difference among fields?
4. Show that if a equals b, then (1) becomes (3).
5. Show that (1) is the same as (4.2.16) when r equals 2.

4.4. MEASURES OF DEPENDENCE

The contingency table is a convenient form for examining data to see if there is some sort of dependence inherent in the data. The particular type of dependence revealed by a contingency table is a row-column dependence. If the different rows represent samples from different populations, and the columns represent different categories of classification of the data from the samples, then a row-column dependence is synonymous with a functional dependence of the probabilities of being in the various categories on the population from which the sample was obtained. Similarly, if the observations from one random sample are classified into rows and columns according to each of two different criteria, then a row-column dependence has an obvious interpretation as a dependence between the two criteria of classification.

Suppose that instead of testing hypotheses, as we have been doing so far in this chapter, we merely wish to express the degree of dependence shown in a particular contingency table. Ideally, we would like to be able to express the degree of dependence in some simple form, and in a form that easily conveys to other people the exact degree of dependence exhibited by the table.

As a first approach, we could use the test statistic of the previous sections

$$(1) \qquad T = \sum_{i=1}^{r} \sum_{j=1}^{c} \frac{(0_{ij} - E_{ij})^2}{E_{ij}}$$

as a measure of dependence, with the philosophy, "If it is good enough to test for dependence, it is good enough to measure dependence." The use of T appears satisfactory as far as convenience and simplicity are concerned.

However, in order to convey the degree of dependence to other people, the number of degrees of freedom should also be stated along with the T value, because without knowing the number of degrees of freedom it is not possible to tell the degree of dependence conveyed by the value of T. Even if the number of degrees of freedom is known, the nonexpert must consult a chi-square table in order to interpret T.

One widely used measure of dependence is the *critical level*, the smallest level of significance which would result in rejection of the null hypothesis in a hypothesis test involving T. Instead of the experimenter reporting the T value, and the number of degrees of freedom, so that the reader may consult a chi-square table to determine how large T is, the experimenter could do all this for the reader and find the quantile x_p (of the appropriate chi-square distribution) closest to T. The experimenter then simply reports

$$(2) \qquad\qquad \hat{\alpha} = 1 - p$$

where T is the pth quantile of the appropriate chi-square distribution. If $\hat{\alpha}$ is small (close to zero) much dependence is shown. If $\hat{\alpha}$ is large (close to 1.0) the interpretation is that an extreme amount of independence is shown. Values of $\hat{\alpha}$ around 0.5 would be commonplace if there were no row-column dependence.

The above measure of dependence is also called simply the *chi-square prob-ability*, but this term may cause confusion because it may create the impression that the number being furnished is p instead of $1 - p$.

The Critical Level as a Measure of Dependence

Example 1. In Example 4.2.1 the contingency table was as follows.

	Scores				
	0–275	276–350	351–425	426–500	Totals
Private school	6	14	17	9	46
Public school	30	32	17	3	82
Totals	36	46	34	12	128

For this contingency table T was found to equal 17.3. Now, 17.3 is approximately the .999 quantile of a chi-square random variable with three degrees of freedom, so p equals .999, and we have

$$\hat{\alpha} = 1 - p$$
$$= .001$$

as computed in the example. Such a small value of the critical level indicates a great degree of dependence between the type of school and the test scores.

Another approach to the problem of providing an easily interpreted measure of dependence consists of modifying the value of T in (1) in such a way that the result does not depend as much on the number of degrees of freedom as T does. One such modification considers dividing T by the maximum value T may attain. We know by now that large values of T arise from contingency tables which have a pronounced unevenness among the cell counts. By examining extremely uneven contingency tables we may find, by trial and error, that T is greatest (for a given number of rows r and columns c and total sample size N) when there are zeros in every cell except for one cell in each row and in each column. (If r does not equal c some rows or columns may be all zeros.) Furthermore, T achieves its maximum value if the nonzero cells are all equal to each other. That is, T is a maximum in a contingency table resembling the following.

Column	1	2	3	4	5	Totals
Row 1	3	0	0	0	0	3
2	0	3	0	0	0	3
3	0	0	3	0	0	3
Totals	3	3	3	0	0	9

For this table, T equals 18, after $0/0$ is defined as 0 or, equivalently, after omitting columns (and rows) with all zero cells.

In general, the maximum value of T is $N(q - 1)$, where q is either r or c whichever is smaller, and N is the total number of observations. Division of T by its maximum gives

$$(3) \qquad R_1 = \frac{T}{N(q - 1)}$$

where q is the smaller of r and c. R_1 is close to 1.0 if the table indicates a strong row-column dependence, and close to 0 if the numbers across each row are in the same proportions to each other as the column totals are to each other. This measure was suggested by Cramér (1946, p. 443).

Cramér's Contingency Coefficient

Example 2. In the previous example, the 2 × 4 contingency table furnished a value of T equal to 17.3. Because N equals 128 and q equals 2, we have R_1

given by

$$R_1 = \frac{T}{N(q-1)} = \frac{17.3}{128}$$
$$= .135$$

This rather small value of R_1 might be interpreted as being indicative of little or no dependence, but such an interpretation is incorrect as we have already seen in the previous example. It is true that the measure of dependence is slight when compared with the possibility of total dependence, but it is large when compared with the possibility of no dependence. In general R_1 has the desirable feature of being between 0 and 1.0 at all times, but it has the undesirable feature of depending on r and c for its interpretation. The larger r and c are, the larger T tends to be, and division by $(q-1)$ only partially offsets this tendency.

Two other coefficients are sometimes used. The first is called *Pearson's coefficient of mean square contingency* by Yule and Kendall (1950, p. 53), and is given as

(4) $$R_2 = \sqrt{\frac{T}{N+T}}$$

We stated that the maximum value of T is $N(q-1)$, and so the maximum value of R_2 is

(5) $$R_2(\max) = \sqrt{\frac{N(q-1)}{N+N(q-1)}} = \sqrt{\frac{q-1}{q}}$$

which is close to 1 in many cases. The smallest possible value of T is zero, and so

(6) $$0 \leq R_2 \leq \sqrt{\frac{q-1}{q}} < 1.0$$

R_2 is also called the *contingency coefficient* by McNemar (1962, p. 198) and Siegel (1956, p. 196).

Pearson's Contingency Coefficient

Example 3. In the contingency table of the two previous examples we have $T = 17.3$ and $N = 128$, so

$$R_2 = \sqrt{\frac{T}{N+T}} = \sqrt{\frac{17.3}{128+17.3}}$$
$$= .345$$

We present a third measure of dependence, R_3, also attributed to Pearson (by Cramér, 1946, p. 282) and also called the *mean-square contingency* (by Yule and Kendall, 1950, p. 53). R_3 is defined as

$$(7) \qquad R_3 = \frac{T}{N}$$

From the above discussions we may conclude

$$0 \le R_3 \le q - 1$$

and we may conclude that knowledge of r and c is necessary in order to accurately interpret the degree of dependence from the value of R_3.

Pearson's Mean-Square Contingency Coefficient

Example 4.   For the same contingency table used in the previous examples we have

$$R_3 = \frac{17.3}{128} = .135$$

Finally, we just mention *Tschuprow's coefficient*, given by Yule and Kendall (1950) as

$$(8) \qquad R_4 = \sqrt{\frac{T}{N\sqrt{(r-1)(c-1)}}}$$

The choice of a measure of dependence is largely a personal decision, motivated primarily by local traditions rather than by statistical considerations. See Stuart (1953) for further discussion.

For the 2×2 contingency table

Column	1	2	
Row 1	a	b	r_1
Row 2	c	d	r_2
	c_1	c_2	N

the above measures simplify somewhat. We know from previous sections that T reduces to

$$(9) \qquad T = \frac{N(ad - bc)^2}{r_1 r_2 c_1 c_2}$$

Therefore R_3 and R_1 (because $q = 2$) reduce to

$$(10) \qquad R_3 = \frac{T}{N} = \frac{(ad - bc)^2}{r_1 r_2 c_1 c_2}$$

and

$$(11) \qquad R_1 = \frac{T}{N(q - 1)} = \frac{T}{N} = \frac{(ad - bc)^2}{r_1 r_2 c_1 c_2} = R_3$$

R_2 may be written as

$$(12) \qquad R_2 = \sqrt{\frac{T}{N + T}} = \sqrt{\frac{(ad - bc)^2}{r_1 r_2 c_1 c_2 + (ad - bc)^2}}$$

In a fourfold contingency table, unlike the general $r \times c$ contingency table, it is sometimes meaningful to distinguish between a positive association and a negative association, such as when the two criteria of classification have corresponding categories.

Example 5. Forty children are classified according to whether their mother has dark hair or light hair, and also according to whether their father has dark or light hair. The results may show a positive association (positive correlation)

		Father Dark	Father Light	
	Dark	28	0	28
Mother	Light	5	7	12
		33	7	40

or a negative association (negative correlation)

		Father Dark	Father Light	
	Dark	21	7	28
Mother	Light	12	0	12
		33	7	40

according to whether $(ad - bc)$ is positive or negative. A lack of association (zero correlation) is indicated by the table

		Father		
		Dark	Light	
Mother	Dark	23	5	28
	Light	10	2	12
		33	7	40

   If the type of association is of interest, then care must be taken to set up the table so that a and d represent the numbers of similar classifications (dark-dark, and light-light), while b and c represent the numbers of unlike classifications (dark-light and light-dark). One measure of association which preserves direction is the *phi coefficient*, given by

(13)
$$R_5 = \frac{ad - bc}{\sqrt{r_1 r_2 c_1 c_2}}$$

which may vary from $+1$, when all items are classified in the "alike" cells (both b and c equal zero) to -1, when all items are classified as "unlike" (both a and d equal zero). The phi coefficient is merely the square root of R_3 [see Equation (10)] with the sign of $(ad - bc)$ being preserved. One reason for the popularity of the phi coefficient is because it is a special case of the *Pearson product moment correlation coefficient* (presented in the next chapter) computed by representing the classes by numbers.

The Phi Coefficient

Example 6.   For the first table in Example 5 we have

$$a = 28 \qquad r_1 = 28$$
$$b = 0 \qquad r_2 = 12$$
$$c = 5 \qquad c_1 = 33$$
$$d = 7 \qquad c_2 = 7$$

so that R_5 is computed as

$$R_5 = \frac{ad - bc}{\sqrt{r_1 r_2 c_1 c_2}} = \frac{(28)(7) - 0}{\sqrt{(28)(12)(33)(7)}}$$

(14)
$$= .703$$

For the second table in Example 5

$$R_5 = \frac{(21)(0) - (7)(12)}{\sqrt{(28)(12)(33)(7)}}$$

(15) $= -.302$

which reflects the negative association of hair types.

Other measures of association for the four-fold contingency table include one proposed by Yule and Kendall (1950, p. 30)

(16) $$R_6 = \frac{ad - bc}{ad + bc}$$

and one proposed by Ives and Gibbons (1967)

(17) $$R_7 = \frac{(a + d) - (b + c)}{a + b + c + d}$$

There is no end to the possible measures which may be defined. One's choice of a coefficient is solely a result of one's personal preferences.

Sometimes the question arises, "How can I test the null hypothesis of independence, using R_1 (or R_2, R_3, etc.) as a test statistic?" The answer is that you can find the exact small sample distribution of any of these measures in the same laborious manner that the exact small sample distribution of T was found in Section 4.2. Therefore, theoretically a test may be devised. But it is much easier, and just as effective, to use the tests presented in Sections 4.1 and 4.2 for the same hypotheses.

In particular, the coefficients

$$R_1 = \frac{T}{N(q - 1)}$$

$$R_2 = \sqrt{\frac{T}{N + T}}$$

$$R_3 = \frac{T}{N}$$

and

$$R_4 = \sqrt{\frac{T}{N\sqrt{(r - 1)(c - 1)}}}$$

will all be "too large" whenever T is "too large," because they increase or decrease when T increases or decreases. The tests of Section 4.2 use T as a test statistic, and when T is significant we may conclude that R is significant.

A one-tailed test, appropriate only for the 2×2 contingency table, may be based on the phi coefficient R_5. Because of the relationship

$$(18) \qquad\qquad R_5 = \pm \sqrt{\frac{T}{N}}$$

where the sign is determined by $(ad - bc)$, we may conclude that

$$(19) \qquad\qquad \pm\sqrt{T} = \sqrt{N} \cdot R_5$$

and therefore $\sqrt{N} \cdot R_5$ is approximately normally distributed. This is because T is approximately chi-square distributed with one degree of freedom (see Theorem 1.6.3). Then the null hypothesis

$$(20) \qquad\qquad H_0: \text{There is no positive correlation}$$

may be rejected if $\sqrt{N} \cdot R_5$ is too large (exceeds $x_{1-\alpha}$ from Table 1, for a level of α), and the null hypothesis

$$(21) \qquad\qquad H_0: \text{There is no negative correlation}$$

may be rejected if $\sqrt{N} \cdot R_5$ is too small (smaller than x_α from Table 1, for a level of α).

As an alternative to the above procedure the one tailed test for four-fold contingency tables, presented in Section 4.1, may be used, even though the model here is more correctly the model in the second test of Section 4.2. The rule in Section 4.1 that says, "If $0_{11}/n_1$ is less than or equal to $0_{21}/n_2$, in agreement with H_0, accept H_0 immediately," may be restated as follows. The inequality

$$(22) \qquad\qquad \frac{0_{11}}{n_1} \leq \frac{0_{21}}{n_2}$$

becomes, in the notation of this section,

$$(23) \qquad\qquad \frac{a}{a+b} \leq \frac{c}{c+d}$$

Because the denominators are positive, the inequality (23) is equivalent to

$$(24) \qquad\qquad a(c+d) \leq c(a+b)$$

or

$$(25) \qquad\qquad ad \leq bc$$

The restatement of the above rule is then, "If ad is less than or equal to bc, accept H_0 immediately," for the H_0 in (20). For the H_0 in (21) the inequality is reversed.

A one-tailed test based on the phi coefficient is illustrated in the following example.

Example 7.   In order to see if seat belts help prevent fatalities, records of the last 100 automobile accidents to occur along a particular highway were examined. These 100 accidents involved 242 persons. Each person was classified as using or not using seat belts when the accident occurred, and as injured fatally or a survivor.

		Injured Fatally?		
		Yes	No	Totals
Wearing Seat Belts?	Yes	7	89	96
	No	24	122	146
	Totals	31	211	242

The statement we wish to prove, "Seat belts help prevent fatalities," becomes the alternative hypothesis. The null hypothesis may be stated as

H_0:Seat belts do not help prevent fatalities

or more correctly as

H_0:There is no negative correlation between wearing seat belts and being killed in an automobile accident.

This example is slightly different than the situation we have been describing in that the two criteria of classification do not have the same corresponding classes. ("yes" and "no" mean one thing in the rows and another in the columns.) Therefore we need to stop and assess the situation. If H_1 were true we would expect b and c to be larger than a and d. Therefore the inequality

$$ad - bc < 0$$

tends to support H_1, and also causes R_5 to be negative. So we reject H_0 if $\sqrt{N} \cdot R_5$ is less than -1.645, the .05 quantile of the standard normal distribution given in Table 1. In this example the test statistic,

$$\sqrt{N} \cdot R_5 = \frac{\sqrt{N}(ad - bc)}{\sqrt{r_1 r_2 c_1 c_2}}$$
$$= \frac{\sqrt{242((7)(122) - (89)(24))}}{\sqrt{(96)(146)(31)(211)}}$$
$$= -2.08$$

is less than -1.645 so H_0 is rejected. We may conclude that the use of seat belts is associated with fewer fatalities. (Whether the relationship is a causal relationship, as stated by H_1, remains an open question.) The critical level is found from Table 1 to be about .02.

To illustrate the alternative method mentioned above, we first check to see if $(ad - bc)$ tends to agree with H_1. Because $(ad - bc)$ is less than zero, in agreement with H_1, we compute T.

$$T = \frac{N(ad - bc)^2}{r_1 r_2 c_1 c_2}$$

$$= 4.34$$

(Note that (19) gives the relationship between T and $\sqrt{N} \cdot R_5$.) At $\alpha = .05$ we reject H_0 if T exceeds $x_{1-2\alpha} = x_{.90} = 2.706$, the .90 quantile of a chi-square random variable with one degree of freedom obtained from Table 2. (Note that 2.706 equals $(-1.645)^2$, where -1.645 is the critical value used above.) H_0 is rejected, and $\hat{\alpha}$ equals about .02 (from interpolation in Table 2) as before.

Several other measures of association between variables that are classified according to two criteria are introduced in classical papers by Goodman and Kruskal (1954, 1959, 1963). A partial coefficient is introduced by Davis (1967).

PROBLEMS

1. One hundred married couples were interviewed, and the husband and the wife were asked separately for their first choice for the next U.S. president, with the following results.

| | | Wife's Choice | | |
		A	B	Other
Husband's Choice	A	12	22	6
	B	25	21	4
	Other	3	7	0

Compute the following.

(a) T (d) R_2

(b) $\hat{\alpha}$ (e) R_3

(c) R_1 (f) R_4

2. Fifty factory workers reported to the nurse complaining of soreness due to arthritis. Twenty-five of them were given aspirin and the rest were given a placebo without their knowledge. One hour later each was asked if the pill they took helped them to feel better. Seventeen in the aspirin group and twelve in the placebo group said it did. Compute the following.

(a) T (d) R_2

(b) $\hat{\alpha}$ (e) R_3

(c) R_1 (f) R_4

3. A traffic study was conducted for a short time on a well-traveled city street. Of the sixty-four cars observed, sixteen were exceeding the speed limit and forty-eight were not. Also, twenty-four of the drivers had passengers and the rest did not. Twelve of the speeders were driving alone. Assume that the observed traffic behaves the same as a random sample of all traffic would.
   (a) Use R_5 to see if there is a positive correlation between speeding and driving alone.
   (b) Compute R_6.
   (c) Compute R_7.

4. Show that T equals $N(q - 1)$ for the following contingency table (here $r < c$).

	1	2	$\cdots$	r	$r + 1$	$\cdots$	c
1	$\dfrac{N}{r}$	0	$\cdots$	0	0	$\cdots$	0
2	0	$\dfrac{N}{r}$	$\cdots$	0	0	$\cdots$	0
$\cdots$	$\cdots$	$\cdots$	$\cdots$	$\cdots$	$\cdots$	$\cdots$	$\cdots$
r	0	0	$\cdots$	$\dfrac{N}{r}$	0	$\cdots$	0

5. Think of another $r \times c$ contingency table, other than the one in Problem 4, which you would suspect of having a large value of T. Compute T for your contingency table. Is it greater than $N(q - 1)$?

6. Prove the identities:

(a) $R_2^2 = \dfrac{R_3}{1 + R_3}$

(b) $R_3 = \dfrac{R_2^2}{1 - R_2^2}$

(c) When $r = 2$ and $c = 2$, $R_3 = R_5^2 = R_1$

4.5. A GOODNESS OF FIT TEST

Often the hypotheses being tested are statements concerning the unknown probability distribution of the random variable being observed. Examples include, "The median is 4.0," and "The probability of being in class 1 is the same for both populations". More comprehensive hypotheses than the ones we've been examining would include, "The unknown distribution function is the normal distribution function with mean 3.0 and variance 1.0," or

"The distribution function of this random variable is the binomial, with parameters $n = 10$ and $p = .2$." These latter hypotheses are more comprehensive because they include statements concerning all of the quantiles simultaneously, rather than just the median, and all of the probabilities simultaneously instead of an isolated statement about some of the probabilities. Hypotheses such as these may be tested with a "goodness of fit" test, that is, with a test designed to compare the sample obtained with the type of sample one would expect from the hypothesized distribution, to see if the hypothesized distribution function "fits" the data in the sample.

The oldest and best known goodness of fit test is the chi-square test for goodness of fit, first presented by Pearson (1900).

The Chi-Square Test for Goodness of Fit

DATA. The data consist of N independent observations of a random variable X. These N observations are grouped into c classes and the numbers of observations in each class are presented in the form of a $1 \times c$ contingency table.

Class	1	2	$\ldots$	c	Total
Observed Frequencies	0_1	0_2	$\ldots$	0_c	N

Let O_j denote the number of observations in class j, for $j = 1, 2, \ldots, c$.

ASSUMPTIONS
1. The sample is a random sample.
2. The measurement scale is at least nominal.

HYPOTHESES. Let $F(x)$ be the true but unknown distribution function of X, and let $F^*(x)$ be some completely specified distribution function, the hypothesized distribution function. (If $F^*(x)$ is completely specified except for several unknown parameters which must be estimated from the sample, then this test requires a slight modification, as described in COMMENT on p. 190.)

$$H_0: F(x) = F^*(x) \quad \text{for all } x$$

$$H_1: F(x) \neq F^*(x) \quad \text{for at least one } x$$

The hypotheses may be stated in words.

H_0: The distribution function of the observed random variable is $F^*(x)$
H_1: The distribution function of the observed random variable is different than $F^*(x)$

TEST STATISTIC. Let p_j^* be the probability of a random observation on X being in class j, under the assumption that $F^*(x)$ is the distribution function of X. Then define E_j as

$$(1) \qquad\qquad E_j = p_j^* N, \quad j = 1, 2, \ldots, c$$

where E_j represents the expected number of observations in class j when H_0 is true The test statistic T is given by

(2)
$$T = \sum_{j=1}^{c} \frac{(O_j - E_j)^2}{E_j}$$

An equivalent expression, more convenient for use with a desk calculator, is

(3)
$$T = \sum_{j=1}^{c} \frac{O_j^2}{E_j} - N$$

If some of the E_j are small, the asymptotic chi-square distribution (described below) may not be appropriate. Cells with small E_j should be combined with other cells in some meaningful way, so that no more than 20% of the E_j are less than 5.0, and none are less than 1.0. This rule may be relaxed somewhat, as in Section 4.2, if all of the E_j are equal.

DECISION RULE. The exact distribution of T is difficult to use, so the large sample approximation is used. The approximate distribution of T, valid for large samples, is the chi-square distribution with $(c - 1)$ degrees of freedom. (If parameters are being estimated from the sample, fewer degrees of freedom may be used. See COMMENT below.) Therefore the critical region of approximate size α corresponds to values of T greater than $x_{1-\alpha}$, the $(1 - \alpha)$ quantile of a chi-square random variable with $(c - 1)$ degrees of freedom, obtained from Table 2. Reject H_0 if T exceeds $x_{1-\alpha}$. Otherwise accept H_0.

We may always be quite sure that the true distribution function is never exactly the same as the hypothesized distribution function. However, in many cases we are looking for a good approximation to the true distribution function and this test provides a means of justifying the use of $F^*(x)$ as the good approximation, by the acceptance of H_0. We realize that in any goodness of fit test H_0 will be rejected if the sample size is large enough. For this reason T is often used as a *measure* of goodness of fit.

Example 1.   The following is a random sample of size twenty, after being ordered from smallest to largest.

16.7	18.8	24.0	35.1	39.8
17.4	19.3	24.7	35.8	42.1
18.1	22.4	25.9	36.5	43.2
18.2	22.5	27.0	37.6	46.2

We wish to test the null hypothesis

H_0: This random sample represents observations on a normally distributed random variable, with mean 30 and variance 100

against the alternative hypothesis

H_1: The distribution function is other than as described by H_0

We arbitrarily decide to form four classes with equal expected cell counts. Four classes are formed from the three quartiles $w_{.25}$, $w_{.50}$, and $w_{.75}$ of the standard normal distribution obtained from Table 1, which are converted to the quartiles x_p of the hypothesized normal distribution by applying Theorem 1.6.1

$$x_p = \mu + \sigma w_p$$
(4)
$$= 30 + 10w_p$$

The quartiles x_p are

$$x_{.25} = 30 + 10(-.6745) = 23.255$$
$$x_{.50} = 30$$
$$x_{.75} = 36.745$$

Class 1 contains all observations less than or equal to 23.255. Class 2 contains those observations between 23.255 and 30 including 30. The other two classes are formed in the same way. The data may now be classified.

	Class 1 $(-\infty, 23.26]$	Class 2 $(23.26, 30]$	Class 3 $(30, 36.75]$	Class 4 $(36.75, \infty)$	Total
Observed Frequency	8	4	3	5	20
Expected Frequency	5	5	5	5	

The decision rule is to reject H_0 at $\alpha = .05$ if T exceeds 7.815, the .95 quantile of a chi-square random variable with $c - 1 = 3$ degrees of freedom, obtained from Table 2. The test statistic is computed using (2).

$$T = \frac{(8-5)^2}{5} + \frac{(4-5)^2}{5} + \frac{(3-5)^2}{5} + \frac{(5-5)^2}{5}$$
(5)
$$= 2.8$$

The decision is to accept H_0. The critical level is well above .25.

The following is an example where $F^*(x)$ is a discrete distribution function.

Example 2. A certain computer program is supposed to furnish random digits. If the program is accomplishing its purpose, the computer prints out digits (2, 3, 7, 4, etc.) that appear to be observations on independent and identically distributed random variables, where each digit 0, 1, 2, . . . , 8, 9 is equally likely (probability 0.1) to be obtained. One way of testing

$$H_0: \text{The numbers appear to be random digits}$$

against the alternative of nonrandomness is by separating a long sequence of the digits into groups of size 10, and then counting the number of even digits in each group. The number of even digits in a group of 10 digits should have a binomial distribution with $n = 10$ and $p = .5$ (because the probability of each digit being even is $5/10$) if H_0 is true.

Suppose the following digits are generated, and are grouped into blocks of ten digits each.

1578748416	4705188926	6936349612
4653843213	0282868892	3928057043
5101259393	9837006785	3011679938
7122863085	6528271107	2956427027
2671728075	9759178719	9373309535
8363265100	2546793732	2212122529
9453087720	3976759377	9593511031
5605373242	1819898287	3872181027
3494768396	9296177240	8620774591
4659773922	9246724287	8326143939

In the first block there are 5 even digits $(8, 4, 8, 4, 6)$. In the second block there are 6 even digits $(4, 0, 8, 8, 2, 6)$. In each block the number of even digits is counted. The results are summarized in the following table.

The number of blocks containing j even digits, for j equal to

0	1	2	3	4	5	6	7	8	9	10	Total
0	4	1	1	3	14	5	1	0	1	0	30

Each block consists of 10 independent trials (digits) where each trial has probability .5 of resulting in "even" rather than "odd." Therefore, as in Example 1.4.5, the probability of obtaining exactly j even numbers is $\binom{10}{j}(1/2)^{10}$. Thirty repetitions of the experiment, represented by the thirty blocks, result in

$$(6) \qquad E_j = p_j^* N = 30 \binom{10}{j} \left(\frac{1}{2}\right)^{10}$$

from Equation (1). The values of p_j^* are obtained from Table 3, $n = 10, p = .5$, by subtracting successive entries to obtain the difference.

The expected number of blocks containing j even digits, for j equal to

	0	1	2	3	4	5	6	7	8	9	10
$F^*(j)$	.0010	.0107	.0547	.1719	.3770	.6230	.8281	.9453	.9893	.9990	1.0000
$p^*(j)$	.0010	.0097	.0440	.1172	.2051	.2460	.2051	.1172	.0440	.0097	.0010
E_j	.030	.291	1.320	3.516	6.153	7.380	6.153	3.516	1.320	.291	.030

Therefore, as a test of randomness, we are actually using a goodness of fit test where $F^*(j)$ is the binomial distribution with $n = 10$ and $p = .5$. If H_0 is true, then the observed numbers should agree well with the E_j. If H_0 is not true, some types of nonrandomness will result in poor agreement with $F^*(x)$, while other types of nonrandomness will not be detectable using this test. No test for nonrandomness is consistent against all types of nonrandomness.

Because some of the E_j are small, the categories $j = 0$, 1, 2, and 3 are combined into one class, and so are the categories $j = 7$, 8, 9, and 10. The result is as follows.

Class:	$j \leq 3$	$j = 4$	$j = 5$	$j = 6$	$j \geq 7$	Total
O_j = observed number	6	3	14	5	2	30
E_j = expected number	5.157	6.153	7.380	6.153	5.157	30

Now all E_j exceed 5.0. The number of classes is now five, so the critical region of approximate size .05 corresponds to values of T greater than 9.488, the .95 quantile of a chi-square random variable with four degrees of freedom, obtained from Table 2. The test statistic T is computed using (2).

$$T = \frac{(6 - 5.157)^2}{5.157} + \frac{(3 - 6.153)^2}{6.153} + \frac{(14 - 7.380)^2}{7.380} + \frac{(5 - 6.153)^2}{6.153}$$
$$+ \frac{(2 - 5.157)^2}{5.157}$$

$$(7) \qquad = 9.84$$

The observed value of T is greater than the critical value 9.488, so H_0 is rejected. The critical level $\hat{\alpha}$ is about .04. We conclude that the digits produced by the computer program do not appear to be random.

COMMENT. If $F^*(x)$ is completely specified except for a number k of parameters, it is first necessary to estimate the parameters and then to proceed with the test as outlined above. The only change is in the distribution of T, which now may be approximated using a chi-square distribution with $c - 1 - k$ degrees of freedom. That is, one degree of freedom is subtracted for each parameter estimated. However, subtraction of degrees of freedom is a privilege accorded only when the parameters are estimated in the proper manner. For example, in a goodness of fit test with four classes, H_0 is usually rejected (at $\alpha = .05$) if T exceeds 7.815 (see Table 2). However, if one parameter is estimated from the data before the test is applied, then the hypothesized distribution has already been modified so that it will fit the data better. (This is true if the estimate is a "good" estimate. A poor estimate may be used deliberately to result in a poor fit, but then the goodness of fit test is no longer valid.)

The goodness of fit test will then be more likely to result in acceptance of H_0; the test becomes conservative, and therefore less powerful. We would like to enlarge the critical region, so that α again becomes .05, and the test regains some (or all) of the power that was lost. If we subtract one degree of freedom, using two degrees of freedom instead of three, the critical region is enlarged, and H_0 is rejected if T exceeds 5.991 instead of 7.815 as before. The question is, "Are we justified in subtracting one degree of freedom, as we did?"

Cramér (1946, p. 424; or see Birnbaum, 1962, p. 258) shows that one degree of freedom may be subtracted for each parameter estimated by the *minimum chi-square method*. The minimum chi-square method simply involves using that value of the parameter which results in the smallest value of the test statistic, for the given observations. From a practical standpoint this means trying all possible values of the parameter, or all possible combinations if several parameters are unknown, computing a set of E_j and then T for each value of the parameter, and selecting that value of the parameter that results in the smallest T. However, such a procedure is impractical. Therefore, Cramér also presents a more usable *modified minimum chi-square method*. Even that procedure is tedious, and so Cramér and Birnbaum, in their examples, actually use a modification of the modified minimum chi-square method, which asymptotically still permits subtracting one degree of freedom for each parameter estimated. The method eventually used consists of estimating the k unknown parameters by computing the first k sample moments of the grouped data. (Each observation is assumed to be at the midpoint of its class interval, where all intervals containing observations are of finite length.) Then the first k population moments are set equal to the first k sample moments of the grouped data, and the resulting k equations are solved simultaneously for the k unknown parameters. The following example should help to clarify the above procedure.

Example 3.   Fifty two-digit numbers were drawn at random from a telephone book, and the chi-square test for goodness of fit is used to see if they could have been observations on a normally distributed random variable. The numbers, after being arranged in order from the smallest to the largest, are as follows.

23	23	24	27	29	31	32	33	33	35
36	37	40	42	43	43	44	45	48	48
54	54	56	57	57	58	58	58	58	59
61	61	62	63	64	65	66	68	68	70
73	73	74	75	77	81	87	89	93	97

The null hypothesis is

H_0: These numbers are observations on a normally distributed random variable.

The normal distribution has two parameters (Definition 1.6.3), both of which are unspecified by H_0, and must be estimated before the goodness of fit test may be applied. For illustration, the procedure is divided into steps.

Step 1. Divide the observations into intervals of finite length. We arbitrarily choose the intervals 20 to 40, 40 to 60, 60 to 80, and 80 to 100, not including the upper limit of each interval.

Interval	20 to 40	40 to 60	60 to 80	80 to 100	Total
Number of observations	12	18	15	5	50

Step 2. Estimate μ and σ with the sample mean $\bar{X}$ and sample standard deviation S of the grouped data. The twelve observations in the interval 20 to 40 are treated as if they all equal the middle point 30. The eighteen observations from 40 to 60 are all considered to be 50, and so on. These are the numbers used for computing $\bar{X}$ and S, using the equations of Definition 2.2.3.

$$\bar{X} = \frac{1}{N} \sum_{i=1}^{N} X_i$$

$$= \tfrac{1}{50}[12(30) + 18(50) + 15(70) + 5(90)]$$

(8) $= 55.2$

$$S = \sqrt{S^2} = \sqrt{\frac{1}{N} \sum_{i=1}^{N} X_i^2 - \bar{X}^2}$$

$$= \{\tfrac{1}{50}[12(30)^2 + 18(50)^2 + 15(70)^2 + 5(90)^2] - (55.2)^2\}^{\frac{1}{2}}$$

(9) $= 18.7$

Therefore our estimates of μ and σ are 55.2 and 18.7 respectively.

Step 3. Using the estimated parameters from Step 3, compute the E_j for the groups in Step 1 and for the "tails."

Class Boundaries b_j	$(b_j - \bar{X})/S = x_p$	$F(x_p)$	Interval	p_j^*
$b_1 = 20$	-1.88	.03	<20	.03
$b_2 = 40$	$- .813$	.21	20 to 40	.18
$b_3 = 60$	$+ .256$	.60	40 to 60	.39
$b_4 = 80$	$+1.33$	.91	60 to 80	.31
$b_5 = 100$	$+2.40$	.99	80 to 100	.08
			≥ 100	.01

To find the hypothesized probabilities of being in the various classes, when the hypothesized distribution is the normal distribution with mean 55.2 and standard deviation 18.7, the class boundaries (column 1 above) are considered to be the quantiles of the hypothesized distribution. These quantiles are converted to quantiles of a standard normal random variable (column 2 above) by Equation (1.6.3) in order to find out *which* quantile the boundaries represent (column 3 above). Subtraction of the items in column 3 then yields the probabilities p_j^* of being in the various intervals under the hypothesized distribution. The E_j equal $50p_j^*$, from Equation (1), and are given below.

Class	<20	20 to 40	40 to 60	60 to 80	80 to 100	≥100
Expected number E_j	1.5	9.0	19.5	15.5	4	0.5
Observed number O_j	0	12	18	15	5	0

Because of the small E_j, the first and last cells are combined with the cells adjacent to them.

Class	<40	40 to 60	60 to 80	≥80
Expected number E_j	10.5	19.5	15.5	4.5
Observed number O_j	12	18	15	5

Step 4. Compute T. The test statistic is now computed using (2).

$$T = \frac{(12 - 10.5)^2}{10.5} + \frac{(18 - 19.5)^2}{19.5} + \frac{(15 - 15.5)^2}{15.5} + \frac{(5 - 4.5)^2}{5}$$

(10) $= .395$

The critical region of size .05 corresponds to values of T greater than 3.841, the .95 quantile of a chi-square random variable with $c - 1 - k = 4 - 1 - 2 = 1$ degree of freedom. Therefore H_0 is accepted, with a critical level $\hat{\alpha}$ well above .25.

Usually a modification called Sheppard's correction is used when the variance is being estimated from grouped data, and when the interior intervals are of equal width, say h. Sheppard's correction consists of subtracting $h^2/12$ from S^2 in order to obtain a better estimate of variance. In this example h equals 20 (the width of each interval) so $(20)^2/12 = 33.33$ could have been subtracted in step 2, before extracting the square root. The result is $S = 17.8$, a smaller estimate for σ. This smaller estimate of σ results in a larger value of T in this example, and since our objective is to obtain the smallest possible value for T the correction was not used. In most situations we can expect a smaller T when the correction is used.

Another peculiarity of this example is the fact that a smaller value of T (.279) may be obtained by using $\bar{X} = 55.04$ and $S = 19.0$ as estimates of μ and σ. These estimates are the sample moments obtained from the original observations, before grouping. No matter how they are obtained, the estimates to use are the estimates that result in the smallest value of T. The procedure described in this example can be relied upon to provide a value of T not far from its minimum value in most cases. Therefore it is the recommended procedure.

In the above example the test statistic just happened to be smaller when μ and σ were estimated using the sample mean and standard deviation based on the original, rather than the grouped, observations. This procedure may be used, and is even recommended by Yule and Kendall (1950), but it is usually inferior in results to the other method which uses the grouped data (Chernoff and Lehmann, 1954).

Theory.   If the null hypothesis completely specifies the hypothesized distribution function, then once the classes are defined there is a known probability p_j associated with each class, if H_0 is true. The probability of any particular arrangement $0_1, 0_2, \ldots , 0_c$ of the N sample values is then given by the multinomial probability distribution,

$$(11) \qquad P(0_1, 0_2, \ldots , 0_c \mid N) = \frac{N!}{0_1! \, 0_2! \ldots 0_c!} \, p_1^{0_1} p_2^{0_2} \cdots p_c^{0_c}$$

which is an immediate extension of the binomial distribution to c classes instead of only two classes. With the probability function given by (11) the probability distribution of T may be determined, although the calculation becomes laborious for large N and c. There appears to be no theory developed to find the exact distribution of T when several parameters are first estimated from the sample. Therefore the large sample approximation is both practical and necessary in order to apply this goodness of fit test. The theory behind the large sample chi-square approximation may be found in Cramér (1946).

Usage of the chi-square goodness of fit test with small but equal expected frequencies is discussed by Slakter (1966, 1968). If the sample is grouped according to time of observation rather than according to numerical value, the usage of the test may require some modification (Putter, 1964). Chernoff (1967) gives a recent discussion of the adjustment of the degrees of freedom.

PROBLEMS

1. Test the following data to see if they could have come from a population whose values are uniformly distributed between .0000 and .9999.

.4755	.5233	.5440	.5456	.9056
.2186	.7500	.2484	.5101	.8283
.5112	.5484	.5758	.3607	.4352
.3826	.6454	.9145	.3943	.5381
.5758	.8620	.6687	.3979	.5646
.4274	.5482	.3007	.4438	.4102
.4295	.5926	.6521	.6328	.5689
.7297	.3768	.8403	.2925	.2113
.8757	.4403	.4993	.3900	.5166
.8230	.8522	.8312	.7979	.4632
.8432	.4004	.4295	.9763	.5590
.4396	.2595	.3003	.3003	.5836
.5337	.8008	.4887	.2172	.9329
.5498	.3686	.4067	.5274	.4579
.9096	.4995	.2172	.6793	

2. A die was cast 600 times with the following results.

Occurrence	1	2	3	4	5	6
Frequency	87	96	108	89	122	98

Is the die balanced?

3. Test the data in Problem 1 to see if they might have come from a normal population.

4. Without the aid of books or tables, attempt to write 300 *random* digits. Then apply the test of randomness described in Example 2 to see if you are a good random digit generator.

4.6. COCHRAN'S TEST FOR RELATED OBSERVATIONS

Sometimes the use of a treatment, or condition, results in one of two possible outcomes. For example, the response to a salesman's technique may be classified as either "sale" or "no sale", or a certain treatment may result in "success" or "failure." Of course, if the several treatments, c in number, are each applied in several different independent trials, the results may be given in the form of a $2 \times c$ contingency table, where one row represents the number of successes and the other row represents the number of failures, and the null hypothesis of no treatment differences may be tested using a chi-square contingency table test as previously described in Section 4.2. However it is often possible to detect more subtle differences between treatments, that is, increase the power of the test, by applying all c treatments independently to the same blocks, such as by trying all c sales techniques on each of several persons in an experimental situation, and then recording for each person the results of each technique. Thus each block, or person, acts as its own control and the treatments are more effectively compared with each other. Such an experimental technique is called "blocking," and the experimental design is called a "randomized complete block design." If the treatment result may be classified into one of two categories, the following test, proposed by Cochran (1950), may be an appropriate method of analysis.

The Cochran Test

DATA. Each of c treatments is applied independently to each of r blocks, or subjects, and the result of each treatment application is recorded as either 1 or 0, to represent "success" or "failure," or any other dichotomization of the possible treatment results. The results are then given in the form of a table with r rows representing the blocks and c columns representing the c treatments, with entries that are either zeros or ones. Let R_i represent the row totals, $i = 1, 2, \ldots, r$, and let C_j represent the column totals, $j = 1, 2, \ldots, c$. Then the data appear as follows, where the X_{ij} are either 0 or 1, and N represents the total number of ones in the table.

Treatment	1	2		c	Row totals
Blocks 1	X_{11}	X_{12}	$\ldots$	X_{1c}	R_1
2	X_{21}	X_{22}	$\ldots$	X_{2c}	R_2
$\ldots$	$\ldots$	$\ldots$	$\ldots$	$\ldots$	$\ldots$
r	X_{r1}	X_{r2}	$\ldots$	X_{rc}	R_r
Column totals	C_1	C_2		C_c	N = grand total

ASSUMPTIONS
1. The blocks were randomly selected from the populations of all possible blocks.
2. The outcomes of the treatments may be dichotomized in a manner common to all treatments within each block, so the outcomes are listed as either "0" or "1."

HYPOTHESES

H_0: The treatments are equally effective

H_1: There is a difference in effectiveness among treatments

In more mathematical terms, let

(1) $p_{ij} = P(X_{ij} = 1); \quad i = 1, \ldots, r; \quad j = 1, \ldots, c$

Then equal effectiveness among treatments implies

(2) $p_{i1} = p_{i2} = \ldots = p_{ic}, \quad$ for each i from 1 to r

That is, for each block the probability of a treatment being a success does not depend on which treatment is being used. Then the hypotheses may be restated as follows.

H_0: $p_{i1} = p_{i2} = \ldots = p_{ic}, \quad$ for each i from 1 to r

H_1: $p_{ij} \neq p_{ik}$ for some j and k, and for some i

TEST STATISTIC. The test statistic T may be written as

(3)
$$T = \sum_{j=1}^{c} \frac{c(c-1)\left(C_j - \dfrac{N}{c}\right)^2}{\sum_{i=1}^{r} R_i(c - R_i)}$$

or, equivalently, as

(4)
$$T = c(c-1) \frac{\sum_{j=1}^{c}\left(C_j - \dfrac{N}{c}\right)^2}{\sum_{i=1}^{r} R_i(c - R_i)}$$

The following form is more suitable for machine computation.

(5)
$$T = \frac{c(c-1)\sum_{j=1}^{c} C_j^2 - (c-1)N^2}{cN - \sum_{i=1}^{r} R_i^2}$$

DECISION RULE. The exact distribution of T is difficult to tabulate, so the large sample approximation is used instead. The number of blocks r is assumed to be large. Then the critical region of approximate size α corresponds to all values of T greater than $x_{1-\alpha}$, the $(1 - \alpha)$ quantile of a chi-square random variable with $(c - 1)$ degrees of freedom, obtained from Table 2. If T exceeds $x_{1-\alpha}$, reject H_0. Otherwise accept the null hypothesis of no differences in the effectiveness of the various treatments.

Example 1.   Each of three basketball enthusiasts had devised his own system for predicting the outcomes of collegiate basketball games. Twelve games were selected at random, and each sportsman presented a prediction of the outcome of each game. After the games were played the results were tabulated, using 1 for successful prediction and 0 for unsuccessful prediction.

	Sportsman	1	2	3	Totals
Game	1	1	1	1	3
	2	1	1	1	3
	3	0	1	0	1
	4	1	1	0	2
	5	0	0	0	0
	6	1	1	1	3
	7	1	1	1	3
	8	1	1	0	2
	9	0	0	1	1
	10	0	1	0	1
	11	1	1	1	3
	12	1	1	1	3
	Totals	8	10	7	25

The assumptions of the Cochran test were met, because the games (blocks) were selected at random from among all college basketball games being played. Therefore the Cochran test was used to test the null hypothesis

H_0: Each sportsman is equally effective in his ability to predict the outcomes of the basketball games

The test statistic is computed using (4).

$$T = c(c-1) \frac{\sum\limits_{j=1}^{c} \left(C_j - \frac{N}{c}\right)^2}{\sum\limits_{i=1}^{r} R_i(c - R_i)}$$

$$= \frac{(3)(2)[(-\frac{1}{3})^2 + (\frac{5}{3})^2 + (-\frac{4}{3})^2]}{2 + 2 + 2 + 2 + 2}$$

(6) $= 2.8$

The critical region of approximate size .05 corresponds to values of T greater than 5.99, the .95 quantile of a chi-square random variable with 2 degrees of freedom obtained from Table 2. Therefore, in this example H_0 is accepted and we conclude that no significant differences among prediction methods were detected. H_0 could have been rejected using an α of about .25, so $\hat{\alpha}$ equals about .25.

Theory.   Each X_{ij}, as defined above, is a random variable that follows the point binomial distribution (the binomial distribution with $n = 1$) with parameter p_{ij}. The column total, C_j, defined by

(7) $$C_j = \sum_{i=1}^{r} X_{ij}$$

is therefore a random variable also. If the p_{ij} in each column were the same, then C_j would follow a binomial distribution, but even if H_0 is assumed to be true the p_{ij} down the column may differ from each other. The hypothesis merely states that the p_{ij} across each row are equal to each other, but may vary from one block (row) to another. However, because the random variable C_j is the sum of r independent random variables, the Central Limit Theorem still applies, and for large r the distribution of C_j is approximately normal. This implies that the distribution function of

$$\frac{C_j - E(C_j)}{\sqrt{\text{Var}(C_j)}}$$

may be approximated by the standard normal distribution function.

The C_j are not mutually independent, because the X_{ij} within each block are related. However, as r gets large (actually, as the number of nonties gets large, as emphasized by Tate and Brown, 1970) the C_j approach independence, and according to Theorem 1.6.3 the sum

$$(8) \qquad \sum_{j=1}^{c} \left[\frac{C_j - E(C_j)}{\sqrt{\text{Var}(C_j)}} \right]^2 = \sum_{j=1}^{c} \frac{[C_j - E(C_j)]^2}{\text{Var}(C_j)}$$

may be approximated by the chi-square distribution with c degrees of freedom. However, the parameters $E(C_j)$ and $\text{Var}(C_j)$ are unknown, and the following method of estimating those parameters is shown by Blomqvist (1951) to result in the loss of one degree of freedom.

The mean of C_j may be estimated by the sample mean

$$(9) \qquad \frac{1}{c} \sum_{j=1}^{c} C_j = \frac{N}{c} = \text{estimate of } E(C_j)$$

The same estimate is used for the mean of every C_j. The variance of C_j equals the sum of the variances of the X_{ij} in the jth column,

$$(10) \qquad \text{Var}(C_j) = \sum_{i=1}^{r} \text{Var}(X_{ij})$$

because of the block to block independence of the X_{ij} (see also Theorem 1.5.3). The variance of X_{ij} is given by Equation (1.5.8) as

$$(11) \qquad \text{Var}(X_{ij}) = p_{ij}(1 - p_{ij})$$

Under H_0 the probability of a "success" p_{ij} is the same for all columns within a block, and therefore it is natural to estimate p_{ij} by the average number of successes in row i, R_i/c. That is,

$$(12) \qquad \text{estimate of } p_{ij} = R_i/c$$

and from (11),

$$(13) \qquad \text{estimate of } \text{Var}(X_{ij}) = \frac{R_i}{c}\left(1 - \frac{R_i}{c}\right)$$

However, such an estimate tends to be too small, and is improved by multiplication by $c/(c-1)$. Then $\text{Var}(X_{ij})$ is estimated by

$$(14) \qquad \text{estimate of } \text{Var}(X_{ij}) = \frac{R_i(c - R_i)}{c(c-1)}$$

and $\text{Var}(C_j)$ is estimated, from (10), as

$$(15) \qquad \text{estimate of } \text{Var}(C_j) = \frac{1}{c(c-1)} \sum_{i=1}^{r} R_i(c - R_i)$$

which does not depend on j and so is used for all C_j. Substitution of the estimates for $E(C_j)$, Equation (9), and Var (C_j), Equation (15), into (8) gives

$$T = \sum_{j=1}^{c} \frac{\left(C_j - \dfrac{N}{c}\right)^2}{\displaystyle\sum_{i=1}^{r} \dfrac{R_i(c - R_i)}{c(c - 1)}}$$

$$(16) \qquad = c(c - 1) \frac{\displaystyle\sum_{j=1}^{c}\left(C_j - \dfrac{N}{c}\right)^2}{\displaystyle\sum_{i=1}^{r} R_i(c - R_i)}$$

which provides some insight into the use of the chi-square distribution with $(c-1)$ degrees of freedom for the test statistic T. For another model in which Cochran's test is valid see Fleiss (1965).

COMMENT. If only two treatments are being considered, such as "before" and "after" observations on the same block, with r blocks, then the experimental situation is the same as that analyzed by the McNemar test for significance of changes. That is, in each situation the null hypothesis is that the proportion of the population in class one is the same using treatment 1 (before) as it is using treatment 2 (after). Thus it appears that if c equals 2 the experimenter has a choice of using the Cochran test or the McNemar test. In fact, there is no choice because if c equals 2 the Cochran test is identical with the McNemar test (Section 3.4) as is shown in the following.

For $c = 2$ the Cochran test statistic reduces to

$$T = 2 \frac{\left(C_1 - \dfrac{C_1 + C_2}{2}\right)^2 + \left(C_2 - \dfrac{C_1 + C_2}{2}\right)^2}{\displaystyle\sum_{i=1}^{r} R_i(2 - R_i)}$$

$$= 2 \frac{\left(\dfrac{C_1}{2} - \dfrac{C_2}{2}\right)^2 + \left(\dfrac{C_2}{2} - \dfrac{C_1}{2}\right)^2}{\displaystyle\sum_{i=1}^{r} R_i(2 - R_i)}$$

$$(17) \qquad = \frac{(C_1 - C_2)^2}{\displaystyle\sum_{i=1}^{r} R_i(2 - R_i)}$$

If a block has ones in both columns, then $R_i = 2$ and $R_i(2 - R_i) = 0$. Similarly if both columns have zeros, then $R_i = 0$ and $R_i(2 - R_i) = 0$. If there is a change from zero to one or one to zero in a given row, then $R_i = 1$ and $R_i(2 - R_i) = 1$. Thus the denominator of (17) is merely the total number of rows that go from 0 to 1 or 1 to 0, which is $b + c$ in the notation of the McNemar test. Also C_1 is the total number of ones in column one, or "before," which is $c + d$ in the notation of the McNemar test. Similarly C_2 equals $b + d$. Therefore we have

$$C_1 - C_2 = c + d - b - d$$

$$= c - b$$

and (17) becomes

$$T = \frac{(c - b)^2}{b + c} = \frac{(b - c)^2}{b + c}$$

which is identical with the form of McNemar's test statistic given in Equation (3.4.6). Both the McNemar test statistic and the Cochran test statistic with $c = 2$ are approximated by a chi-square random variable with one degree of freedom.

PROBLEMS

1. The relative effectiveness of two different sales techniques was tested on twelve volunteer housewives. Each housewife was exposed to each sales technique and asked to buy a certain product, the same product in all cases. At the end of each exposure each housewife rated the technique with a "1" if she felt she would have agreed to buy the product, and a "0" if she probably would not have bought the product.

Housewife 1	2	3	4	5	6	7	8	9	10	11	12
Technique 1 1	1	1	1	1	0	0	0	1	1	0	1
Technique 2 0	1	1	0	0	0	0	0	1	0	0	1

(a) Use Cochran's test.
(b) Rearrange the data and use McNemar's test in the large sample form suggested by Equation (3.4.6).
(c) Ignore the blocking effect in the above experiment, and treat the data as if 24 different housewives were used. Analyze the data using the test for differences in probabilities given in Section 4.1.

2. On a ship, twelve groups with three sailors in each group were chosen in a random manner, where the sailors in each group did similar work and were in the same division aboard ship. In a random manner the sailors in each group were given treatment 1, 2, or 3, no two sailors from the same group receiving the same treatment. Treatment 1 was a "flu shot", treatment 2 was a "flu pill," and

treatment 3 was a promise of two weeks extra leave if they did not catch the flu. As each sailor reported to sick bay with the flu, a report to the experimenter was made. At the end of the winter, these were the results.

Group	Sailors with the flu (by treatment number)
1	2
2	1, 2
3	1, 2, 3
4	2, 3
5	2
6	none
7	1, 2
8	1, 2
9	1
10	2
11	1, 2, 3
12	2

Do these results indicate a significant difference between the various treatments?

3. In an attempt to compare the relative power of three statistical tests, one hundred sets of artificial data were generated using a computer. On each set of data the three statistical tests were used, with $\alpha = .05$, and the decision to "accept" or "reject" H_0 was recorded. The results were as follows.

Test 1	Test 2	Test 3	Number of Sets of Data
accept	accept	accept	26
accept	accept	reject	6
accept	reject	accept	12
reject	accept	accept	4
reject	reject	accept	18
reject	accept	reject	5
accept	reject	reject	7
reject	reject	reject	22

Is there a difference in the power of the three tests when applied to populations from which the simulated data were obtained?

CHAPTER 5

The Use of Ranks

5.1. THE ONE-SAMPLE OR MATCHED PAIRS CASE

Most of the statistical procedures introduced in the previous chapters can be used on data that have a nominal scale of measurement. In Chapter 3 several statistical methods were presented for analyzing data that were naturally dichotomous, that is, the zero-one or success-failure type of data. In Chapter 4 the discussion centered around the analysis of data which may be classified according to two or more different criteria, and into two or more separate classes by each criterion. All of those procedures may also be used where more than nominal information concerning the data is available but for various reasons, such as speed and ease of calculation, abundance of data, or the particular interpretation desired of the data, some of the information contained in the data is disregarded and the data are reduced to nominal type data for analysis. Such a loss of information usually results in a corresponding loss of power. In this chapter several statistical methods are presented which utilize more of the information contained in the data, if the data have at least an ordinal scale of measurement.

Data may be nonnumeric (such as "good, better, best") or numeric (such as 7.36, 4.91, etc.). If data are nonnumeric, but ranked as in ordinal type data, then the methods of this chapter are often the most powerful methods available. If data are numeric, and furthermore are observations on random variables which have the normal distribution so that all of the assumptions of the usual parametric tests are met, then the loss of efficiency caused by using the methods of this chapter is surprisingly small. In those situations the relative efficiency of tests using only the ranks of the observations is frequently about .95, depending on the particular situation.

The first rank test of this chapter deals with the single random sample and the random sample of matched pairs which is reduced to a single sample by considering differences. A matched pair (X_i, Y_i) is actually a single observation on a bivariate random variable. The sign test of Section 3.4 analyzed matched pairs of data by reducing each pair to a plus, a minus, or a tie and applying the binomial test to the resultant single sample. The test of this section also reduces the matched pair (X_i, Y_i) to a single observation, by considering the difference

$$(1) \qquad D_i = Y_i - X_i \qquad \text{for} \qquad i = 1, 2, \ldots, n$$

The analysis is then performed on the D_i's as a sample of single observations. Whereas the sign test merely noted whether D_i was positive, negative, or zero, the test of this section notes the sizes of the positive D_i's relative to the negative D_i's. The model of this section resembles the model used in the sign test. Also the hypotheses resemble the hypotheses of the sign test. The important difference between the sign test and this test is an additional assumption of *symmetry* of the distribution of differences. Before we introduce the test we should clarify the meaning of the adjective *symmetric* as it applies to a distribution, and discuss the influence of symmetry on the scale of measurement.

Symmetry is easy to define if the distribution is discrete. A discrete distribution is symmetric if the left half of the graph of the probability function is the mirror image of the right half. For example, the binomial distribution is symmetric if p equals $1/2$ (see Figure 1), and the discrete uniform distribution is always symmetric (see Figure 2). The dotted lines in the figures represent the lines about which the distributions are symmetric.

For other than discrete distributions we are not able to draw a graph of the probability function. Therefore, a more abstract definition of symmetry is required, such as the following.

Definition 1.   The distribution of a random variable X is *symmetric* about a line $x = c$, for some constant c, if the probability of $X \le c - x$ equals the probability of $X \ge c + x$ for each possible value of x.

In Figure 1, c equals 2 and the definition is easily verified for all real numbers x. In Figure 2, c equals 3.5. Even though we may not know the exact distribution of a random variable, often we are able to say, "It is reasonable to assume that the distribution is symmetric." Such an assumption is not as strong as the assumption of a normal distribution, because while all normal distributions are symmetric, not all symmetric distributions are normal.

If a distribution is symmetric, then the mean (if it exists) coincides with the median, because both are located exactly in the middle of the distribution, at

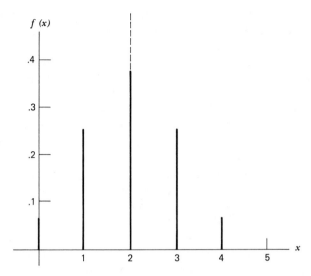

Figure 1. Symmetry in a binomial distribution.

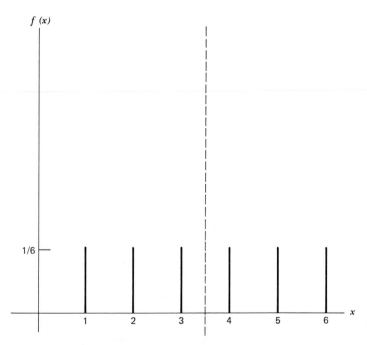

Figure 2. Symmetry in a discrete uniform distribution.

the line of symmetry. One consequence of adding the assumption of symmetry to the model is that any inferences concerning the median are also valid statements for the mean.

A second consequence of adding the assumption of symmetry to the model is that the required scale of measurement is changed from ordinal to interval. With an ordinal scale of measurement, two observations of the random variable need only to be distinguished on the basis of which is larger and which is smaller. It is not necessary to know which one is farthest from the median, such as when the two observations are on opposite sides of the median. If the assumption of symmetry is a meaningful one, the distance from the median is a meaningful measurement, and therefore the distance between two observations is a meaningful measurement. As a result the scale of measurement is more than just ordinal, it is interval.

One other assumption usually included in the model of this section and, in fact, of this entire chapter is the assumption that the random variables are continuous. The continuity assumption is made so that we may consider the distribution of the test statistics as if no ties (two or more observations exactly equal to each other) occur. In practice ties do occur, so we shall include a modification for ties along with each test whenever possible.

A test presented by Wilcoxon (1945) is designed to test whether a particular sample came from a population with a specified median. It may also be used in situations where observations are paired, such as "before" and "after" observations on each of several subjects, to see if the second random variable in the pair has the same median as the first.

The Wilcoxon Signed Ranks Test

DATA. The data consist of n' observations (x_1, y_1), (x_2, y_2), . . . , $(x_{n'}, y_{n'})$ on the respective bivariate random variables (X_1, Y_1), (X_2, Y_2), . . . , $(X_{n'}, Y_{n'})$. (For use as a median test with a single sample see Example 2.) The absolute differences (without regard to sign)

$$(2) \qquad |D_i| = |Y_i - X_i| \qquad i = 1, 2, \ldots, n'$$

are then computed for each of the n' pairs (X_i, Y_i).

Omit from further consideration all pairs with a difference of zero (i.e., where X_i equals Y_i, or $D_i = 0$). Let the number of pairs remaining be denoted by n, $n \leq n'$. Ranks from 1 to n are assigned to these n pairs according to the relative size of the absolute difference, as follows. The rank 1 is given to the pair (X_i, Y_i) with the smallest absolute difference $|D_i|$; the rank 2 is given to the pair with the second smallest absolute difference; and so on, with the rank n being assigned to the pair with the largest absolute difference.

If several pairs have absolute differences which are equal to each other, assign to each of these several pairs the *average* of the ranks that would have otherwise been assigned. (That is, if the ranks 3, 4, 5, and 6 belong to four pairs, but we don't know which rank to assign to which pair because all four absolute differences are exactly equal to each other, assign the average rank $(1/4)(3 + 4 + 5 + 6) = 4.5$ to each of the four pairs.)

ASSUMPTIONS
1. Each D_i is a continuous random variable.
2. The distribution of each D_i is symmetric.
3. The D_i's are mutually independent.
4. The D_i's all have the same median.
5. The measurement scale of the D_i's is at least interval.

HYPOTHESES. Let the common median of the D_i's be denoted by $d_{.50}$. Then the hypotheses may be stated in several ways, depending on whether the test is one tailed or two tailed.

A. (one-tailed test)
$$H_0 : d_{.50} \leq 0$$
$$H_1 : d_{.50} > 0$$

This alternative hypothesis may be loosely stated as, "The values of the X_i's tend to be smaller than the values of the Y_i's."

B. (one-tailed test)
$$H_0 : d_{.50} \geq 0$$
$$H_1 : d_{.50} < 0$$

This alternative hypothesis may be loosely stated as, "The values of the X_i's tend to be larger than the values of the Y_i's."

C. (two-tailed test)
$$H_0 : d_{.50} = 0$$
$$H_1 : d_{.50} \neq 0$$

If the model is changed slightly, the above hypotheses may be broadened considerably. The change consists of adding the assumption, "The (X_i, Y_i), for $i = 1, 2, \ldots, n$, constitute a random (bivariate) sample." This new assumption actually includes assumptions 3 and 4 of the model. The hypotheses may then be stated as

A.
$$H_0 : E(X) \geq E(Y)$$
$$H_1 : E(X) < E(Y)$$

B.
$$H_0 : E(X) \leq E(Y)$$
$$H_1 : E(X) > E(Y)$$

C.
$$H_0 : E(X) = E(Y)$$
$$H_1 : E(X) \neq E(Y)$$

in place of A, B, and C originally given, if $E(X)$ and $E(Y)$ exist.

TEST STATISTIC. The test statistic T equals the sum of the ranks assigned to those pairs (X_i, Y_i) where Y_i exceeds X_i. That is, let R_i be defined for each pair (X_i, Y_i) as follows:

(3)
$$R_i = 0 \quad \text{if } X_i > Y_i \ (D_i \text{ is negative})$$
$$R_i = \text{the rank assigned to } (X_i, Y_i), \text{ if } X_i < Y_i \ (D_i \text{ is positive})$$

Then the test statistic T may be written as

(4)
$$T = \sum_{i=1}^{n} R_i$$

DECISION RULE. The decision rule may be expressed three different ways, corresponding to the three sets of hypotheses, A, B, and C. Let w_p be the pth quantile, obtained from Table 7.

A. (one-tailed test) Large values of T indicate that H_0 is false, so reject H_0 at the level of significance α if T exceeds $w_{1-\alpha}$. Accept H_0 if T is less than or equal to $w_{1-\alpha}$.

B. (one-tailed test) Small values of T indicate that H_0 is false, so reject H_0 at the level of significance α if T is less than w_α. Accept H_0 if T is greater than or equal to w_α.

C. (two-tailed test) Reject H_0 at the level of significance α if T exceeds $w_{1-\alpha/2}$ or if T is less than $w_{\alpha/2}$. If T is between $w_{\alpha/2}$ and $w_{1-\alpha/2}$, or equal to either quantile, accept H_0.

Example 1. Twelve sets of identical twins were given psychological tests to determine whether the first-born of the twins tends to be more aggressive than the other. The results were as follows, where the higher score indicates more aggressiveness.

Set	1	2	3	4	5	6	7	8	9	10	11	12
First-born X_i	86	71	77	68	91	72	77	91	70	71	88	87
Second-born Y_i	88	77	76	64	96	72	65	90	65	80	81	72
Difference D_i	+2	+6	−1	−4	+5	0	−12	−1	−5	+9	−7	−15
Rank of $|D_i|$	3	7	1.5	4	5.5	—	10	1.5	5.5	9	8	11
R_i	3	7	0	0	5.5	—	0	0	0	9	0	0

The hypotheses are:

H_0: The first-born twin does not tend to be more aggressive than the other $(d_{.50} \geq 0)$.

H_1: The first-born twin tends to be more aggressive than the second-born twin $(d_{.50} < 0)$.

These correspond to the set B of hypotheses. We are assuming that the test scores are accurate measures of the aggressiveness of the individuals. The test statistic is

$$T = \sum_{i=1}^{11} R_i$$

(5)
$$= 24.5$$

The critical region (see decision rule B) of size $\alpha = .05$ corresponds to values of T less than 14 (from Table 7, with $n = 11$). Therefore H_0 is readily accepted.

To estimate $\hat{\alpha}$, note that the quantiles $w_{.20}$ and $w_{.30}$ are given in Table 7 as 23 and 27 respectively. The observed value 24.5 falls between 23 and 27, and so $\hat{\alpha}$ is estimated as

$$(6) \qquad \hat{\alpha} \cong .20 + (.30 - .20) \frac{24.5 - 23}{27 - 23} = .2375$$

The exact $\hat{\alpha}$ is found from Owen (1962) p. 326 as .232, which is exact, strictly speaking, only if there are no ties.

The Wilcoxon signed ranks test is equally appropriate as a median test, where the data consist of a single random sample of size n', $X_1, X_2, \ldots, X_{n'}$. Let X be a random variable with the same distribution as the X_i, and let m be a specified constant. The hypotheses, corresponding to the above sets A, B, and C of hypotheses, are as follows.

A. (one-tailed test)

$\qquad\qquad H_0$: The median of X is $\geq m$

$\qquad\qquad H_1$: The median of X is $< m$

B. (one-tailed test)

$\qquad\qquad H_0$: The median of X is $\leq m$

$\qquad\qquad H_1$: The median of X is $> m$

C. (two-tailed test)

$\qquad\qquad H_0$: The median of X equals m

$\qquad\qquad H_1$: The median of X is not m

The word "mean" may be substituted for "median" in the above hypotheses because of the assumption of symmetry of the distribution of X.

Pairs $(X_1, m), (X_2, m), \ldots, (X_{n'}, m)$ are formed, and the pairs are treated exactly the same as described in the Wilcoxon signed ranks test. The rest of the Wilcoxon test procedure remains unchanged. The following example illustrates the procedure.

Example 2. Thirty observations on a random variable X were obtained in order to test the hypothesis that $E(X)$, the mean of X, was no larger than 30 (set B of hypotheses).

$\qquad\qquad H_0: E(X) \leq 30$

$\qquad\qquad H_1: E(X) > 30$

The observations, the differences $m - X_i$, and the ranks of the pairs are as follows. (The random sample was ordered first, for convenience.)

X_i	$D_i = 30 - X_i$	Rank of $|D_i|$	X_i	$D_i = 30 - X_i$	Rank of $|D_i|$
23.8	+6.2	17	35.9	−5.9	15
26.0	+4.0	11	36.1	−6.1	16
26.9	+3.1	8	36.4	−6.4	18
27.4	+2.6	6	36.6	It is not necessary to compute	
28.0	+2.0	5	37.2	the remaining differences and	
30.3	−0.3	1	37.3	ranks.	
30.7	−0.7	2	37.9		
31.2	−1.2	3	38.2		
31.3	−1.3	4	39.6		
32.8	−2.8	7	40.6		
33.2	−3.2	9	41.1		
33.9	−3.9	10	42.3		
34.3	−4.3	12	42.8		
34.9	−4.9	13	44.0		
35.0	−5.0	14	45.8		

The approximate .05 quantile is obtained from Table 7:

$$w_{.05} \cong \frac{n(n+1)}{4} + x_{.05}\sqrt{n(n+1)(2n+1)/24}$$

$$= \frac{(30)(31)}{4} + (-1.645)\sqrt{(30)(31)(61)/(24)}$$

$$= 232.5 - (1.645)(48.6)$$

(7) $$= 152.6$$

Therefore the critical region of approximate size .05 corresponds to values of the test statistic less than 152.6.

The test statistic T is defined by (4). In this case T equals the sum of the ranks associated with the positive D_i.

(8) $$T = 47$$

The small value of T results in rejection of H_0. We conclude that the mean of X is greater than 30.

The approximate critical level is found by solving the equation

(9) $$47 = 232.5 + x_{\hat{\alpha}}(48.6)$$

to get

(10) $$x_{\hat{\alpha}} = -3.82$$

Table 1 shows that $\hat{\alpha}$ is smaller than .0001.

Theory. The model states that all of the differences D_i share a common median, say $d_{.50}$. By the definition of median, and because the D_i are assumed to have continuous distributions, each D_i has probability .5 of exceeding $d_{.50}$, the median, and probability .5 of being less than $d_{.50}$. If $d_{.50}$ equals zero (H_0 in set C) then each D_i has probability .5 of being positive and the same of being negative. Because of the symmetric distribution of each D_i, the size of each difference and the resulting rank are independent of whether the difference is above or below the median. (Without symmetry it would be possible for the positive differences to tend to be much larger than the negative differences, or vice versa.)

The purpose of the above considerations is to find the distribution of the test statistic T when H_0 is true. First we shall consider the null hypothesis of the two-tailed test. The resulting distribution applies equally well in the one-tailed tests.

Consider n chips numbered from 1 to n, corresponding to the n ranks if there are no ties. Suppose each chip has its number written on one side, and a zero on the other side. Each chip is tossed into the air in such a way that it is equally likely to land with either face showing, corresponding to the ranks which are equally likely to correspond to a positive difference, in which case R_i of (3) equals the rank, or a negative difference, in which case R_i equals zero. Let T equal the sum of the numbers showing after all n chips are tossed, corresponding to the definition of T in (4). The probability distribution of T is the same in the game with the chips as it is when H_0 is true, but the game with the chips is easier to imagine.

The sample space in the game with the chips consists of points like $(1, 2, 3, 0, 0, 6, 7, \ldots 0, n)$ where the zeros represent chips that landed with the zero showing. The tosses are independent of each other and so each of the 2^n points has probability $(1/2)^n$. The test statistic T equals the sum of the numbers in the sample point. Therefore the probability that T equals any number x is found by counting the points whose numbers add to x, then multiplying that count by the probablity $(1/2)^n$.

For example, if n equals 8, T equals 0 one way (all chips land face down) and so $P(T = 0)$ equals $(1/2)^8$. T equals 1 only one way, T equals 2 only one way, but T equals 3 two ways, points $(0, 0, 3, 0, 0, 0, 0, 0)$ and $(1, 2, 0, 0, 0, 0, 0, 0)$. Also T equals 4 two ways. That is,

$$P(T = 0) = (\tfrac{1}{2})^8 = \tfrac{1}{256} \qquad\qquad P(T \leq 0) = .0039$$
$$P(T = 1) = \tfrac{1}{256} \qquad\qquad P(T \leq 1) = .0078$$
$$P(T = 2) = \tfrac{1}{256} \qquad\qquad P(T \leq 2) = .0117$$
$$P(T = 3) = \tfrac{2}{256} \qquad\qquad P(T \leq 3) = .0195$$
$$P(T = 4) = \tfrac{2}{256} \qquad\qquad P(T \leq 4) = .0273$$
$$\text{etc.} \qquad\qquad\qquad\qquad \text{etc.}$$

The distribution function of T, for $n \le 20$, is tabulated in Owen (1962). A table of selected quantiles for $n \le 100$ is given by McCornack (1965). That table is more extensive than we need here and so the more useful quantiles were selected and are given in Table 7. The use of Table 7 will generally result in a slightly conservative test, because the probability of being less than the pth quantile may be less than p. For example, if n equals 8 as in the above illustration, the .025 quantile of T is given in Table 7 as 4, while the actual size of the critical region corresponding to values of T less than 4 is .0195.

For the one-tailed tests, the probability of getting a point in the critical region is a maximum when $d_{.50}$ equals zero, and so this is the situation to be considered. Thus the above distribution of T is equally valid when H_0 is true in the one-tailed tests.

To find the approximate distribution of T when n is large, consider the n random variables R_i, for $i = 1, 2, \ldots, n$, where

(11) $R_i = i$ if the chip with the number i on it lands face up

$\quad\ = 0$ if the chip with the number i on it lands face down

Then

(12) $$E(R_i) = i(\tfrac{1}{2}) + 0(\tfrac{1}{2}) = \tfrac{1}{2}i$$
$$E(R_i^2) = i^2(\tfrac{1}{2}) + 0(\tfrac{1}{2}) = \tfrac{1}{2}i^2$$

and

$$\mathrm{Var}\,(R_i) = E(R_i^2) - E^2(R_i)$$
$$= \tfrac{1}{2}i^2 - \tfrac{1}{4}i^2$$
(13) $$= \tfrac{1}{4}i^2$$

Now

$$T = \text{the sum of the numbers showing on the chips}$$
$$= \sum_{i=1}^{n} R_i$$

Therefore

$$E(T) = E\left(\sum_{i=1}^{n} R_i\right) = \sum_{i=1}^{n} E(R_i) = \tfrac{1}{2}\sum_{i=1}^{n} i$$

Lemma 1 of Chapter 1 enables $E(T)$ to be written as

(14) $$E(T) = \frac{n(n+1)}{4}$$

Similarly, because the X_i are independent of one another, Theorem

1.5.3 and Lemma 2 may be used to give

$$\text{Var}\,(T) = \sum_{i=1}^{n} \text{Var}\,(R_i) = \tfrac{1}{4}\sum_{i=1}^{n} i^2$$

(15)
$$= \frac{n(n+1)(2n+1)}{24}$$

Because T is the sum of n random variables, use of the Central Limit Theorem (Theorem 1.6.2, Set B) allows the distribution of

(16)
$$\frac{T - E(T)}{\sqrt{\text{Var}\,(T)}} = \frac{T - [n(n+1)/4]}{n(n+1)(2n+1)/24}$$

to be approximated by the standard normal distribution, for large n. Therefore the pth quantile of T, w_p, may be approximated by

$$w_p = E(T) + x_p\sqrt{\text{Var}\,(T)}$$

(17)
$$= \frac{n(n+1)}{4} + x_p\sqrt{\frac{n(n+1)(2n+1)}{24}}$$

where x_p is obtained from Table 1, as stated at the bottom of Table 7.

The relative efficiency of the Wilcoxon signed ranks test, and other tests appearing later in this chapter, is discussed thoroughly by Noether (1967a). Unfortunately for us, the mathematical treatment requires calculus and so only the results will be mentioned here. The usual parametric counterpart to the Wilcoxon nonparametric test is the "paired t-test." The A.R.E. of the Wilcoxon test, relative to the paired t-test, is computed under the following restrictions.

1. The bivariate random variables $(X_1, Y_1), \ldots, (X_n, Y_n)$ constitute a random sample.

2. The distribution of X_i is identical with the distribution of Y_i, except for a possible difference in means.

Under these conditions the A.R.E. may range from $108/125 = .864$ up to infinity, but with the surprising insurance feature of never being less than .864. Therefore the Wilcoxon test never can be too bad, but can be infinitely good as compared with the usual parametric test under these circumstances.

Under the further restriction that

3a. The differences D_i have the normal distribution,
the A.R.E. is $3/\pi = .955$. If instead we assume that

3b. The differences D_i have the uniform distribution.

the A.R.E. is 1.0. For a distribution known as the double exponential distribution,

3c. The differences D_i have the double exponential distribution, the A.R.E. is 1.5.

Under the above restrictions the sign test (Section 3.4) may be used to test the same hypotheses as the Wilcoxon test. Then the A.R.E. of the sign test, relative to the Wilcoxon test, is as follows.

Restriction	A.R.E.
3a. Normal	$\frac{2}{3}$
3b. Uniform	$\frac{1}{3}$
3c. Double exponential	$\frac{4}{3}$

It may be surprising to some that the sign test may be more powerful than the Wilcoxon test under some circumstances, namely 3c above. The Wilcoxon A.R.E., compared to the paired t-test, was $3/2$, and therefore multiplication gives the A.R.E. of the sign test, relative to the paired t-test as

$$(\tfrac{4}{3})(\tfrac{3}{2}) = 2$$

Under restriction 3c the sign test has twice the asymptotic efficiency as the paired t-test. For other investigations of power and efficiency, see Klotz (1963, 1965) and Arnold (1965).

The method presented here for handling zero differences and ties was suggested by Wilcoxon (1949). Another method of handling zeros and ties is thoroughly discussed by Pratt (1959). The problem of how to handle zeros and ties in the best way involves finding the null distribution of T when the continuity assumption is dropped from the model, and appears to be unresolved. Recent work on this is reported by Vorlickova (1970).

COMMENT. A test introduced by Walsh (1949) uses the same model and tests the same hypotheses as the Wilcoxon test. For smaller sample sizes the tests are identical, but differences between the two tests appear when n is 7 or greater. Because of the many similarities with the Wilcoxon test, the Walsh test is not presented here. However, the testing procedure used by Walsh is interesting and different in that the critical region remains the same, but the test statistic is defined differently for each different α and n. The resemblance between the Walsh test and the Wilcoxon signed ranks test is well disguised.

The Wilcoxon signed ranks test is sometimes called a test of symmetry. Symmetry tests for bivariate random variables are discussed by Bell and Haller (1969). Extensions to multivariate random variables are examined by Bennett (1965) and Sen and Puri (1967). Hollander (1970) adapts the Wilcoxon test to test for parallelism of two regression lines. The important application to circular distributions, as in Example 3.4.7 and in examples furnished by Batschelet (1965), is presented by Schach (1969).

PROBLEMS

1. Test the data of Example 3.4.5 to see if there is a tendency for the observations in the second year to be less than the observations in the first year.
2. Find the probability distribution of the test statistic T of this section for $N = 5$. (Assume H_0 is true in the two-tailed test.)
3. A random sample consisting of twenty people who drove automobiles was selected to see if alcohol affected reaction time. Each driver's reaction time was measured in a laboratory before and after drinking a specified amount of a beverage containing alcohol. The reaction times in seconds were as follows.

Subject	Before	After	Subject	Before	After
1	.68	.73	11	.65	.72
2	.64	.62	12	.59	.60
3	.68	.66	13	.78	.78
4	.82	.92	14	.67	.66
5	.58	.68	15	.65	.68
6	.80	.87	16	.76	.77
7	.72	.77	17	.61	.72
8	.65	.70	18	.86	.86
9	.84	.88	19	.74	.72
10	.73	.79	20	.88	.97

Does alcohol affect reaction time?
4. A grocer wishes to see whether the median number of items bought on each sale could be considered to be 10, so he observes twelve customers at the check out counter.

Customer	Number of Items	Customer	Number of Items
1	22	7	15
2	9	8	26
3	4	9	47
4	5	10	8
5	1	11	31
6	16	12	7

Test $H_0 : d_{.50} = 10$ using the Wilcoxon test. Which assumptions of the model are violated in this problem?

5.2. A CONFIDENCE INTERVAL FOR THE MEDIAN

A confidence interval for the unknown median of the D_i's may be obtained. The method for finding the confidence interval was devised by Tukey (1949) and also appears in Walker and Lev (1953), p. 445. This method is based on the Wilcoxon test, and therefore the confidence interval for the median difference will usually be shorter (better) than the one obtained using the method of Section 3.2, because of the added assumption of symmetry. Of course, this confidence interval for the median is also a confidence interval for the mean difference,

$$E(D_i) = E(Y_i) - E(X_i)$$

if (X_i, Y_i) or D_i, $i = 1, 2, \ldots, n$, is a random sample and if the mean difference exists.

Confidence Interval for the Median Difference

DATA. The data consist of n observations $(x_1, y_1), (x_2, y_2), \ldots, (x_n, y_n)$ on the bivariate random variables $(X_1, Y_1), (X_2, Y_2), \ldots, (X_n, Y_n)$ respectively. Compute the differences

$$D_i = Y_i - X_i$$

for each pair and arrange them in order from the smallest (the most negative) to the largest (the most positive), denoted as follows.

$$D^{(1)} \le D^{(2)} \le \ldots \le D^{(n-1)} \le D^{(n)}$$

Or, in the usual situation, the data consist of a single random sample $D_1, D_2, \ldots, D_n$, arranged in order as above. We wish to find a confidence interval for the common median of the D_i's, or for the common mean of the D_i's if the D_i's constitute a random sample.

ASSUMPTIONS

1. Each D_i is a continuous random variable.
2. The distribution of each D_i is symmetric.
3. The D_i's are mutually independent.
4. The D_i's all have the same median.
5. The measurement scale of the D_i's is at least interval.

METHOD A (graphic). Denote the differences D_i by heavy dots along a vertical scale, as in Figure 1. The largest difference is denoted by the top dot, called dot A, and the smallest (most negative) difference is denoted by dot B. Find the point midway between A and B and call it C. Draw a horizontal line through the point C. Somewhere along the horizontal line, it is not important just where, select a point and call it D (see Figure 2). Draw line segments from A to D and from B to D, forming an isosceles triangle. Draw a line segment from each dot slanting upward, parallel to the line segment AD. Draw another line segment downward from each dot, parallel to AD, to the line segment BD. Place a heavy dot at each intersection of an upward line with a downward line, as in Figure 2. A draftsman's triangle may be used for convenience in drawing.

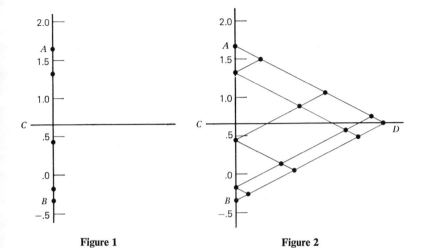

Figure 1   Figure 2

Let $1 - \alpha$ be the approximate confidence coefficient desired. From Table 7 obtain the $\alpha/2$ quantile of T, $w_{\alpha/2}$. Count down from the top dot in the figure to the $w_{\alpha/2}$th dot, including the top dot and the other dots on the vertical line. Draw a horizontal line through the $w_{\alpha/2}$th dot. The intersection of that line with the vertical line is marked U, and represents the upper bound for the confidence interval. Draw another horizontal line, this time through the $w_{\alpha/2}$th dot from the bottom. This line intersects the vertical axis at a point called L, which represents the lower bound. (If $w_{\alpha/2}$ equals zero no confidence interval can be found for that value of α.) Then the interval from L to U, including the end points L and U, is a $1 - \alpha$ confidence interval for the unknown median of the D_i's. See Figure 3 for an illustration, where $w_{\alpha/2}$ equals 3.

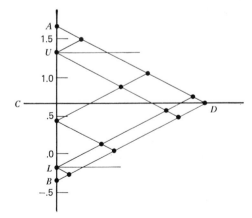

Figure 3

If two or more differences are equal, each resulting dot in the graph needs to be counted a corresponding number of times. This procedure is illustrated in Example 1. To be specific, if there are t differences exactly equal to some number d, each dot on the line segments extending from d is counted t times, and the dot on the vertical axis at d itself is counted $t(t + 1)/2$ times. These are the counts that would result if the t values actually were tiny distances apart and delicate graphical methods were used to distinguish between the different values.

METHOD B (algebraic). To obtain a $1 - \alpha$ confidence interval, obtain the $\alpha/2$ quantile $w_{\alpha/2}$ from Table 7. (If w_α equals zero no confidence interval may be obtained for that value of α.) Then consider the $n(n + 1)/2$ possible averages $(D_i + D_j)/2$ for all i and j, including $i = j$. The $w_{\alpha/2}$th largest of these averages, and the $w_{\alpha/2}$th smallest of these averages constitute the upper and lower bounds for the $1 - \alpha$ confidence interval. It is not necessary to compute all $n(n + 1)/2$ averages; only those averages near the largest and the smallest need to be computed to obtain a confidence interval.

Example 1.   An environmental control study was conducted to determine the median temperature at which people become uncomfortably warm. Several subjects were selected and given standard instructions concerning clothing to wear, exercises to perform, and so on. One by one the subjects were observed in a controlled environment room, as the temperature was slowly raised. Each subject indicated the time at which the room became uncomfortably warm, and the experimenter noted the temperature at that time. It was assumed that the humidity and air movement were representative of average conditions, and it was assumed that the responses of the subjects would resemble the responses of a random sample of people who lived in that climate.

The recorded temperatures for eight subjects were as follows.

Subject	1	2	3	4	5	6	7	8
Temperature (°F)	83	87	86	79	86	80	82	89

The temperatures were represented by heavy dots on the vertical axis in Figure 4. The midline was drawn through 84, an arbitrary point D was selected, and the line segments were drawn. Heavy dots were placed at the intersections. Circles were placed around those dots that were to be counted twice because of the two readings of 86 degrees, and a square was placed around the dot that was to be counted three times.

For a 90% confidence interval, α equals .10. From Table 7, $w_{.05}$ for $n = 8$ is found to equal 6. The sixth dot from the top of Figure 4 is at 86.5, the junction of line segments from 87 and 86. (A higher dot is counted twice.) The sixth dot from the bottom is at 81. Therefore the interval from 81 to 86.5 (inclusive) is a 90% confidence interval for the median temperature for the threshold of discomfort.

The algebraic method of finding the confidence interval consists of finding

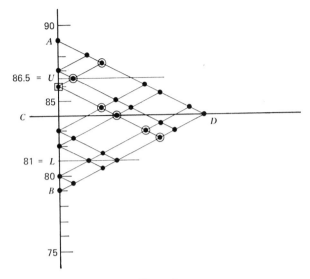

Figure 4

the sixth largest and the sixth smallest averages. The greatest averages are

$$89, 88, 87.5, 87.5, 87, \underline{86.5}, 86.5, \ldots$$

and the smallest averages are

$$79, 79.5, 80, 80.5, 81, \underline{81}, 82, \ldots$$

The sixth largest and sixth smallest averages are underlined. The confidence interval is the same as before.

Theory. It is easy to see that the algebraic method and the graphic method of obtaining confidence intervals are equivalent. Each dot in the graphic method represents an average of two D_i's; the ordinate of each dot is the average of the ordinates of the two dots on the vertical axis, representing D_i's, as in Figure 5, because each triangle is an isosceles triangle similar to the original large isosceles triangle. Therefore the nth dot from the top represents the nth largest average $(D_i + D_j)/2$, so the two methods are equivalent.

It is not as easy to see the relationship between the graphic method of finding confidence intervals and the Wilcoxon test statistic. With the aid of a numerical illustration we shall describe first the relation between counting dots in the graphic method, and the rank of the absolute

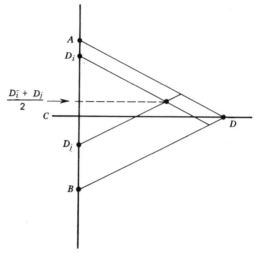

Figure 5

difference. Consider some hypothetical D_i's arranged in order, and the $|D_i|$'s arranged in order.

i	D_i	D_i's in order	i	$|D_i|$'s in order	Ranks of $|D_i|$	R_i
1	−3	+7	5	$|+7|$	6	6
2	0	+5	6	$|+5|$	5	5
3	+2	+2	4	$|-4|$	4	0
4	−4	0	1	$|-3|$	3	0
5	+7	−3	3	$|+2|$	2	2
6	+5	−4	2	$|0|$	1	0

The graphical method of analysis results in the following diagram.

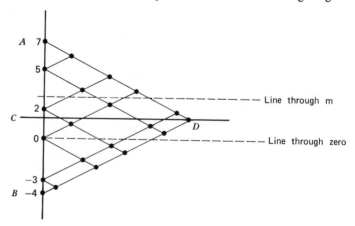

Figure 6

Consider the largest D_i, $D_5 = 7$. There are six heavy dots on the line segment extending downward from D_5, and above the line through zero. This value 6 equals the rank of $|D_5|$. Consider any other positive D_i, such as $D_3 = 2$. The line segment extending downward from D_3 contains two heavy dots above the line through zero, and "two" is also the rank of $|D_3|$. The number of heavy dots on the downward line segment from any positive D_i, and above the zero line, equals the rank of $|D_i|$. This is because the ordinate of each dot on the downward line segment from a positive D_i equals the average $(D_i + D_j)/2$ of D_i with some D_j smaller or equal to D_i; that average is above the zero line (positive) if and only if $|D_i|$ is greater than $|D_j|$; and the rank of $|D_i|$ equals the number of absolute differences less than or equal to $|D_i|$, counting itself. Therefore the Wilcoxon test statistic T may be computed in either of two ways; by counting all of the heavy dots above the line through zero, or by summing the ranks of the positive D_i's.

That is, the two ways of computing T are equivalent if there are no zero differences and if there are no ties of positive with negative differences. The method of handling zeros and ties in this section is equivalent to retaining the zeros in the Wilcoxon test, and assigning ranks to the ties so as to minimize T.

A confidence interval for the median $d_{.50}$ of D_i is found by using the Wilcoxon test to test

$$H_0 : d_{.50} = m$$

for various values of m. This procedure is equivalent to subtracting the value m from each D_i and testing to see if the median of the new D_i's equals zero. The test statistic may be found either by counting the number of heavy dots above the line through m (see Figure 6), or by subtracting m from each D_i and then counting the number of dots above zero. The values of m that result in accepance of H_0 collectively form the confidence interval. This method of forming a confidence interval is similar in principle to the method used in Section 3.1 to find a confidence interval for the unknown parameter p in the binomial distribution.

Intervals formed in this way have probability $1 - \alpha$ of containing the true unknown parameter, which may be seen as follows. Suppose m' is the true median. A random sample is drawn, and the null hypothesis $d_{.50} = m'$ is tested. There is (at most) probability α that H_0 will be rejected (because H_0 is actually true). Suppose H_0 is rejected. Then the confidence interval, formed by collecting all values of m for which $H_0 : d_{.50} = m$ is acceptable, will not contain the true value m'. That is, the probability that the confidence interval does *not* contain m' is (at most) α.

Similarly, suppose the random sample is such that $H_0: d_{.50} = m'$ is accepted. Then m' is part of the resulting confidence interval. This happens with probability (at least) $1 - \alpha$. The method of this section therefore results in an interval which has probability $1 - \alpha$ of containing the true parameter.

A recent paper by Noether (1967b) shows that if the continuity assumption is not true, then the confidence interval with its endpoints (U and L) has a confidence coefficient of at least $1 - \alpha$, while the interval without its endpoints has a confidence coefficient of at most $1 - \alpha$. Therefore we recommend inclusion of the endpoints, and a statement of the form

$$P(L \leq d_{.50} \leq U) \geq 1 - \alpha$$

A recent discussion of this method of finding confidence intervals is given by Moses (1965). If the sampling is stratified rather than random, see the article by McCarthy (1965). For another type of dependency in the sample see Høyland (1968). Confidence regions for the case of multivariate random variables are given by Puri and Sen (1968).

PROBLEMS

1. A candidate for political office realizes that he will maximize his vote-getting ability if he adopts the median position of his constituents. Therefore he devises a questionnaire and distributes it to fifteen voters (who hopefully resemble a random sample). The results of the questionnaire are scored from one extreme (zero) to the other (ten).

Voter	Score	Voter	Score	Voter	Score
1	6.7	6	9.3	11	8.8
2	4.2	7	8.9	12	5.4
3	4.1	8	7.4	13	6.1
4	2.3	9	7.4	14	6.0
5	6.1	10	9.3	15	4.9

Find a 90% confidence interval for the median score. Based on the procedure of this section, what is your point estimate of the median score?

2. An emergency rescue squad is responsible for accidents occurring in a long narrow lake. They wish to build a permanent station in the spot that will minimize the total distance they will have to travel going to accidents in the future. That spot should be at the median (by distance from some point of reference) spot at which accidents will occur. Assuming that the accidents occurred thus far

resemble a random sample of all accidents yet to occur, the distances (from the dam) are measured.

Accident	Distance (mi.)	Accident	Distance (mi.)
1	7.1	8	6.1
2	4.4	9	2.2
3	3.9	10	6.7
4	2.2	11	4.9
5	4.2	12	7.3
6	3.4	13	0.3
7	1.1	14	7.6

What is a 95% confidence interval for the optimal distance from the dam for the station?

5.3. TWO INDEPENDENT SAMPLES

The first test presented in this section is known as the Mann-Whitney test and also as the Wilcoxon test. Equivalent forms of the same test appeared in the literature under various names, probably due in part to the intuitive appeal of the test procedure. Although primarily a two-sample test, the Mann-Whitney test may be applied in many different situations other than the usual two-sample situation.

The usual two-sample situation is one in which the experimenter has obtained two samples from possibly different populations, and he wishes to use a statistical test to see if he can reject the null hypothesis that the two populations are identical. That is, he wishes to detect differences between the two populations on the basis of random samples from those populations. If the samples consist of ordinal type data, the difference which is usually the most interesting is a difference in the locations of the two populations. Does one population tend to yield larger values than the other population? Are the two medians equal? Are the two means equal?

An intuitive approach to the two-sample problem is to combine both samples into a single ordered sample and then assign ranks to the sample values from the smallest value to the largest, without regard to which population each value came from. Then the test statistic might be the *sum* of the ranks assigned to those values from one of the populations. If the sum is too small, or too large, there is some indication that the values from that population tend to be smaller, or larger as the case may be, than the values from the other population. Hence the null hypothesis of no differences between populations may be rejected if the ranks associated with one sample tend to be larger than those of the other sample.

Ranks may be considered preferable to the actual data for several reasons

First, if the numbers assigned to the observations have no meaning by themselves, but rather attain meaning only in an ordinal comparison with the other observations, then the numbers contain no more information than the ranks contain. Such is the nature of ordinal data. Second, even if the numbers have meaning but the distribution function is not a normal distribution function, the probability theory is usually beyond our reach when the test statistic is based on the actual data. The probability theory of statistics based on ranks is relatively simple and does not depend on the distribution in many cases. A third reason for preferring ranks is that the A.R.E. of the Mann-Whitney test is never too bad when compared with the two-sample t-test, the usual parametric counterpart. And yet the contrary is not true; the A.R.E. of the t-test compared to the Mann-Whitney test may be as small as zero, or "infinitely bad." So the Mann-Whitney test is a safer test to use.

The Mann-Whitney Test

DATA. The data consist of two random samples. Let $X_1, X_2, \ldots, X_n$ denote the random sample of size n from population 1 and let $Y_1, Y_2, \ldots, Y_m$ denote the random sample of size m from population 2. Assign the ranks 1 through $n + m$ to the two samples in the usual way. That is, assign the rank 1 to the smallest value in the combined sample of X's and Y's, the rank 2 to the next smallest, and so on to the largest, which receives the rank $n + m$. Let $R(X_i)$ and $R(Y_j)$ denote the rank assigned to X_i and Y_j, for all i and j.

If several sample values are exactly equal to each other (tied), assign to each the average of the ranks that would have been assigned to them had there been no ties (see Example 1).

ASSUMPTIONS
1. Both samples are random samples from their respective populations.
2. In addition to independence within each sample, there is mutual independence between the two samples.
3. Both samples consist of continuous random variables. (A moderate number of ties is tolerable.)
4. The measurement scale is at least ordinal.

HYPOTHESES
A. (two-tailed test) Let $F(x)$ and $G(x)$ be the distribution functions corresponding to populations 1 and 2 respectively, and of X and Y respectively. Then the hypotheses may be stated as follows.

$$H_0: F(x) = G(x) \quad \text{for all } x$$
$$H_1: F(x) \neq G(x) \quad \text{for some } x$$

In many real situations any difference between distributions implies that $P(X < Y)$

is no longer equal to 1/2. Therefore the following set of hypotheses is often used instead of the above.

$$H_0:P(X < Y) = \tfrac{1}{2}$$
$$H_1:P(X < Y) \neq \tfrac{1}{2}$$

B. (one-tailed test)

$$H_0:P(X < Y) \leq \tfrac{1}{2}$$
$$H_1:P(X < Y) > \tfrac{1}{2}$$

C. (one-tailed test)

$$H_0:P(X < Y) \geq \tfrac{1}{2}$$
$$H_1:P(X < Y) < \tfrac{1}{2}$$

The Mann-Whitney test is unbiased and consistent when testing the above hypotheses involving $P(X < Y)$. However, the same is not always true for the following hypotheses, which are sometimes stated instead of the above forms. In order to insure that the test remains consistent and unbiased for the following hypotheses, it is sufficient to add another assumption to the model above.

Assumption 5. If there is a difference between population distribution functions, that difference is a difference in the location of the distribution. That is, if $F(x)$ is not identical with $G(x)$, then $F(x)$ is identical with $G(x + c)$, where c is some constant.

Then the hypotheses are stated in terms of the means of X and Y, if they exist.

A. (two-tailed test)

$$H_0:E(X) = E(Y)$$
$$H_1:E(X) \neq E(Y)$$

B. (one-tailed test)

$$H_0:E(X) \geq E(Y)$$
$$H_1:E(X) < E(Y)$$

C. (one-tailed test)

$$H_0:E(X) \leq E(Y)$$
$$H_1:E(X) > E(Y)$$

TEST STATISTIC. The test statistic T may be found by first finding S, the sum of the ranks assigned to the observations from population 1. That is,

(1)
$$S = \sum_{i=1}^{n} R(X_i)$$

The test statistic is then given by

(2)
$$T = S - \frac{n(n + 1)}{2}$$

DECISION RULE. Use decision rule A, B, or C depending on whether the hypothesis of interest is classified as A, B, or C above. The quantiles w_p of T are given in Table 8 for p equal to .001, .005, .01, .025, .05, and .10. The upper quantiles are not given, but may be computed by subtraction from nm. That is,

(3) $$w_{1-p} = nm - w_p$$

As an alternative to using upper quantiles, the statistic T', defined as

(4) $$T' = nm - T$$

may be used with the lower quantiles whenever an upper tailed test is desired.

A. (two-tailed test) Reject H_0 at the level of significance α if T is less than the $\alpha/2$ quantile $w_{\alpha/2}$ or if T is greater than the $1 - \alpha/2$ quantile $w_{1-\alpha/2}$. Accept H_0 if T is between or equal to the two quantiles.

B. (one-tailed test) Small values of T indicate that H_1 is true. Therefore reject H_0 at a level of significance α if T is less than the αth quantile w_α. Accept H_0 if T is greater than or equal to w_α.

C. (one-tailed test) Large values of T, or small values of T, indicate that H_1 is true. Therefore reject H_0 at the level of significance α if T is greater than $w_{1-\alpha}$, or (equivalently) if T' is less than w_α. Accept H_0 if T is less than or equal to $w_{1-\alpha}$.

Example 1. The senior class in a particular high school had forty-eight boys. Twelve boys lived on farms and the other thirty-six lived in town. A test was devised to see if farm boys in general were more physically fit than town boys. Each boy in the class was given a physical fitness test in which a low score indicates poor physical condition. The scores of the farm boys (X_i) and the town boys (Y_i) are as follows.

X_i: Farm Boys				Y_i: Town Boys			
14.8	10.6	12.7	16.9	7.6	2.4	6.2	9.9
7.3	12.5	14.2	7.9	11.3	6.4	6.1	10.6
5.6	12.9	12.6	16.0	8.3	9.1	15.3	14.8
6.3	16.1	2.1	10.6	6.7	6.7	10.6	5.0
9.0	11.4	17.7	5.6	3.6	18.6	1.8	2.6
4.2	2.7	11.8	5.6	1.0	3.2	5.9	4.0

Neither group of boys is a random sample from any population. However, it is reasonable to assume that these scores resemble hypothetical random samples from the populations of farm and town boys in that age group, at least for similar localities. The other assumptions of the model appear to be reasonable, such as independence between groups. Therefore the Mann-Whitney test is selected to test

H_0: Farm boys do not tend to be more fit, physically, than town boys

H_1: Farm boys tend to be more fit than town boys

The null hypothesis could also be stated as $H_0: P(X < Y) \geq 1/2$, or $H_0: E(X) \leq E(Y)$, according to set C of hypotheses.

The scores are ranked as follows.

X	Y	Rank	X	Y	Rank	X	Y	Rank
	1.0	1		6.2	17		11.3	33
	1.8	2	6.3		18	11.4		34
	2.1	3		6.4	19		11.8	35
	2.4	4		6.7	20.5⎤	12.5		36
	2.6	5		6.7	20.5⎦		12.6	37
2.7		6	7.3		22		12.7	38
	3.2	7		7.6	23	12.9		39
	3.6	8		7.9	24		14.2	40
	4.0	9		8.3	25		14.8	41.5⎤
4.2		10	9.0		26	14.8		41.5⎦
	5.0	11		9.1	27		15.3	43
	5.6	13⎤		9.9	28		16.0	44
	5.6	13⎥		10.6	30.5⎤	16.1		45
5.6		13⎦		10.6	30.5		16.9	46
	5.9	15	10.6		30.5		17.7	47
	6.1	16		10.6	30.5⎦		18.6	48

There are four groups of tied scores, as indicated by the square brackets above. Within each group the ranks that should have been assigned are averaged, and the average rank is assigned instead, as illustrated.

The test is one tailed. The critical region corresponds to large values of T. Because m, the number of Y values, exceeds 20, the large sample approximation at the end of Table 8 is used. The critical region of approximate size .05 corresponds to values of T greater than $w_{.95}$, where $w_{.95}$ is found as follows.

$$w_{.95} = \frac{nm}{2} + x_{.95}\sqrt{nm(n + m + 1)/12}$$

$$= \frac{(12)(36)}{2} + (1.645)\sqrt{(12)(36)(49)/12}$$

$$= 285.1$$

The sum of the ranks assigned to the X's is

$$S = \sum_{i=1}^{n} R(X_i)$$

$$= 6 + 10 + 13 + 18 + 22 + 26$$

$$+ 30.5 + 34 + 36 + 39 + 41.5 + 45$$

$$= 321$$

Equation (2) is used to find T.

$$T = S - \frac{n(n + 1)}{2}$$
$$= 321 - 78$$
$$= 243$$

Because T is less than 285.1, the null hypothesis is accepted. The critical level $\hat{\alpha}$ turns out to be 0.26 in this example. It has not been shown by these data that farm boys are more physically fit than town boys.

The next example illustrates a situation in which no random variables are defined explicitly. The pieces of flint are ranked according to hardness by direct comparison with each other. A random variable which assigns a measure of hardness to each piece of flint is conceivable, but unnecessary in this case.

Example 2.   A simple experiment was designed to see if flint in area A tended to have the same degree of hardness as flint in area B. Four sample pieces of flint were collected in area A and five sample pieces of flint were collected in area B. To determine which of two pieces of flint was harder, the two pieces were rubbed against each other. The piece of flint sustaining less damage was judged the harder piece of the two. In this manner all nine pieces of flint were ordered according to hardness. The rank 1 was assigned to the softest piece, rank 2 to the next softest, and so on.

Origin of piece	A	A	A	B	A	B	B	B	B
Rank	1	2	3	4	5	6	7	8	9

The hypothesis to be tested is

H_0: The flints from areas A and B are of equal hardness

against the alternative

H_1: The flints are not of equal hardness

The Mann-Whitney two-tailed test was used, where

S = sum of the ranks of pieces from area A
$= 1 + 2 + 3 + 5$
$= 11$

and

$$T = S - \frac{(4)(5)}{2}$$
$$= 1$$

The two-tailed critical region of approximate size .05 corresponds to values of T less than 2, and values of T greater than $20 - 2 = 18$. Because T in this example is less than 2, the null hypothesis is rejected and it is concluded that flints from the two areas differ in degree of hardness. Because of the direction of the difference the further conclusion that the flint in area B is harder than the flint in area A may also be drawn.

The critical level $\hat{\alpha}$ may be considered to be .05, because of the values for α given in Table 8, .05 is the smallest that results in rejection of H_0.

COMMENT. It is sometimes simpler to compute the value of T directly, without first computing S. This may be accomplished by letting T equal the sum of the number of Y values less than X_i, for each X_i. Mathematically this may be stated as

$$(5) \qquad\qquad T = \sum_{i=1}^{n} U_i$$

where

(6) U_i = the number of values of Y which are smaller than the ith smallest value of X in the combined sample.

In Example 1 the smallest X was preceded by five smaller Y values, so $U_1 = 5$. The next value of X was preceded by a total of eight Y values, so $U_2 = 8$. Similarly, U_3 equals 10 (counting each tied value as one-half), U_4 equals 14, and so on.

The Mann-Whitney test may be modified slightly to test

$$H_0 : \text{Var } (X) \leq \text{Var } (Y)$$

against

$$H_1 : \text{Var } (X) > \text{Var } (Y)$$

The principal modification is in the method of assigning ranks. In the ordered combined sample, assign rank 1 to the smallest value, rank 2 to the largest value, rank 3 to the second largest value, rank 4 to the second smallest value, rank 5 to the third smallest value, and so on, alternately assigning ranks to the end values two at a time (after the first) and proceeding toward the middle. The ordered combined sample then has the ranks

$$1, 4, 5, 8, 9, \ldots, (n + m), \ldots, 7, 6, 3, 2$$

If H_0 is false the values of X will tend to be in the tails of the combined sample (if the two sample means are approximately equal), and will thereby be assigned the smaller ranks. The result is that T will tend to be small if H_0 is false. The test consists of rejecting H_0 if T is too small, as defined by Table 8. The resemblance of this test, introduced by Siegel and Tukey (1960), to the test of Mann and Whitney is obvious. Equally obvious are the modifications required to convert this test into a two-tailed test.

The Siegel-Tukey test requires a slightly different model than that given for the Mann-Whitney test. Here we want the differences to show as differences in variances, rather than as differences in location as implied by Assumption 5 above. However, if differences between populations appear as both differences in location and differences in scale, the Siegel-Tukey test is not likely to detect the difference. If the population means or medians differ by a known amount it is legitimate first to adjust the data by subtracting that known amount and then to test the hypothesis. If the difference in location is unknown the logical step is to adjust the data on the basis of the sample means or medians, that is, by subtracting the sample median from each observation in the respective samples, and then using the test as described. However, such an adjustment affects the null distribution of T so that Table 8 is no longer accurate, and therefore the test is not valid, strictly speaking. For large sample sizes the error does not appear to be serious. An illustration of the Siegel-Tukey test follows.

Example 3.   Suppose we want to test

$$H_0: \text{Var}(X) \leq \text{Var}(Y)$$

against

$$H_1: \text{Var}(X) > \text{Var}(Y)$$

for the data in Example 1. Then the ranks are assigned to the observations as follows.

X	Y	Rank	X	Y	Rank	X	Y	Rank
	1.0	1		6.2	33		11.3	31
	1.8	4	6.3		36	11.4		30
	2.1	5		6.4	37		11.8	27
	2.4	8		6.7	40.5⎤			26
	2.6	9		6.7	40.5⎦ 12.5		12.6	23
2.7		12	7.3		44		12.7	22
	3.2	13		7.6	45	12.9		19
	3.6	16		7.9	48		14.2	18
	4.0	17		8.3	47		14.8	14.5⎤
4.2		20	9.0		46	14.8		14.5⎦
	5.0	21		9.1	43		15.3	11
	5.6	25⅔⎤		9.9	42		16.0	10
	5.6	25⅔		10.6	36.5⎤ 16.1		16.9	7
5.6		25⅔⎦		10.6	36.5		16.9	6
	5.9	29	10.6	10.6	36.5		17.7	3
	6.1	32		10.6	36.5⎦		18.6	2

The ranks are assigned to tied values, in the same manner as before, by assigning the average of the ranks that would have been assigned to untied values. For instance, one X and two Y's are tied at 5.6. The ranks that would have been assigned to those values, had they not been tied, are 24, 25, and 28. The average of those three ranks is $25\frac{2}{3}$, which is then assigned as a rank to the X and the two Y's.

The sum of the ranks of the X values is $316\frac{2}{3}$. That is,

$$S = 316\frac{2}{3}$$

and

$$T = S - \frac{n(n + 1)}{2}$$

$$= 316\frac{2}{3} - \frac{(12)(13)}{2}$$

$$= 238\frac{2}{3}$$

The critical region of approximate size .05 corresponds to values of T smaller than the quantile $w_{.05}$ which is approximated by

$$w_{.05} \cong \frac{nm}{2} + x_{.05}\sqrt{\frac{nm(n + m + 1)}{12}}$$

$$= \frac{(12)(36)}{2} + (-1.645)\sqrt{\frac{(12)(36)(49)}{12}}$$

$$= 146.9$$

The observed value of T is greater than 146.9, so H_0 is accepted.

The observed value of T is larger than even the mean of T, $nm/2 = 216$. Therefore $\hat{\alpha}$ is larger than .50. In fact $\hat{\alpha}$ may be found to equal about .70.

Theory. The null distribution of T is found by assuming that X_i and Y_j are identically distributed. Strictly speaking, this is true only when H_0 is true in the two-tailed test. However, in the one-tailed tests α is found by maximizing the probability of T falling into the appropriate rejection region, and that probability is a maximum, under H_0, when the two populations have identical distributions. Therefore the distribution of T is found by assuming that the X_i and Y_j are identically distributed, no matter which of the three forms of H_0 is being tested.

If the X_i and the Y_j are independent and identically distributed, then every arrangement of the X's and Y's in the ordered combined sample is equally likely. This is the basic principle behind many rank tests. A formal proof of the above statement requires calculus and is therefore beyond the scope of this book. However, the truth of the statement may seem to be intuitively obvious after one attempts to furnish a reason for some arrangements being more probable than others. There is no valid reason for this, and therefore we can accept the fact that all ordered arrangements are equally likely as an intuitively obvious but unproved (here) statement.

If the X_i and Y_j are independent and identically distributed, then the ranks assigned to the X_i in the combined sample should resemble a random selection of n of the integers from 1 to $n + m$. That is, there is no reason why any particular rank should have a better chance than any other rank of being assigned to a value of X_i. Because each number from 1 to $n + m$ is equally likely to be assigned to X_i as its rank, and because n different numbers are selected as ranks for the X's, the probability distribution of S, the sum of the ranks, may be obtained by considering the probability distribution of the sum of n integers selected at random, without replacement, from among the integers from 1 to $n + m$. Once the distribution of S is found, the distribution of T follows easily.

The number of ways of selecting n integers from a total number of $n + m$ integers is $\binom{n + m}{n}$, and each way has equal probability according to the basic premise stated above. Hence the probability that S equals k may be found by counting the number of different sets of n integers from 1 to $n + m$ that add up to the value k, and then by dividing that number by $\binom{n + m}{n}$.

For example, if the sample sizes are $n = 3$ and $m = 4$, then the number of ways of selecting 3 out of 7 ranks is

$$\binom{n + m}{n} = \frac{7!}{3!\,4!} = 35$$

The smallest value S may assume is 6, which occurs if the three ranks, 1, 2, 3, are selected. The next value S may assume is 7, which occurs only one way: 1, 2, 4. The value $S = 8$ may be assumed two ways, with the ranks 1, 2, 5 or with 1, 3, 4. Each value of S represents a single value of T; from Equation (2) we have

$$T = S - \frac{n(n + 1)}{2}$$

$$= S - \frac{(3)(4)}{2}$$

(7) $$= S - 6$$

Therefore,

$P(S = 6) = \frac{1}{35}$	$P(T = 0) = \frac{1}{35}$	$P(T \le 0) = .029$
$P(S = 7) = \frac{1}{35}$	$P(T = 1) = \frac{1}{35}$	$P(T \le 1) = .057$
$P(S = 8) = \frac{2}{35}$	$P(T = 2) = \frac{2}{35}$	$P(T \le 2) = .114$
etc.	etc.	etc.

Because S is the sum of the ranks of the nX's, for large n and m the Central Limit Theorem may be applied to obtain an approximate distribution for S. This was done in Example 1.6.7. (The requirement, in Example 1.6.7, that n be less than $(n + m)/2$ is not needed if both n and m are large.) The results of Example 1.6.7 state that S is approximately normal with mean and variance given by Theorem 1.5.5 as

$$(8) \qquad E(S) = \frac{n(n + m + 1)}{2}$$

and

$$(9) \qquad \mathrm{Var}\,(S) = \frac{n(n + m + 1)m}{12}$$

Because T equals S less a constant, the variance of T equals the variance of S, the mean of T equals the mean of S less a constant, and T is approximately normally distributed.

$$E(T) = E\left(S - \frac{n(n + 1)}{2}\right)$$

$$= E(S) - \frac{n(n + 1)}{2}$$

$$(10) \qquad = \frac{nm}{2}$$

and

$$\mathrm{Var}\,(T) = \mathrm{Var}\,(S)$$

$$(11) \qquad = \frac{nm(n + m + 1)}{12}$$

Therefore the quantiles w_p of T may be approximated with the aid of Theorem 1.6.1,

$$w_p \cong E(T) + x_p\sqrt{\mathrm{Var}\,(T)}$$

$$(12) \qquad = \frac{nm}{2} + x_p\sqrt{\frac{nm(n + m + 1)}{12}}$$

where x_p is the pth quantile of a standard normal random variable.

The Mann-Whitney test was first introduced for the case $n = m$ by Wilcoxon (1945), who used S as a test statistic. Wilcoxon's test was extended to the case of unequal sample sizes by White (1952) and van der Reyden (1952). A test equivalent to Wilcoxon's was also developed independently and introduced by Festinger (1946). Mann and Whitney

(1947) appear to be the first to consider unequal sample sizes and to furnish tables suitable for use with small samples. Their test statistic was in the form used here, given by (2), rather than the sum of ranks as used by Wilcoxon. It is largely the work of Mann and Whitney that led to widespread use of the test. Because the test is attributed to various authors, it is the user's perogative as to which name to call the test.

The modification of the Mann-Whitney test to examine differences in dispersion or variance or scale, introduced by Siegel and Tukey in 1960, is similar in principle to an earlier test devised by Freund and Ansari (1957). The Freund-Ansari test assigns the rank 1 to both the largest and the smallest values in the combined sample, the rank 2 to both the second largest and second smallest values, and so on toward the middle. The test statistic is the sum of the ranks of the X values, and enjoys large sample normality with

$$(13) \qquad \text{mean} = \frac{n(n + m + 2)}{4}$$

$$(14) \qquad \text{variance} = \frac{nm(n + m - 2)(n + m + 2)}{48(n + m - 1)}$$

if $n + m$ is even. However, special tables are needed for small samples. In this respect the Siegel-Tukey test is preferred because the Mann-Whitney tables may be used for small samples. Under the null hypothesis that the two samples are identically distributed, the distribution of the Siegel-Tukey test statistic is found in the same way, and therefore has the same distribution, as the Mann-Whitney test statistic. The relationship between the two tests is described more explicitly on page 126 of Hájek and Šidák (1967).

Mood (1954) suggested the use of

$$(15) \qquad T = \sum_{i=1}^{n} \left[R(X_i) - \frac{n + m + 1}{2} \right]^2$$

as a statistic for testing dispersion, where $R(X_i)$ refers as before to the rank of X_i as assigned in the Mann-Whitney test. If n and m are large, Mood's test statistic approaches normality with

$$(16) \qquad \text{mean} = \frac{n(n + m + 1)(n + m - 1)}{12}$$

and

$$(17) \qquad \text{variance} = \frac{nm(n + m + 1)(n + m + 2)(n + m - 2)}{180}$$

so that Table 1 may be used to determine the crtical region. Exact tables are given by Laubscher, Steffens, and De Lange (1968). The null distribution tables of a related statistic

$$(18) \qquad T = \sum_{i=1}^{n} [R(X_i) - \bar{R}_x]^2 \qquad \text{where} \qquad \bar{R}_x = \frac{1}{n} \sum_{i=1}^{n} R(X_i)$$

are given by Hollander (1963).

The usual parametric test for testing the equality of two variances is called the F test. However the F test is known to be quite sensitive to departures from normality, as pointed out by Siegel and Tukey (1960). This suggests using one of the above nonparametric tests when the populations may be nonnormal. In turn, the above nonparametric tests rapidly lose power in the presence of unlike location parameters, a factor by which the F test is unaffected. The A.R.E. of the various tests relative to the F test is given as follows (Ansari and Bradley, 1960).

True Distribution	A.R.E. of Siegel-Tukey	A.R.E. of Freund-Ansari	A.R.E. of Mood
Normal	.61	.61	.76
Uniform	.60	.60	1.00
Double Exponential	.94	.94	1.08

Of the above tests, Mood's has the highest A.R.E. in the three situations examined.

The usual two-sample parametric test to detect differences between means is the two-sample t test, which assumes normally distributed random variables in addition to the other assumptions of the model. The Mann-Whitney test is almost as powerful as the t test if the normality assumption is true. That is, if the underlying distribution is normal, the A.R.E. of the Mann-Whitney test is $3/\pi$ or .955, compared to the t test. The comparable A.R.E. is 1.00 if the distribution is uniform, and may be as high as infinity for other distributions. However, an interesting safety feature of the Mann-Whitney test is that the A.R.E. is never less than .864, a valuable property shown by Hodges and Lehmann (1956). These results assume that the distribution functions of the two populations differ only by a location parameter.

More extensive tables for the Mann-Whitney test are given by Verdooren (1963) for n and $m \le 25$, and by Milton (1964) for $n \le 20$ and $m \le 40$. Other tables and a bibliography are found in Jacobson (1963). The distribution of the Mann-Whitney test statistic is discussed by Klotz (1966) and Buckle, Kraft, and van Eeden (1969). Other recent articles are by Zaremba (1965) and Serfling (1968).

The efficiency of the Mann-Whitney test and other closely related tests is the subject of papers by Chanda (1963), Noether (1963), Haynam and Govindarajulu (1966), McNeil (1967), R. A. Shorack (1967), and Stone (1967). Modifications for sequential testing are given by Alling (1963), Woinsky and Kurz (1969), Bradley, Martin, and Wilcoxon (1965), and Bradley, Merchant, and Wilcoxon (1966). The problem of testing circular distributions, as discussed by Batschelet (1965), is approached by Beran (1969) and Schach (1969b).

If the two samples are censored, that is, if some of the largest and/or smallest sample values are not observable, the data may be analyzed with modifications of the Mann-Whitney test, such as those discussed by Gastwirth (1965a), Gehan (1965a, 1965b), Gehan and Thomas (1969), Saw (1966), Basu (1968), Hettmansperger (1968), and Shorack (1968). A rank test for the bivariate two-sample problem is given by Mardia (1967a, 1968). Other two-sample nonparametric tests are presented and discussed by Hudimoto (1959), Haga (1960), Tamura (1963), Potthoff (1963), Wheeler and Watson (1964), Gastwirth (1965b), and Bhattach-aryya and Johnson (1968). The efficiency of some of these tests is examined by Mikulski (1963), Basu (1967a), Hollander (1967a), and Gibbons and Gastwirth (1970).

The correlation between rank tests for scale and rank tests for location is studied by Gibbons (1967) and Hollander (1968). General discussions of rank tests for scale are given by Moses (1963), van Eeden (1964) and Basu and Woodworth (1967). If the location parameters are unknown and possibly unequal, see Raghavachari (1965a), Puri (1968), and Nemenyi (1969) for modified tests. Other tests for dispersion are examined by Sen (1963), Puri (1965), Mielke (1967), Duran and Mielke (1968), and Shorack (1969).

PROBLEMS

1. Test the following data to determine whether the mean of X is larger than the mean of Y.

X: 83,   91,   94,   89,   89,   96,   91,   92,   90

Y: 78,   82,   81,   77,   79,   81,   80,   81

2. In a controlled environment laboratory, ten men and ten women were tested to determine the room temperature they found to be the most comfortable. The results were as follows.

Men: 74,   72,   77,   76,   76,   73,   75,   73,   74,   75

Women: 75,   77,   78,   79,   77,   73,   78,   79,   78,   80

Assuming that these temperatures resemble a random sample from their respective populations, is the average comfortable temperature the same for men and women?

3. In the case where $n = 3$ and $m = 2$, and H_0 is true, find the exact distribution of the test statistic used in

   (a) the Mann-Whitney test
   (b) the Siegel-Tukey test
   (c) the Freund-Ansari test
   (d) the Mood test

4. Show that the Mann-Whitney test statistic may also be defined as the number of pairs (X_i, Y_j) in which X_i exceeds Y_j (counting ties as one-half). This method of computation illustrates the value of the Mann-Whitney test statistic divided by the total number of pairs nm as an estimator for $P(X > Y)$ if there are no ties.

5. Prove that the mean of Mood's test statistic is in fact given by equation (16). (The derivation is similar to the derivation in Example 1.5.7.)

6. At the beginning of the year a first grade class was randomly divided into two groups. One group was taught to read using a uniform method, where all students progressed from one stage to the next at the same time, following the teacher's direction. The second group was taught to read using an individual method, where each student progressed at his own rate according to a pro-grammed workbook, under supervision of the teacher. At the end of the year each student was given a reading ability test with the following results.

First Group				Second Group			
227	55	184	174	209	271	63	19
176	234	147	194	14	151	184	127
252	194	88	248	165	235	53	151
149	247	161	206	171	147	228	101
16	99	171	89	292	99	271	179

   (a) Use the Mann-Whitney test to test for differences between means.
   (b) Use the Mood test to test for differences between variances.
   (c) Use the Siegel-Tukey test to test for differences between variances.
   (d) Use the Freund-Ansari test to test for differences between variances.

7. Using methods similar to those of Example 1.5.7, show that the mean of the Freund-Ansari test statistic is given by

$$\text{mean} = \frac{n(n + m + 1)^2}{4(n + m)}$$

if $n + m$ is odd.

5.4. A CONFIDENCE INTERVAL FOR THE DIFFERENCE BETWEEN TWO MEANS

There is actually very little technical difference between testing the hypoth-esis of equal means and forming a confidence interval for the difference between two means. If the confidence interval includes the point zero the test results in acceptance of H_0. If zero is not in the confidence interval, H_0 is rejected. However, the confidence interval provides additional information that is much more useful than that furnished by the usual hypothesis test. The confidence interval provides an estimate of the difference between means in addition to stating whether that difference might be considered to be zero.

The method for finding a confidence interval as presented in this section is based on the Mann-Whitney test statistic, introduced in the previous section. The situation involves two mutually independent random samples. In the presence of some qualifying assumptions, this method furnishes a confidence interval for the difference between two location parameters, which may be considered to be the two population means if they exist.

Confidence Interval for the Difference Between Two Means

DATA. The data consist of two random samples $X_1, \ldots, X_n$ and $Y_1, \ldots, Y_m$ of sizes n and m respectively. Let X and Y denote random variables with the same distribution as the X_i and the Y_j respectively.

ASSUMPTIONS
1. Both samples are random samples from their respective populations.
2. In addition to independence within each sample, there is mutual independence between the two samples.
3. The two population distribution functions are identical except for a possible difference in location parameters. That is, there is a constant d (say) such that $X + d$ has the same distribution function as Y.

Note that no assumption of continuity need be made here. Noether (1967b) shows that if the confidence coefficient of a confidence interval is $1 - \alpha$ when sampling from a continuous population, then for general populations the true confidence coefficient of the same interval including its endpoints is at least $1 - \alpha$, and without its end points is at most $1 - \alpha$. We shall include the endpoints.

METHOD A (graphic). Denote the X_i by lines parallel to the vertical axis, and denote the Y_j by lines parallel to the horizontal axis, as illustrated in Figure 1 for $n = 4$, $m = 6$. It is important that the same scale be used along both axes. Also it is necessary that the two axes intersect at the origin $(0, 0)$ as usual. It is permissible to subtract a constant from all X_i and Y_j, as long as the same constant is subtracted from both the X's and the Y's, in order to bring the graph closer to the origin. Such a transformation merely moves the resulting graph toward the origin at a $45°$ angle and does not affect the resulting confidence interval.

Place a dot at each intersection determined by a pair (X_i, Y_j), as indicated by Figure 2. There will be nm such dots. If the confidence coefficient is to be $1 - \alpha$,

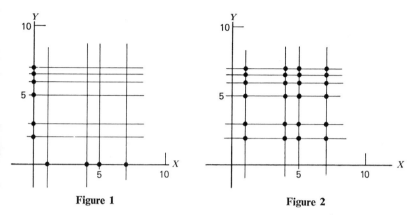

Figure 1   **Figure 2**

find the $\alpha/2$ quantile $w_{\alpha/2}$ of the Mann-Whitney test statistic from Table 8. Take a 45° angle drawing triangle and draw a 45° angle line through the graph in such a way that $w_{\alpha/2}$ of the dots are above or on the line, counting from the left. Draw a similar 45° angle line in the lower right corner of the graph so that $w_{\alpha/2}$ dots are below or on the second line, counting from the right (see Figure 3). The points L and U where the two 45° lines intersect the horizontal axis are the endpoints of the confidence interval. The interval from L to U, inclusive, contains the difference between means, $E(X) - E(Y)$, with probability $1 - \alpha$. That is,

$$(1) \qquad P(L \leq E(X) - E(Y) \leq U) \geq 1 - \alpha$$

If confusion results because several points are on the same 45° line (tied differences) or because $w_{\alpha/2}$ is not an integer (from the large sample approximation), use the rule of selecting that 45° line such that the number of points outside the line is *less than* $w_{\alpha/2}$, but nearest to $w_{\alpha/2}$.

METHOD B (algebraic). This method is more tedious than the graphic method described above, but may be preferred if good drawing equipment is not available.

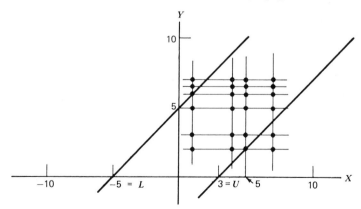

Figure 3. The .025 quantile for $n = 4$ and $m = 6$ is 3. The confidence coefficient is .95. The confidence interval is from L to U, inclusive.

Also it may be used in conjunction with the graphic method to determine precisely the values of L and U after the approximate values have been graphically determined.

For each of the pairs (X_i, Y_j), compute the difference $X_i - Y_j$. Arrange the nm differences in order from smallest to the largest. (Actually it is necessary only to arrange the largest differences and the smallest differences in order.) Determine the $\alpha/2$ quantile $w_{\alpha/2}$ for n and m from Table 8, where $1 - \alpha$ is the desired confidence coefficient. The $w_{\alpha/2}$th smallest difference is the lower limit L, and the $w_{\alpha/2}$th largest difference is U. That is, counting toward the middle of the ordered array of differences, the $w_{\alpha/2}$th differences from each end of the array are the points L and U. Then the confidence interval is given by

$$(2) \qquad\qquad P(L \leq E(X) - E(Y) \leq U) \geq 1 - \alpha$$

Example 1.   A certain type of batter is to be mixed until it reaches a specified level of consistency. Five batches of the batter are mixed using mixer A, and another five batches are mixed using mixer B. The required times for mixing are given as follows (in minutes).

Mixer A: 7.3,   6.9,   7.2,   7.8,   7.2

Mixer B: 7.4,   6.8,   6.9,   6.7,   7.1

A 95% confidence interval is sought for the difference in mixing times, more specifically for $E(X) - E(Y)$ where X refers to mixer A and Y refers to mixer B.

Before using the graphic method the constant 7.0 is subtracted from all of the times to get the following.

$(X_i - 7)$:   .3, $-.1$,   .2,   .8,   .2

$(Y_j - 7)$:   .4, $-.2$, $-.1$, $-.3$,   .1

The numbers are then plotted on the axes of a graph, horizontal and vertical lines are drawn through each number, and dots placed at the 25 intersections

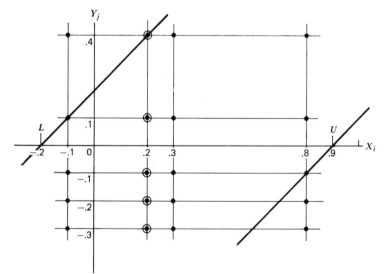

Figure 4

in Figure 4. The circled dots represent two points each. The .025 quantile is 3, from Table 8 for $n = 5$, $m = 5$. A $45°$ line is drawn through the point (actually two points) at (.2, .4) because it is the third point from the upper left corner. This line intersects the horizontal axis at $-.2$, which is now established as the lower limit of the confidence interval. Another $45°$ line is drawn through the point (.8, $-.1$), which establishes the upper limit as .9. The confidence interval is given by

$$P(-.2 \leq E(X) - E(Y) \leq .9) \geq .95$$

The algebraic method for obtaining a confidence interval involves finding the third largest and the third smallest differences $X_i - Y_j$, because $w_{.025}$ equals 3 as before. It is convenient to order the samples first.

$$X_i: \quad 6.9, \quad 7.2, \quad 7.2, \quad 7.3, \quad 7.8$$
$$Y_j: \quad 6.7, \quad 6.8, \quad 6.9, \quad 7.1, \quad 7.4$$

Then the differences are arranged in order.

Smallest Differences	Largest Differences
$6.9 - 7.4 = -.5$	$7.8 - 6.7 = 1.1$
$6.9 - 7.1 = -.2$	$7.8 - 6.8 = 1.0$
$7.2 - 7.4 = -.2 = L$	$7.8 - 6.9 = \quad .9 = U$
$7.2 - 7.4 = -.2$	$7.8 - 7.1 = \quad .7$
$7.3 - 7.4 = -.1$	$7.3 - 6.7 = \quad .7$
etc.	etc.

The lower limit is $-.2$ and the upper limit is .9, as before in the graphic method.

Theory.   A $45°$ angle line through a point (a, b) has the equation $x - y = a - b$. All points on that line have the same difference between coordinate values. That is, if (c, d) is another point on the same line, then $c - d$ equals $a - b$. If the y coordinate equals zero, the $45°$ line is intersecting the x axis, and the x coordinate equals the difference common to all points on the line. Therefore each difference $X_i - Y_j$ may be determined graphically by drawing a $45°$ line through the point (X_i, Y_j) down to the horizontal axis, where the y (vertical) coordinate equals zero. The $45°$ line that has, say, k points above it passes through the point (X_i, Y_j) which has k differences smaller than its own difference $X_i - Y_j$. Thus the graphic method of obtaining the $w_{\alpha/2}$th difference is equivalent to the algebraic method.

Now we need to show that one of the methods gives a correct confidence interval. One way of forming a confidence interval is to let the confidence interval consist of all numbers d, say, that would result in acceptance of the null hypothesis

(3) $$H_0: E(X) - E(Y) = d$$

when using the two-tailed Mann-Whitney test. The Mann-Whitney test described in the previous section was presented for testing H_0 when d equals zero.

(4) $$H_0: E(X) = E(Y)$$

A simple modification enables the test to be used for any d. The modification consists of subtracting the constant d from each observation on the X_i's, and then performing the usual test on the modified data. This is because the equation

(5) $$E(X) - E(Y) = d$$

is equivalent to the equation

(6) $$E(X) - d = E(Y)$$

which in turn is equivalent to

(7) $$E(X - d) = E(Y)$$

Therefore the test of whether the difference between means equals d as in Equation (5) is obviously equivalent to the test of whether the mean of $X - d$ equals the mean of Y.

The confidence interval with a confidence coefficient of $1 - \alpha$ is then found by obtaining the two-sided critical value $w_{\alpha/2}$. We could then collect all the values of d that can be subtracted from the X values and still result in acceptance of H_0. However, repeated hypothesis testing for various values of d is at best a tedious chore. Therefore it is more convenient to regard the Mann-Whitney test statistic as the number of pairs (X_i, Y_j) in which X_i exceeds Y_j (see Problem 4 in the previous section). If each pair (X_i, Y_j) is represented as a point on the coordinate axes, then it is easy to see that X_i exceeds Y_j only when the 45° line through (X_i, Y_j) intersects the horizontal axis to the right of the origin $(0, 0)$. Furthermore, $X_i - d$ exceeds Y_j only when the 45° line through (X_i, Y_j) passes to the right of d on the horizontal axis. All points to the right of this line have differences larger than d so that $X_i - d$ exceeds Y_j for those points. The Mann-Whitney test statistic equals the number of those points. Therefore the largest value of d that results in acceptance of

$$H_0: E(X) - E(Y) = d$$

is found by moving the 45° line toward the right until there are fewer than $w_{\alpha/2}$ points to the right of the line. This is accomplished by the graphic method. A lower value for d is obtained in a similar manner, by moving the 45° line to the left. In this way all values of d that result in acceptance of H_0 are collected into a confidence interval. The two endpoints of the interval actually represent threshold values of d with the property that a slightly smaller value of d (for the upper limit) gives acceptance of H_0 and a slightly larger value of d gives rejection of H_0. Therefore the confidence coefficient is at least $1 - \alpha$ if the endpoints are included.

The graphic method above is described by Lincoln Moses in Walker and Lev (1953), while the algebraic method with further discussion is given by Noether (1967a). Neither method is very manageable when sample sizes are large.

Additional discussion of this procedure may be found in Moses (1965). A related paper is by Govindarajulu (1968). Other estimates of differences in location are discussed by Hodges and Lehmann (1963) and Høyland (1965).

PROBLEMS

1. Seven students were taught algebra using the former method, and six students learned algebra according to a new method. Find a 90% confidence interval for the difference in achievement scores expected from the two methods.

Method	Students' Achievement Scores						
Former	68	72	79	69	84	80	78
New	64	60	68	73	72	70	

2. Diet A was given to four overweight girls and diet B was given to five other overweight girls, with the following weight losses. Find a 90% confidence interval for mean difference in effectiveness of the two diets.

Diet	Weight Losses				
A	7,	2,	-1,	4	
B	6,	5,	2,	8,	3

5.5. MEASURES OF RANK CORRELATION

A measure of correlation is a random variable which is used in situations where the data consist of pairs of numbers, such as the type of data described in Section 5.1. Suppose a *bivariate* random sample of size n is represented by $(X_1, Y_1), (X_2, Y_2), \ldots, (X_n, Y_n)$. We shall use (X, Y) when referring to the (X_i, Y_i) in general. That is, the (X_i, Y_i) for $i = 1, 2, \ldots, n$ have identical bivariate distributions, the same bivariate distribution as (X, Y) has.

Examples of bivariate random variables include one where X_i represents the height of the ith man and Y_i represents his father's height, or where X_i represents a test score of the ith individual and Y_i represents his amount of training. The random variables X and Y may even be independent, as might be the case if X_i represented the scoring average of a basketball player and Y_i his current girlfriend's present grade point average.

By tradition, a measure of correlation between X and Y should satisfy the following requirements in order to be acceptable.

1. The measure of correlation should assume only values between -1 and $+1$.

2. If the larger values of X tend to be paired with the larger values of Y, and hence the smaller values of X and Y tend to be paired together, then the measure of correlation should be positive, and close to $+1.0$ if the tendency is strong. Then we would speak of a positive correlation between X and Y.

3. If the larger values of X tend to be paired with the smaller values of Y, and vice versa, then the measure of correlation should be negative and close to -1.0 if the tendency is strong. Then we say that X and Y are negatively correlated.

4. If the values of X appear to be randomly paired with the values of Y, the measure of correlation should be fairly close to zero. This should be the case when X and Y are independent, and possibly some cases where X and Y are not independent. We then say that X and Y are uncorrelated, or have no correlation, or have correlation zero.

The most commonly used measure of correlation is Pearson's product moment correlation coefficient, denoted by r and defined as

$$(1) \qquad r = \frac{\sum_{i=1}^{n}(X_i - \bar{X})(Y_i - \bar{Y})}{\left(\sum_{i=1}^{n}(X_i - \bar{X})^2 \sum_{i=1}^{n}(Y_i - \bar{Y})^2\right)^{\frac{1}{2}}}$$

where $\bar{X}$ and $\bar{Y}$ are the sample means

$$(2) \qquad \bar{X} = \frac{1}{n}\sum_{i=1}^{n} X_i$$

$$(3) \qquad \bar{Y} = \frac{1}{n}\sum_{i=1}^{n} Y_i$$

as defined in Section 2.2. If the numerator and denominator in (1) are divided by n, then r becomes

$$(4) \qquad r = \frac{\dfrac{1}{n}\sum_{i=1}^{n}(X_i - \bar{X})(Y_i - \bar{Y})}{\left[\dfrac{1}{n}\sum_{i=1}^{n}(X_i - \bar{X})^2\right]^{\frac{1}{2}}\left[\dfrac{1}{n}\sum_{i=1}^{n}(Y_i - \bar{Y})^2\right]^{\frac{1}{2}}}$$

which may be easily remembered as the sample covariance in the numerator, and the product of the two sample standard deviations in the denominator.

This measure of correlation may be used with any data of a numeric nature without any requirements concerning the scale of measurement or the type of underlying distribution, although it is difficult to interpret unless the scale of measurement is at least interval. It meets the necessary requirements of an acceptable measure of correlation. However, r is a random variable and as such r has a distribution function. Unfortunately, the distribution function of r depends on the bivariate distribution function of (X, Y). Therefore r has no value as a test statistic in nonparametric tests, or for forming confidence intervals, unless of course the distribution of (X, Y) is known.

In addition to the widely accepted r above, many other measures of correlation have been invented which satisfy the above requirements for acceptability. An excellent and readable survey article by Kruskal (1958) discusses many of these. Some measures of correlation possess distribution functions which do not depend on the bivariate distribution function of (X, Y) if X and Y are independent, and therefore they may be used as test statistics in nonparametric tests of independence. The measures of correlation selected for presentation here are functions of only the ranks assigned to the observations. They possess distribution functions which are independent of the bivariate distribution function of (X, Y) if X and Y are independent and continuous. They may even be used as measures of correlation on certain types of nonnumeric data.

Spearman's Rho

DATA. The data may consist of a bivariate random sample of size n, (X_1, Y_1), $(X_2, Y_2), \ldots, (X_n, Y_n)$. Let $R(X_i)$ be the rank of X_i as compared with the other X values, for $i = 1, 2, \ldots, n$. That is, $R(X_i) = 1$ if X_i is the smallest of $X_1, X_2, \ldots, X_n$, $R(X_i) = 2$ if X_i is the second smallest, and so on, with rank n being assigned to the largest of the X_i. Similarly let $R(Y_i)$ equal $1, 2, \ldots,$ or n depending on the relative magnitude of Y_i as compared with $Y_1, Y_2, \ldots, Y_n$, for each i.

Or the data may consist of nonnumeric observations occurring in n pairs, if the observations are such that they can be ranked in the manner just described. The ranking may be based on the quality of the observations ("worst" observation to "best" observation), or according to the degree of preference attached to the observations, and so on.

In case of ties, assign to each tied value the average of the ranks that would have been assigned if there had been no ties, as was done in the Wilcoxon and Mann-Whitney tests.

MEASURE OF CORRELATION. The measure of correlation as given by Spearman (1904) is usually designated by ρ (rho) and, if there are no ties, is defined as

$$(5) \quad \rho = \frac{\sum_{i=1}^{n} \left[R(X_i) - \frac{n+1}{2} \right] \left[R(Y_i) - \frac{n+1}{2} \right]}{n(n^2 - 1)/12}$$

An equivalent but computationally easier form is given by

$$(6) \quad \rho = 1 - \frac{6 \sum_{i=1}^{n} [R(X_i) - R(Y_i)]^2}{n(n^2 - 1)} = 1 - \frac{6T}{n(n^2 - 1)}$$

where T represents the entire sum in the numerator. The above forms are equivalent only if there are no ties. If there are many ties Pearson's r, Equation (4), should be used on the ranks as described below. If a moderate number of ties is present in the data, then Equation (6) is recommended for computational simplicity, as the difference between the two coefficients obtained from Equations (4) and (6) will be very slight.

   If there are no ties in the data, Spearman's ρ is merely what one obtains by replacing the observations by their ranks and then computing Pearson's r on the ranks. This may be seen as follows. If the data are replaced by their ranks, then $\bar{X}$ and $\bar{Y}$ correspond to

$$\overline{R(X)} = \frac{1}{n} \sum_{i=1}^{n} R(X_i) = \frac{1}{n} \sum_{i=1}^{n} i = \frac{1}{n} \frac{n(n+1)}{2}$$

$$(7) \quad = \frac{n+1}{2}$$

and

$$(8) \quad \overline{R(Y)} = \frac{n+1}{2}$$

Also $\sum_{i=1}^{n} (X_i - \bar{X})^2$ and $\sum_{i=1}^{n} (Y_i - \bar{Y})^2$ correspond to

$$\sum_{i=1}^{n} [R(X_i) - \overline{R(X)}]^2 = \sum_{i=1}^{n} \left(i - \frac{n+1}{2} \right)^2$$

$$= \sum_{i=1}^{n} \left[i^2 - i(n+1) + \left(\frac{n+1}{2} \right)^2 \right]$$

$$= \frac{n(n+1)(2n+1)}{6} - \frac{n(n+1)^2}{2} + \frac{n(n+1)^2}{4}$$

$$(9) \quad = \frac{n(n^2 - 1)}{12}$$

and

$$(10) \quad \sum_{i=1}^{n} [R(Y_i) - \overline{R(Y)}]^2 = \frac{n(n^2 - 1)}{12}$$

so that (1) becomes (5), and Pearson's r reduces to Spearman's ρ if the data are replaced by their ranks.

Example 1.   This example uses the data of Example 5.1.1 where twelve sets of identical twins were given psychological tests to determine whether the first-born of the twins tended to be more aggressive than the other. In that example the emphasis was on examination of difference between two people who are identical twins. In this example the emphasis is on examination of the degree of similarity between twins within the same set as contrasted with differences between sets of twins.

The data were measures of aggressiveness, and are repeated below.

Twin set	i	1	2	3	4	5	6	7	8	9	10	11	12
First-born	X_i	86	71	77	68	91	72	77	91	70	71	88	87
Second-born	Y_i	88	77	76	64	96	72	65	90	65	80	81	72

The first-born twins were ranked among themselves, and the second-born twins were ranked among themselves, with the following results.

Twin set, i	1	2	3	4	5	6	7	8	9	10	11	12
$R(X_i)$	8	3.5	6.5	1	11.5	5	6.5	11.5	2	3.5	10	9
$R(Y_i)$	10	7	6	1	12	4.5	2.5	11	2.5	8	9	4.5
$[R(X_i) - R(Y_i)]^2$	4	12.25	0.25	0	0.25	0.25	16	0.25	0.25	20.25	1	20.25

The five pairs of ties were given the average ranks for each pair.

First the statistic T in Equation (6) is computed.

$$T = \sum_{i=1}^{12} [R(X_i) - R(Y_i)]^2 = 75$$

Then ρ is obtained from (6).

$$\rho = 1 - \frac{6T}{n(n^2 - 1)} = 1 - \frac{6(75)}{12(143)}$$
$$= .7378$$

As a point of interest, the value of ρ obtained using Equation (5) is .7290, and the ρ obtained using Pearson's r on the ranks is .7354, with the differences being due to the use of average ranks for the ties.

HYPOTHESIS TEST. The Spearman rank correlation coefficient above is often used as a test statistic to test for independence between two random variables. Actually Spearman's ρ is insensitive to some types of dependence, so it is better to be specific as to what types of dependence may be detected. Therefore, the hypotheses takes the following form.

A. (two-tailed test)

H_0: The X_i and Y_i are mutually independent.

H_1: Either (a) there is a tendency for the larger values of X to be paired with the larger values of Y, or (b) there is a tendency for the smaller values of X to be paired with the larger values of Y.

B. (one-tailed test for positive correlation)
H_0: The X_i and Y_i are mutually independent.
H_1: There is a tendency for the larger values of X and Y to be paired together.

C. (one-tailed test for negative correlation)
H_0: The X_i and Y_i are mutually independent.
H_1: There is a tendency for the smaller values of X to be paired with the larger values of Y, and vice versa.

The alternative hypotheses given above and throughout this section state the existence of correlation between X and Y, so that a null hypothesis of "no correlation between X and Y" would be more accurate than the statement of independence between X and Y as given above. Nevertheless, we shall persist in using the null hypothesis of independence because it is in widespread usage and it is easier to interpret.

Spearman's ρ may be used as a test statistic for the above hypotheses. Table 10 gives the quantiles of ρ under the assumption of independence, the null hypothesis. Then H_0 in B is rejected if ρ is too large (at a level α if ρ exceeds the $1 - \alpha$ quantile), H_0 in C is rejected if ρ is too small, and the two-tailed test involves rejecting H_0 if ρ exceeds the $1 - \alpha/2$ quantile or if ρ is less than the $\alpha/2$ quantile.

Instead of Spearman's ρ, it is usually more convenient to use directly the statistic T in Equation (6), where T is defined explicitly as

$$(11) \qquad\qquad T = \sum_{i=1}^{n} [R(X_i) - R(Y_i)]^2$$

The use of T eliminates some of the arithmetic involved in computing ρ. The test in this form is called the Hotelling-Pabst test, probably because of the paper by Hotelling and Pabst (1936) which emphasized the nonparametric nature of Spearman's ρ. Quantiles of T are given in Table 9. Note however that T is large when ρ is small, and vice versa. Therefore the H_0 in B is rejected at the level α if T is less than its α quantile. Also H_0 in C is rejected if T exceeds its $1 - \alpha$ quantile.

Example 2.   Let us continue with Example 1. Suppose we want to test

H_0: The measures of aggressiveness of two identical twins are mutually independent

against the two-sided alternative

H_1: There is either a positive correlation or a negative correlation between the two measures of aggressiveness

at the .05 level of significance. The .025 (half of .05) quantile of T is given by Table 9, for $n = 12$, as

$$w_{.025} = 120$$

Because T equals 75 in Example 1, and because 75 is less than 120, the null hypothesis is rejected. H_0 would have been rejected if T had exceeded the .975

quantile also, given by

$$w_{.975} = \tfrac{1}{3}n(n^2 - 1) - w_{.025}$$
$$= \tfrac{1}{3}(12)(143) - 120$$
$$= 452$$

as explained by the heading of Table 9.

Since ρ had already been computed, it would have been easier to use ρ as a test statistic. The ρ in Example 1 equals .7378, which exceeds the .975 quantile given by Table 10 as .5804.

The approximate critical level $\hat{\alpha}$, the smallest level at which H_0 could have been rejected, is seen from Tables 9 and 10 to be about .01.

The next measure of correlation we are presenting resembles Spearman's ρ in that it is based on the order (ranks) of the observations rather than the numbers themselves, and the distribution of the measure does not depend on the distribution of X and Y if X and Y are independent and continuous. This measure, called Kendall's tau (τ) is usually considered to be more difficult to compute than Spearman's ρ. The chief advantage of Kendall's τ is that its the distribution approaches the normal distribution quite rapidly so that the normal approximation is better for Kendall's τ than it is for Spearman's ρ, when the null hypothesis of independence between X and Y is true. Another advantage of Kendall's τ is its direct and simple interpretation in terms of probabilities of observing concordant and discordant pairs, as defined below.

Kendall's Tau

DATA. The data may consist of a bivariate random sample of size n, (X_i, Y_i) for $i = 1, 2, \ldots, n$. Two observations, for example $(1.3, 2.2)$ and $(1.6, 2.7)$, are called *concordant* if both members of one observation are larger than their respective members of the other observation. Let N_c denote the number of concordant pairs of observations, out of the $\binom{n}{2}$ total possible pairs. A pair of observations, like $(1.3, 2.2)$ and $(1.6, 1.1)$, are called *discordant* if the two numbers in one observation differ in opposite directions (one negative and one positive) from the respective members in the other observation. Let N_d be the total number of discordant pairs of observations. Pairs with ties between respective members are neither concordant nor discordant. Because the n observations may be paired $\binom{n}{2} = n(n-1)/2$ different ways, the number of concordant pairs N_c plus the number of discordant pairs N_d plus the number of pairs with ties should add up to $n(n-1)/2$.

The data may also consist of nonnumeric observations occurring in n pairs, if the observations are such that N_c and N_d described above may be computed.

MEASURE OF CORRELATION. The measure of correlation proposed by Kendall (1938) is

(12)
$$\tau = \frac{N_c - N_d}{n(n-1)/2}$$

If all pairs are concordant, Kendall's τ equals 1.0. If all pairs are discordant, the value is -1.0. As a measure of correlation Kendall's τ satisfies the requirements stated at the beginning of this section.

The computation of τ is simplified if the observations (X_i, Y_i) are arranged in a column according to increasing values of X. Then each Y may be compared only with those below it and the number of concordant and discordant comparisons is easily determined. Also each pair of observations is considered only once. The procedure is illustrated in the following example.

Example 3. Again we shall use the data in Example 1 for purposes of illustration. Arrangement of the data (X_i, Y_i) according to increasing values of X gives the following.

(X_i, Y_i)	Concordant Pairs Below (X_i, Y_i)	Discordant Pairs Below (X_i, Y_i)
(68, 64)	11	0
(70, 65)	9	0
tie $\begin{cases}$ (71, 77)	4	4
 (71, 80) $\end{cases}$	4	4
(72, 72)	5	1
tie $\begin{cases}$ (77, 65)	5	0
 (77, 76) $\end{cases}$	4	1
(86, 88)	2	2
(87, 72)	3	0
(88, 81)	2	0
tie $\begin{cases}$ (91, 90)	0	0
 (91, 96) $\end{cases}$	0	0
	$N_c = 49$	$N_d = 12$

Kendall's τ is given by

$$\tau = \frac{N_c - N_d}{n(n-1)/2} = \frac{49 - 12}{(12)(11)/2}$$
$$= .5606$$

There is a positive rank correlation between aggression scores in the twins we observed, as measured by Kendall's τ.

HYPOTHESIS TEST. Kendall's τ may also be used as a test statistic to test the null hypothesis of independence between X and Y, with possible one-tailed or two-tailed alternatives as described above with Spearman's ρ. Some arithmetic may be saved, however, by using $N_c - N_d$ as a test statistic, without dividing by

$n(n-1)/2$ to obtain τ. Therefore we use T as the Kendall test statistic, where T is defined as

$$(13) \qquad\qquad T = N_c - N_d$$

Quantiles of T are given in Table 11. If T exceeds the $1 - \alpha$ quantile reject H_0 in favor of the one-sided alternative of positive correlation, at level α. Values of T less than the α quantile lead to acceptance of the alternative of negative correlation.

Example 4.   In Example 3 Kendall's τ was computed by first finding the value of

$$T = N_c - N_d = 49 - 12$$
$$= 37$$

In Table 11 the quantiles for a two-tailed test of size $\alpha = .05$ are found, for n equal to 12, to be

$$w_{.975} = 28$$

and

$$w_{.025} = -28$$

For $T = 37$ the null hypothesis of independence is rejected. The critical level is estimated from Table 11 to be about

$$\hat{\alpha} \cong 2(.005) = .01$$

The same data were used for both Spearman's ρ and Kendall's τ in order to better compare the two statistics. It was seen that Spearman's ρ ($\rho = .7378$) was a larger number than Kendall's τ ($\tau = .5606$). However, the two tests using the two statistics (or their equivalents) produced nearly identical results. Both of the above statements hold true in most, but not all, situations. Spearman's ρ tends to be larger than Kendall's τ, in absolute value. However, as a test of significance there is no strong reason to prefer one over the other, because both will produce nearly identical results in most cases.

Daniels (1950) proposed the use of Spearman's ρ to test for trend by pairing measurements, called X_i, with the time (or order) at which the measurements were taken. The assumption is that the X_i are mutually independent, and the null hypothesis is that they are identically distributed. The alternative hypothesis is that the distribution of the X_i's is related to time so that as time goes on, the X measurements tend to become larger (or smaller). The idea of trend was discussed more fully in Section 3.4, where the Cox and Stuart test for trend was presented. Tests of trend based on Spearman's ρ or Kendall's τ are generally considered to be more powerful than the Cox and Stuart test. It was mentioned in Section 3.4 that the A.R.E. of the Cox and Stuart test for trend, when applied to random variables known to be normally distributed, is about .78 with respect to the test based on the regression coefficient, while the A.R.E. of these tests using Spearman's ρ or Kendall's τ is about .98 under the same conditions, according to Stuart

(1956). However, these tests are not as widely applicable as the Cox and Stuart test. For instance, these tests would be inappropriate in Example 3.4.5. These tests are appropriate in Example 3.4.4, and so we use that example to illustrate the use of Spearman's ρ as a test for trend. The procedure using Kendall's τ is similar.

Example 5.   In Example 3.4.4, nineteen years of annual precipitation records are given as follows.

Precipitation X_i	Year Y_i	$R(X_i)$	$R(Y_i)$	$[R(X_i) - R(Y_i)]^2$
45.25 inches	1950	12	1	121
45.83	1951	15	2	169
41.77	1952	11	3	64
36.26	1953	6	4	4
45.27	1954	13	5	64
52.25	1955	17	6	121
35.37	1956	2.5	7	20.25
57.16	1957	18	8	100
35.37	1958	2.5	9	42.25
58.32	1959	19	10	81
41.05	1960	9	11	4
33.72	1961	1	12	121
45.73	1962	14	13	1
37.90	1963	7	14	49
41.72	1964	10	15	25
36.07	1965	4	16	144
49.83	1966	16	17	1
36.24	1967	5	18	169
39.90	1968	8	19	121
			Total	1421.5

The two-tailed test for trend involves rejection of the null hypothesis of no trend if the total of the last column above is too large or too small. The test statistic is given by

$$T = \sum_{i=1}^{19} [R(X_i) - R(Y_i)]^2$$

$$= 1421.5$$

and the quantiles of T, for $\alpha = .05$, are given in Table 9, for $n = 19$, as

$$w_{.025} = 618$$

and

$$w_{.975} = \tfrac{1}{3}(19)(360) - 618$$

$$= 1662$$

As before, H_0 is readily accepted.

Theory. The exact distributions of ρ and τ are quite simple to obtain in principle, although in practice the procedure is most tedious for even moderate sized n. The exact distributions are found under the assumption that X_i and Y_i are independent and identically distributed. Then each of the $n!$ arrangements of the ranks of the X_i's paired with the ranks of the Y_i's is equally likely. As in the previous sections of this chapter, the distribution functions are obtained simply by counting the number of arrangements that give a particular value of ρ, or τ, and by dividing that number by $n!$ to get the probability of that value of ρ, or τ.

A form of the Central Limit Theorem is applied to obtain the large sample approximate distributions, because both ρ and τ are based on the sum of random variables. Both ρ and τ have probability distributions symmetric about zero, so the means are zero for both. The variances are more difficult to obtain and will not be derived here. Division of ρ and τ by their respective variances thus results in a random variable that is approximately distributed as a standard normal random variable for large n. The approximation is considered quite good when used to find the quantiles of τ for $n \geq 8$, but not nearly as good when used to find the quantiles of ρ.

If (X_i, Y_i), $i = 1, \ldots, n$ are independent and identically distributed bivariate normal random variables, both ρ and τ have an asymptotic relative efficiency of $9/\pi^2 = .912$, relative to the parametric test which uses Pearson's r as a test statistic (Stuart, 1954).

Kendall's Partial Correlation Coefficient

The concept of partial correlation is not an easy one to grasp. However, in order to illustrate the manner in which Kendall's τ may be extended to partial correlation, a brief attempt to describe partial correlation will be made.

In a multivariate random variable $(X_1, X_2, \ldots, X_k)$ there may be correlation between X_1 and X_2, between X_2 and X_5, etc., and a measure of this correlation might be any of the above measures already described. Those measures estimate the total influence (correlation) of one random variable on the other, including the indirect influence felt because the second random variable is correlated not only with the first random variable, but perhaps with a third random variable which is in turn correlated with the first random variable and hence acts as a carrier of indirect influence between the first and second random variables. Sometimes it is desirable to measure the correlation between two random variables, under the condition that the indirect influence due to the other random variables is somehow eliminated. An estimate of this "partial" correlation between X_1 and X_2, say, while the indirect correlation due to $X_3, X_4, \ldots$, and X_n is eliminated, is denoted by $r_{12.34\ldots n}$ when using the extension of Pearson's r, or by $\tau_{X_1 X_2 . X_3 X_4 \ldots X_n}$ when using the extension of Kendall's τ. In the simple case where $n = 3$, the partial correlation may be estimated by Pearson's partial correlation coefficient

(14)
$$r_{12.3} = \frac{r_{12} - r_{13}r_{23}}{\sqrt{(1 - r_{13}^2)(1 - r_{23}^2)}}$$

where r_{ij} is the ordinary Pearson r computed between X_i and X_j, and by Kendall's partial correlation coefficient

(15)
$$\tau_{X_1 X_2 \cdot X_3} = \frac{\tau_{X_1 X_2} - \tau_{X_1 X_3} \tau_{X_2 X_3}}{\sqrt{(1 - \tau_{X_1 X_3}^2)(1 - \tau_{X_2 X_3}^2)}}$$

where $\tau_{X_i X_j}$ is the ordinary Kendall's τ computed between X_i and X_j. There is no apparent reason why Spearman's ρ could not be extended to measure partial correlation in the same way as above.

The distribution of $r_{12.3}$ depends on the multivariate distribution function of (X_1, X_2, X_3) and therefore may not be used as a test statistic in a nonparametric test. The distribution of $\tau_{X_1 X_2 \cdot X_3}$ depends only on the ranks from 1 to N, but it also is not distribution free. Therefore at the present time no nonparametric tests of hypotheses may be made using Kendall's partial correlation coefficient as a test statistic. This fact severely limits the use of this statistic.

Another measure of correlation proposed by Kendall for use in another situation, is the "coefficient of concordance." This may be used to measure total correlation when more than two variates are involved. However, the close relationship between Kendall's coefficient of concordance and a test statistic proposed by Friedman makes it advisable to present both of these statistics at the same time, which will be done in Section 5.7.

See Kendall (1942) for his discussion of partial rank correlation. A comprehensive study of rank correlation is in Kendall's (1955) book on the subject. Knight (1966) gives a computer method of calculating Kendall's τ. Spearman's ρ for contingency tables is explained by Stuart (1963).

Usage of rank correlation methods in regression is discussed by Konijn (1961), Adichie (1967a, 1967b), and Sen (1968a). Other recent papers on rank correlation and the concept of dependence are by Aitkin and Hume (1965), Lehmann (1966), Bell and Doksum (1967), and Gokhale (1968).

PROBLEMS

1. A husband and wife who go bowling together kept their scores for ten lines to see if there was a correlation between their scores. The scores were:

Line	Husband's Score	Wife's Score	Line	Husband's Score	Wife's Score
1	147	122	6	151	120
2	158	128	7	196	108
3	131	125	8	129	143
4	142	123	9	155	124
5	183	115	10	158	123

(a) Compute ρ.

(b) Compute τ.

(c) Test the hypothesis of independence using a two-tailed test based on ρ.

(d) Do the same as (c) for τ.

2. Show that (5) and (6) are equivalent expressions for ρ.

3. For $n = 5$, what pairing of ranks results in

(a) $\rho = 1$?

(b) $\tau = 1$?

(c) $\rho = -1$?

(d) $r = -1$?

4. Generalize the result of Problem 3 to any value of n in general, and show that ρ and τ do in fact assume the values indicated.

5. Suppose someone suggests using

$$R = 1 - \frac{\sum_{i=1}^{n} |R(X_i) - R(Y_i)|}{(1/4)n^2}$$

(a) Under what conditions will $R = 1$?

(b) Under what conditions will $R = -1$?

6. Find the exact distribution of ρ, τ, and R from Problem 5, for the case where $n = 3$, under the usual assumption of independence.

7. The following is an example of a situation in which τ and ρ yield widely varying estimates of correlation.

X_i	Y_i	X_i	Y_i	X_i	Y_i
−8.7	−0.6	−1.9	−4.7	2.2	3.8
−8.3	−0.8	−1.6	−5.5	4.0	3.5
−8.2	−1.3	−1.3	−5.6	5.6	3.1
−7.2	−1.9	−0.2	−6.0	5.9	2.6
−6.1	−2.0	0.7	4.6	6.2	2.0
−6.0	−2.1	1.3	4.4	6.6	1.2
−4.1	−4.0	1.6	4.2	6.7	0.6
−2.0	−4.6	2.1	3.9	8.1	0.4

(a) Make a rough scatter diagram.

(b) Compute τ.

(c) Compute ρ.

(d) Does either ρ or τ lead to rejection of the null hypothesis that X and Y are independent?

(e) Compute R defined in Problem 5.

5.6. SEVERAL INDEPENDENT SAMPLES

The Mann-Whitney test for two independent samples, presented in Section 5.3, was extended to the problem of analyzing k independent samples, for $k \geq 2$, by Kruskal and Wallis (1952). The experimental situation is one where k random samples have been obtained, one from each of k possibly different populations, and we want to test the null hypothesis that all of the populations are identical against the alternative that some of the populations tend to furnish greater observed values than other populations. The term "greater" applies to observations on random variables, but actually any observation which may be arranged in increasing order according to some property such a quality, value, etc., may be analyzed using the Kruskal-Wallis test, in a manner analogous to the analysis of nonnumeric data using the Mann-Whitney test as in Example 5.3.2.

The experimental design described above is called the "completely random-ized" design, and was already introduced in Section 4.3, following Example 4.3.1, where the median test was introduced as a possible method of analysis. The Kruskal-Wallis test uses more information contained in the observations than does the median test. That is, the Kruskal-Wallis test statistic is a function of the ranks of the observations in the combined sample, as was true with the Mann-Whitney test, while the median test statistic was dependant only on the knowledge of whether the observations were below or above the grand median. For this reason the Kruskal-Wallis test is usually more power-ful than the median test. However, computation of the test statistic involves ranking all of the observations, and therefore involves more effort than the median test. If ranks may be assigned to the observations in many different ways because of many apparently equal observations, then it may be better to use the median test despite the possible loss of power, because in such a situation the Kruskal-Wallis test may furnish an estimated level of significance much different than the true level of significance. However, such advice should not be followed unless ties are a major problem.

The Kruskal-Wallis Test

DATA. The data consist of k random samples of possibly different sizes. Denote the ith random sample of size n_i by $X_{i1}, X_{i2}, \ldots, X_{in_i}$. Then the data may be arranged into columns;

Sample 1	Sample 2	$\ldots$	Sample k
$X_{1,1}$	$X_{2,1}$		$X_{k,1}$
$X_{1,2}$	$X_{2,2}$		$K_{k,2}$
$\ldots$	$\ldots$		$\ldots$
$X_{1,n}$	X_{2,n_2}		X_{k,n_k}

Let N denote the total number of observations;

$$(1) \qquad N = \sum_{i=1}^{k} n_i$$

Assign rank 1 to the smallest of the totality of N observations, rank 2 to the second smallest, and so on to the largest of all N observations, which receives rank N. No attention is given to the sample to which the observation belongs. Let $R(X_{ij})$ represent the rank assigned to X_{ij}. Let R_i be the sum of the ranks assigned to the ith sample.

$$(2) \qquad R_i = \sum_{j=1}^{n_i} R(X_{ij}) \qquad i = 1, 2, \ldots, k$$

Compute R_i for each sample.

If the ranks may be assigned in several different ways because several observations are equal to each other, assign the average rank to each of the tied observations, as was done in the previous tests of this chapter.

ASSUMPTIONS

1. All samples are random samples from their respective populations.
2. In addition to independence within each sample, there is mutual independence among the various samples.
3. All random variables X_{ij} are continuous. (A moderate number of ties is tolerable.)
4. The measurement scale is at least ordinal.
5. Either the k population distribution functions are identical, or else some of the populations tend to yield larger values than other populations do.

HYPOTHESES

H_0: All of the k population distribution functions are identical.
H_1: At least one of the populations tends to yield larger observations than at least one of the other populations.

Because the Kruskal-Wallis test is designed to be sensitive against differences among means in the k populations, the alternative hypothesis is sometimes stated as follows.

H_1: The k populations do not all have identical means.

TEST STATISTIC. The test statistic T is defined as

$$(3) \qquad T = \frac{12}{N(N+1)} \sum_{i=1}^{k} \frac{[R_i - (1/2)n_i(N+1)]^2}{n_i}$$

where N and R_i are defined by (1) and (2) respectively. An equivalent form for T, and one which is usually more convenient to use, is

$$(4) \qquad T = \frac{12}{N(N+1)} \sum_{i=1}^{k} \frac{R_i^2}{n_i} - 3(N+1)$$

DECISION RULE. If k equals 3 and all three sample sizes are 5 or less, then the critical region of exact size α may be determined from Table 12, and corresponds to all values of T *equal to or greater than* the critical value of T given in Table 12 for the corresponding α.

Otherwise the approximate distribution of T is used, the chi-square distribution with $k - 1$ degrees of freedom. Reject H_0 at the level α if T exceeds the $1 - \alpha$ quantile of a chi-square random variable with $k - 1$ degrees of freedom, obtained from Table 2.

Although the chi-square approximation is justified only for reasonably large sample sizes n_i, in practice the approximation is used in all situations not covered by Table 12, and appears to be fairly good even in adverse situations.

Example 1.   Data from a completely randomized design were given in Example 4.3.1, where four different methods of growing corn resulted in various yields per acre on various plots of ground where the four methods were tried. Ordinarily only one statistical analysis is used, but here we shall use the Kruskal-Wallis test so that a rough comparison may be made with the median test, which previously furnished a critical level $\hat{\alpha}$ of slightly less than .001.

The hypotheses may be stated as follows:

H_0:The four methods are equivalent.
H_1:Some methods of growing corn tend to furnish higher yields than others.

The observations are ranked from the smallest, 77, of rank 1 to the largest, 101, of rank $N = 34$. Tied values receive the average ranks. The ranks of the observations, with the sums R_i, are given below.

Method	1		2		3		4	
	Obs.	Rank	Obs.	Rank	Obs.	Rank	Obs.	Rank
	83	11	91	23	101	34	78	2
	91	23	90	19.5	100	33	82	9
	94	28.5	81	6.5	91	23	81	6.5
	89	17	83	11	93	27	77	1
	89	17	84	13.5	96	31.5	79	3
	96	31.5	83	11	95	30	81	6.5
	91	23	88	15	94	28.5	80	4
	92	26	91	23			81	6.5
	90	19.5	89	17				
			84	13.5				
R_i:		196.5		153.0		207.0		38.5
n_i:		9		10		7		8
$N = 34$								

The critical region of approximate size $\alpha = .05$ corresponds to values of T greater than the .95 quantile of a chi-square random variable with $k - 1 = 3$ degrees of freedom, which is given in Table 2 as 7.815. (Note that the median test also uses the chi-square distribution with $k - 1$ degrees of freedom, so the critical regions of the two tests will appear to be the same although the test statistics are different.)

The value of T obtained using (3) or (4) is

$$T = 25.46$$

which clearly leads to rejection of H_0. A rough idea of the power of the Kruskal-Wallis test as compared with the median test may be obtained by comparing the value of the test statistics in both tests. Both test statistics have identical asymptotic distributions, the chi-square distribution with three degrees of freedom. However, the value 25.46 attained in the Kruskal-Wallis test is somewhat larger than the value 17.6 computed in the median test.

COMMENT. If analysis of the data leads to rejection of the null hypothesis, then any group of two or more samples may be further analyzed using the Kruskal-Wallis test, until the differences between populations have been satisfactorily detected. However, the level of significance in all but the first test is distorted and almost completely devoid of meaning, except possibly to aid one in ordering the differences from smallest to largest.

Theory. The exact distribution of T is found under the assumption that all observations were obtained from the same or identical populations. The method is that of randomization, which was also used in finding the distribution of the Mann-Whitney test statistic. That is, under the above assumption, each arrangement of the ranks 1 to N into groups of sizes $n_1, n_2, \ldots, n_k$, respectively, is equally likely, and occurs with probability $n_1! \, n_2! \ldots n_k!/N!$, which is the reciprocal of the number of ways the N ranks may be divided into groups of sizes $n_1, n_2, \ldots, n_k$. The value of T is computed for each arrangement. The probabilities associated with equal values of T are then added to give the probability distribution of T.

For example, if $n_1 = 2$, $n_2 = 1$, $n_3 = 1$ in the three sample cases, then there are twelve equally likely arrangements of the four ranks; thus each arrangement has probability 1/12. The twelve arrangements, with the associated values of T, are given as follows.

	Sample	*1*	*2*	*3*	*T*
Arrangement	1	1,2	3	4	2.7
	2	1,2	4	3	2.7
	3	1,3	2	4	1.8
	4	1,3	4	2	1.8
	5	1,4	2	3	0.3
	6	1,4	3	2	0.3
	7	2,3	1	4	2.7
	8	2,3	4	1	2.7
	9	2,4	1	3	1.8
	10	2,4	3	1	1.8
	11	3,4	1	2	2.7
	12	3,4	2	1	2.7

Therefore the probability function $f(x)$ and the distribution function $F(x)$ are given as follows, for $n_1 = 2$, $n_2 = 1$, and $n_3 = 1$.

x	$f(x) = P(T = x)$	$F(x) = P(T \leq x)$
0.3	$2/12 = 1/6$	1/6
1.8	$4/12 = 1/3$	1/2
2.7	$6/12 = 1/2$	1.0

This result may be checked against Table 12, where $P(T \geq 2.7)$ is given as .5000.

The large sample approximation for the distribution of T is based on the fact that R_i in (2) is the sum of n_i random variables, and for large n_i the Central Limit Theorem may be used. Thus

$$\frac{R_i - E(R_i)}{\sqrt{\text{Var}(R_i)}}$$

is approximately distributed as a standardized normal random variable when H_0 is true. From Theorem 1.5.5 the mean and variance of R_i may be expressed by

$$(5) \qquad E(R_i) = \frac{n_i(N + 1)}{2}$$

and

$$(6) \qquad \text{Var}(R_i) = \frac{n_i(N + 1)(N - n_i)}{12}$$

Therefore

$$(7) \qquad \left[\frac{R_i - E(R_i)}{\sqrt{\text{Var}(R_i)}}\right]^2 = \frac{\{R_i - [n_i(N + 1)/2]\}^2}{n_i(N + 1)(N - n_i)/12}$$

is approximately distributed as a chi-square random variable with one degree of freedom. If the R_i were independent of each other the distribution of the sum

$$(8) \qquad T' = \sum_{i=1}^{k} \frac{\{R_i - [n_i(N + 1)/2]\}^2}{n_i(N + 1)(N - n_i)/12}$$

could be approximated using the chi-square distribution with k degrees of freedom. However, the sum of the R_i's is N, so there is dependence among the R_i's. Kruskal (1952) showed that if the ith term in T' is multiplied by $(N - n_i)/N$, for $i = 1, 2, \ldots, k$, then the result

$$(9) \qquad T = \sum_{i=1}^{k} \frac{\{R_i - [n_i(N + 1)/2]\}^2}{n_i(N + 1)N/12}$$

is asymptotically distributed as a chi-square random variable with $k - 1$ degrees of freedom. Equation (9) is merely a rearrangement of the terms in Equation (3) which originally defined the test statistic T. Therefore we have rationalized the use of the chi-square approximation for the distribution of the Kruskal-Wallis test statistic.

Kruskal and Wallis (1952) found that for small α (less than about .10) and for selected small values of n_1, n_2, and n_3, the true level of significance is smaller than the stated level of significance associated with the chi-square distribution, which indicates that the chi-square approximation furnishes a conservative test in many if not most situations. Gabriel and Lachenbruch (1969) show that the chi-square approximation is good even though the sample sizes may be small.

For two samples the Kruskal-Wallis test is equivalent to the Mann-Whitney test. Recall that in the Mann-Whitney test, Section 5.3, one sample was called $X_1, \ldots, X_n$ while the other was $Y_1, \ldots, Y_m$. The statistic S was defined by (5.3.1) as

$$(10) \qquad S = \sum_{i=1}^{n} R(X_i)$$

the sum of the ranks of the X's in the combined sample corresponding to R_1 in the Kruskal-Wallis test. The Mann-Whitney two-tailed test consisted of rejecting H_0 if the quantity $S - n(n + 1)/2$ was either too large or too small. Because S is approximately normal for large sample sizes, one could reject H_0 if the quantity

$$(11) \qquad \frac{S - E(S)}{\sqrt{\text{Var}(S)}}$$

is above or below the appropriate standardized normal quantiles, or if the square of the above,

$$(12) \qquad \frac{[S - E(S)]^2}{\text{Var}(S)}$$

is above the $1 - \alpha$ quantile in a chi-square distribution with one degree of freedom, according to Theorem 1.6.3. So the chi-square distribution with one degree of freedom could have been used in the Mann-Whitney two-tailed test, with the quantity in (12) as a test statistic. The Kruskal-Wallis test, with two samples, also uses the chi-square distribution with one degree of freedom to test the same hypothesis as in the Mann-Whitney two-tailed test, and in fact the Kruskal-Wallis test statistic is identical to the form of the Mann-Whitney test statistic given in (12). We shall now prove this.

An algebraic rearrangement is all that is required to convert the two-sample Kruskal-Wallis test statistic, given by (3) as

$$(13) \qquad T = \frac{12}{N(N+1)} \sum_{i=1}^{2} \frac{[R_i - (1/2)n_i(N+1)]^2}{n_i}$$

into the form suggested by (12). First the R_2 in (13) is replaced with $N(N+1)/2 - R_1$, which is possible because the sum of all ranks, $R_1 + R_2$, adds up to $N(N+1)/2$, as shown in Lemma 1, Section 1.5. Then (13) becomes

$$(14) \quad T = \frac{12}{N(N+1)} \left\{ \frac{[R_1 - \frac{1}{2}n_1(N+1)]^2}{n_1} \right.$$

$$\left. + \frac{\left[\frac{N(N+1)}{2} - R_1 - \frac{1}{2}n_2(N+1) \right]^2}{n_2} \right\}$$

Next the n_2 is replaced by $N - n_1$, and the entire expression is simplified.

$$T = \frac{12}{N(N+1)} \left\{ \frac{[R_1 - \frac{1}{2}n_1(N+1)]^2}{n_1} \right.$$

$$\left. + \frac{\left[\frac{N(N+1)}{2} - R_1 - \frac{1}{2}(N - n_1)(N+1) \right]^2}{N - n_1} \right\}$$

$$= \frac{12}{N(N+1)} \left\{ \frac{[R_1 - \frac{1}{2}n_1(N+1)]^2}{n_1} + \frac{[\frac{1}{2}n_1(N+1) - R_1]^2}{N - n_1} \right\}$$

$$= \frac{12}{N(N+1)} \left[R_1 - \frac{n_1(N+1)}{2} \right]^2 \left[\frac{1}{n_1} + \frac{1}{N - n_1} \right]$$

$$(15) \qquad = \frac{[R_1 - \frac{1}{2}n_1(N+1)]^2}{(N+1)n_1(N - n_1)/12}$$

Equation (15) is the same as (12), because R_1 is the sum of ranks in the first sample, denoted by S in the Mann-Whitney test. The expressions for $E(S)$ and Var (S) were given by Equations (5.3.8) and (5.3.9), and match exactly with the corresponding terms in (15) above. Therefore the Kruskal-Wallis test statistic for two samples is identical with a form of the Mann-Whitney test statistic. The Kruskal-Wallis test is thus shown to be an extension of the Mann-Whitney test to two or more samples.

The A.R.E. of the Kruskal-Wallis test relative to the usual parametric F test is $3/\pi$ or .955, in situations where the latter test is considered appropriate. Hodges and Lehmann (1956) proved that if the distribution functions have identical shapes, but differ only in location, the A.R.E. of the Kruskal-Wallis test relative to the F test is never less than .864, and may exceed 1 for certain types of distributions.

Rank sum tests similar to the Kruskal-Wallis test have been adapted for making multiple comparisons by Steel (1960), Sherman (1965), and McDonald and Thompson (1967). Some tables for making multiple comparisons are provided by Tobach, Smith, Rose, and Richter (1967). Procedures for selecting the best populations are described by Rizvi and Sobel (1967), Sobel (1967), Rizvi, Sobel, and Woodworth (1968), and Puri and Puri (1969). Rank tests are presented for censored data by Basu (1967b), for testing against ordered alternatives by G. R. Shorack (1967), and for analysis of covariance by Puri and Sen (1969a). For other recent work concerned with rank tests and several independent samples see Sen (1962, 1966), Matthes and Truax (1965), Quade (1966), Crouse (1966), Sen and Govindarajulu (1966), Odeh (1967), Deshpande (1970), and Bhapkar and Deshpande (1968). Analysis of covariance is discussed by Quade (1967).

PROBLEMS

1. Test the null hypothesis that the three different brands of lightbulbs have equal mean life, against the appropriate alternative, where the lives of several bulbs, randomly selected, were found to be as follows:

Brand	A	B	C
Life length	73	84	82
	64	80	79
	67	81	71
	62	77	75
	70		

2. Four job training programs were tried on twenty new employees, where five employees were randomly assigned to each training program. The twenty employees were then placed under the same supervisor, and at the end of a certain specified period the supervisor ranked the employees according to job ability, with lowest ranks being assigned to those employees with the lowest job ability.

			Ranks		
Program	1	4,	6,	7,	2,	10
	2	1,	8,	12,	3,	11
	3	20,	19,	16,	14,	5
	4	18,	15,	17,	13,	9

Is there a difference in the effectiveness of the various training programs?

3. Show that equations (3) and (4) are equivalent to each other.
4. Find the exact distribution of the Kruskal-Wallis test statistic when H_0 is true and $n_1 = 2$, $n_2 = 2$, $n_3 = 1$. Compare your results with the critical value given in Table 12.
5. In the two-sample case, what are some of the reasons why we might prefer to use the Mann-Whitney test rather than the Kruskal-Wallis test?

5.7. SEVERAL RELATED SAMPLES

In the previous section we examined the Kruskal-Wallis rank test for several independent samples which is an extension of the Mann-Whitney test for two independent samples introduced in Section 5.3. In this section we consider the problem of analyzing several *related* samples, which may be considered as an extension of the problem of matched pairs, or two related samples, examined in the first section of this chapter on rank tests. However, unlike the problem involving independent samples, this problem of several related samples is not handled by extending the matched pairs ranks test. Instead we shall introduce a different type of rank test, called the Friedman test, which does not use the assumption of symmetry needed for the matched pairs Wilcoxon signed ranks test, and which reduces to a simple sign test of Section 3.4 if there are only two samples. But first we shall describe the type of experiment that results in several related samples.

An experiment might be conducted to detect differences in k different treatments, $k \geq 2$, in the following manner. A "block" (group) consists of k experimental units which are similar to each other in some important respects, such as k puppies which are litter mates. The k experimental units within a block are paired randomly with the k treatments being scrutinized, so that each treatment is administered once and only once within each block. In this way the treatments may be compared with each other without an excess of unwanted effects confusing the results of the experiment. More than one block is used; the total number of blocks used is denoted by b. Within each block it is not necessary to obtain numerical measurements of the effects of each treatment; rather, it is sufficient to be able to rank the treatment results from 1 to k within each block. No comparison is made between blocks, under the assumption that the differences among blocks are sufficiently serious so as to invalidate a comparison of treatment results from one block to another.

The experimental arrangement described above is usually called a *randomized complete block design*. This design may be compared with the *incomplete* block design described in the next section, in which the blocks do not contain enough experimental units to enable all the treatments to be applied in all the blocks, and so each treatment appears in some blocks but not in others. Examples of randomized complete block designs are as follows.

1. *Psychology*. Five litters of mice, with four mice per litter, are used to examine the relationship between environment and aggression. Each litter is considered to be a block. Four different environments are designed. One mouse from each litter is placed in one environment, so that the four mice from each litter are in four different environments. After a suitable length of time the mice are regrouped with their litter mates and are ranked according to degree of agressiveness.

2. *Home economics*. Six different types of bread dough are compared to see which bakes the fastest by forming three loaves with each type of dough. Three different ovens are used, and each oven bakes the six different types of bread at the same time. The ovens are the blocks and the doughs are the treatments.

3. *Environmental engineering*. One experimental unit may form a block if the different treatments may be applied to the same unit without leaving residual effects. Seven different men are used in a study of the effect of color schemes on work efficiency. Each man is considered to be a block, and spends some time in each of three rooms, each with its own type of color scheme. While in the room each man performs a work task and is measured for work efficiency. The three rooms are the treatments.

By now the reader should have some idea of the nature of a randomized complete block design. The usual parametric method of testing the null hypothesis of no treatment differences is called the *two-way analysis of variance*. The following nonparametric method depends only on the ranks of the observations within each block, and is therefore sometimes called the two-way analysis of variance by ranks. It is also known by the name of the man who introduced the test, Friedman (1937).

The Friedman Test

DATA. The data consist of b mutually independent k-variate random variables $(X_{i1}, X_{i2}, \ldots, X_{ik})$, called b blocks, $i = 1, 2, \ldots, b$. The random variable X_{ij} is in block i and is associated with treatment j. The b blocks are arranged as follows.

	Treatment			
	1	2	...	k
Block: 1	X_{11}	X_{12}	...	X_{1k}
2	X_{21}	X_{22}	...	X_{2k}
3	X_{31}	X_{32}	...	X_{3k}
...	...	...	...	...
b	X_{b1}	X_{b2}	...	X_{bk}

Let $R(X_{ij})$ be the rank, from 1 to k, assigned to X_{ij} within block (row) i. That is, for block i the random variables $X_{i1}, X_{i2}, \ldots, X_{ik}$ are compared with each other and the rank 1 is assigned to the smallest observed value, the rank 2 to the second smallest, and so on to the rank k which is assigned to the largest observation in block i. Ranks are assigned in all of the b blocks.

Let R_j denote the sum of the ranks in the jth column (treatment). That is,

$$(1) \qquad\qquad R_j = \sum_{i=1}^{b} R(X_{ij})$$

for $j = 1, 2, \ldots, k$.

In the event that ranks may be assigned in several different ways because of observations that are apparently equal to each other, we recommend assigning the average rank to the tied values, as was done in the previous rank tests of this chapter.

ASSUMPTIONS

1. The b k-variate random variables are mutually independent. (The results within one block do not influence the results within the other blocks.)

2. Within each block the observations may be arranged in increasing order according to some criterion of interest. (A moderate number of ties is tolerable.)

HYPOTHESES

H_0: Each ranking of the random variables within a block is equally likely (i.e., the treatments have identical effects).

H_1: At least one of the treatments tends to yield larger observed values than at least one other treatment.

TEST STATISTIC. The Friedman test statistic is defined as

$$(2) \qquad\qquad T = \frac{12}{bk(k+1)} \sum_{j=1}^{k} \left[R_j - \frac{b(k+1)}{2} \right]^2$$

which is equivalent to

$$(3) \qquad\qquad T = \frac{12}{bk(k+1)} \sum_{j=1}^{k} R_j^2 - 3b(k+1)$$

The latter equation is more convenient for use on a desk calculator or computer.

DECISION RULE. Reject the null hypothesis at the level α if the Friedman test statistic T exceeds the $1 - \alpha$ quantile of a chi-square random variable with $k - 1$ degrees of freedom, obtained from Table 2. Actually the chi-square distribution only approximates the distribution of T, but the approximation is reasonably close. The approximation improves as b gets larger. The exact distribution of T is given by Owen (1962) for $k = 2, b \leq 15$, and for $k = 3, b \leq 8$. The chi-square approximation in those cases tends to be slightly conservative.

Example 1.   Twelve homeowners were randomly selected to participate in an experiment with a plant nursery. Each homeowner was asked to select four fairly identical areas in his yard and to plant four different types of grasses, one in each area. At the end of a specified length of time each homeowner was asked to rank the grass types in order of preference, weighing such important criteria as expense, maintenance and upkeep required, beauty, hardiness, wife's preference, and so on. The rank 1 was assigned to the least preferred grass and the rank 4 to the favorite.

The null hypothesis was that there is no difference in preferences of the grass types, and the alternative was that some grass types tend to be preferred over others. Each of the twelve blocks consists of four fairly identical plots of land, each receiving care of approximately the same degree of skill because the four plots are presumably cared for by the same homeowner. The results of the experiment are as follows.

The ranks assigned to the grasses by the homeowners

Grass		1	2	3	4
Homeowner	1	4	3	2	1
	2	4	2	3	1
	3	3	1	2	4
	4	3	1	2	4
	5	4	2	1	3
	6	3	1	2	4
	7	1	3	2	4
	8	2	4	1	3
	9	3	1	2	4
	10	4	1	3	2
	11	4	2	3	1
	12	3	1	2	4
R_j (totals):		38	22	25	35

The Friedman test statistic is computed using Equation (2):

$$T = \frac{12}{(12)(4)(5)} [(38 - 30)^2 + (22 - 30)^2 + (25 - 30)^2 + (35 - 30)^2]$$
$$= 8.9$$

The critical region of approximate size $\alpha = .05$ corresponds to all values of T greater than 7.815, the .95 quantile of a chi-square random variable with $k - 1 = 3$ degrees of freedom, obtained from Table 2. Therefore the null hypothesis is rejected. We may conclude that there is a tendency for some types of grass to be preferred over others. The critical level is about .03, which is obtained by interpolation in Table 2. This means that the null hypothesis could have been rejected at a significance level as small as $\alpha = .03$.

COMMENT. If the null hypothesis is rejected, then further comparisons between treatments may be made by repeatedly applying the Friedman test to the reduced number of treatments. However, as was previously mentioned in Sections 4.3 and 5.6, very little meaning may be attached to the value of α in the subsequent tests. Miller (1966) discusses the multiple comparisons aspects of the Friedman test.

Theory.   The exact distribution of T is found under the assumption that each ranking within a block is equally likely, which is the null hypothesis. There are $k!$ possible arrangements of ranks within a block, and therefore $(k!)^b$ possible arrangements of ranks in the entire array of b blocks. The above statements imply that each of these $(k!)^b$ arrangements is equally likely under the null hypothesis. Therefore the probability distribution of T may be found for a given number of samples k and blocks b, merely by listing all possible arrangements of ranks and by computing T for each arrangement.

For example, if k equals 2 and b equals 3, there are $(2!)^3 = 8$ equally likely arrangements of the ranks, which are listed below along with the associated values of T.

	2 1	2 1	2 1	1 2	1 2	1 2	2 1	1 2
Arrangements	2 1	2 1	1 2	2 1	2 1	1 2	1 2	1 2
	2 1	1 2	2 1	2 1	1 2	2 1	1 2	1 2
Probability	1/8	1/8	1/8	1/8	1/8	1/8	1/8	1/8
Value of T	3	$\frac{1}{3}$	$\frac{1}{3}$	$\frac{1}{3}$	$\frac{1}{3}$	$\frac{1}{3}$	$\frac{1}{3}$	3

Therefore $P(T = 1/3) = 3/4$ and $P(T = 3) = 1/4$. It should be noted that in this example the ranks 1 and 2 correspond to "not preferred" and "preferred," a situation similar to that of the sign test. In fact, for two treatments $(k = 2)$ it may be shown that the Friedman test and the two-tailed sign test are equivalent.

If b is large, then the Central Limit Theorem may be applied to R_j, because R_j is the sum of the b independent random variables $R(X_{ij})$, $i = 1, 2, \ldots, b$. Under the null hypothesis $R(X_{ij})$ is a randomly selected integer between 1 and k inclusive. Theorem 1.5.5 gives the mean and variance of $R(X_{ij})$. (Let n equal 1 and replace N by k.)

$$(4) \qquad\qquad E[R(X_{ij})] = \frac{k+1}{2}$$

$$(5) \qquad\qquad \text{Var}\,[R(X_{ij})] = \frac{(k+1)(k-1)}{12}$$

Therefore the mean and variance of R_j may be obtained quite easily, because of the independence between different blocks. The mean is given by

$$E(R_j) = E\left[\sum_{i=1}^{b} R(X_{ij})\right] = \sum_{i=1}^{b} E[R(X_{ij})]$$

(6)
$$= \frac{b(k+1)}{2}$$

The variance of R_j is given by

$$\text{Var}\,(R_j) = \text{Var}\left[\sum_{i=1}^{b} R(X_{ij})\right] = \sum_{i=1}^{b} \text{Var}\,[R(X_{ij})]$$

(7)
$$= \frac{b(k+1)(k-1)}{12}$$

For large b the Central Limit Theorem justifies using the standard normal distribution to approximate the distribution of the random variable

$$\frac{R_j - E(R_j)}{\sqrt{\text{Var}\,(R_j)}}$$

because the set A conditions of Theorem 1.6.2 hold true. Theorem 1.6.3 stated that the sum of squares of *independent* standard normal random variables has the chi-square distribution. Therefore we could use the chi-square distribution with k degrees of freedom to approximate the distribution of

(8)
$$T' = \sum_{j=1}^{k} \frac{[R_j - E(R_j)]^2}{\text{Var}\,(R_j)}$$

if the R_j were mutually independent. But the R_j are dependent because their sum is fixed:

$$\sum_{j=1}^{k} R_j = \sum_{j=1}^{k}\sum_{i=1}^{b} R(X_{ij}) = \sum_{i=1}^{b}\sum_{j=1}^{k} R(X_{ij})$$

(9)
$$= \sum_{i=1}^{b} \frac{k(k+1)}{2} = b\,\frac{k(k+1)}{2}$$

Knowledge of $k - 1$ of the R_j automatically determines the remaining R_j. Friedman (1937) suggests multiplying both sides of (8) by the factor $(k-1)/k$ to obtain

(10)
$$T = \frac{k-1}{k}\,T' = \sum_{j=1}^{k} \frac{(k-1)}{k}\,\frac{[R_j - E(R_j)]^2}{\text{Var}\,(R_j)}$$

which is shown by Wilks (in Friedman's 1937 paper) to be asymptotically equivalent to a chi-square random variable with $k - 1$ degrees of freedom, as b becomes large. The form of T in (10) may be made more explicit by substituting (6) and (7) for the mean and variance of R_j, with the result

$$T = \sum_{j=1}^{k} \frac{(k - 1)}{k} \frac{\{R_j - [b(k + 1)/2]\}^2}{b(k + 1)(k - 1)/12}$$

$$(11) \qquad = \frac{12}{bk(k + 1)} \sum_{j=1}^{k} \left(R_j - \frac{b(k + 1)}{2}\right)^2$$

which is the Friedman test statistic as defined by Equation (2).

Thus the use of the chi-square distribution is justified by the above appeal to intuition. A rigorous proof is not within the intended scope of this book.

The above distribution of T was obtained, as always, under the assumption that H_0 is true. If H_0 is false, and the treatments are effectively different, then the column rank sums R_j may be expected to vary widely from their mean value $b(k + 1)/2$, and Friedman's test statistic will tend to be larger than it would be if H_0 were true. Therefore, the decision rule is to reject H_0 if T is large.

For two samples ($k = 2$), the A.R.E. of the Friedman test relative to the usual parametric t test is the same as that of the sign test, or $2/\pi = .637$, in situations where the t test is the most powerful test. For k samples the A.R.E. of the Friedman test relative to the usual parametric F test is dependent on the number of samples k, and equals $(.955)k/(k + 1)$ if the populations are normal, $k/(k + 1)$ if the populations are uniform, and $3k/2(k + 1)$ if the populations are double exponential. In fact the A.R.E. of the Friedman test relative to the popular F test never falls below $(.864)k/(k + 1)$ under purely translation type alternative hypotheses. The A.R.E. of the Friedman test is discussed more fully in Noether (1967).

THE RELATIONSHIP WITH KENDALL'S COEFFICIENT OF CONCORDANCE. A statistic W called Kendall's coefficient of concordance was introduced independently by Kendall and Babington-Smith (1939) and Wallis (1939). It may be used in the same situation where Friedman's test statistic is applicable although it was probably intended primarily as a measure of "agreement in rankings" in the b blocks, rather than as a test statistic. Using the same notation as above, Kendall's W is defined as

$$(12) \qquad W = \frac{12}{b^2 k(k + 1)(k - 1)} \sum_{j=1}^{k} \left(R_j - \frac{b(k + 1)}{2}\right)^2$$

If there is perfect agreement in the rankings in all b blocks, then treatment 1 receives the same rank in all b blocks, treatment 2 receives the same rank in all b blocks, and so on, and the resulting value of W is 1.0. If there is "perfect disagreement" among rankings, then the values of R_j will either be equal or very nearly equal to each other and their mean, and W will be 0 or very close to 0.

A comparison of Kendall's W with Friedman's T of Equation (2) reveals the relationship

$$(13) \qquad W = \frac{T}{b(k-1)}$$

Thus W is a simple modification of the Friedman test statistic, and any hypothesis test which uses W as a test statistic may be conducted by computing T instead of W. If T exceeds its $1 - \alpha$ quantile, then W exceeds its own $1 - \alpha$ quantile.

THE RELATIONSHIP WITH SPEARMAN'S RHO. Spearman's ρ defined by Equation (5.5.5), may be computed between any two blocks, say block i and block m, by considering the two blocks as two samples and the two ranks under each treatment as being a pair of related ranks. If Spearman's ρ computed between blocks i and m is denoted by ρ_{im}, then conversion of the equation for Spearman's ρ, Equation (5.5.5), into the above notation gives

$$(14) \qquad \rho_{im} = \frac{\sum\limits_{j=1}^{k} \{R(X_{ij}) - [(k+1)/2]\}\{R(X_{mj}) - [(k+1)/2]\}}{k(k+1)(k-1)/12}$$

The average value of Spearman's ρ, averaged over all pairs of blocks, bears a direct relationship with Friedman's test statistic as we shall now demonstrate.

Let ρ_a denote the average value of Spearman's ρ. There are $b(b-1)$ values of ρ_{im} to be averaged, counting both ρ_{im} and ρ_{mi} even though ρ_{im} equals ρ_{mi} by symmetry. To compute the average ρ we shall sum over all i and m, and then subtract those ρ_{im} where i equals m; that is, we shall subtract those values of ρ where a block is paired with itself. In those b cases ρ_{im} equals 1. Thus the average ρ may be expressed as

$$(15) \qquad \rho_a = \frac{1}{b(b-1)}\left(\sum_{i=1}^{b}\sum_{m=1}^{b} \rho_{im} - b\right)$$

The random variable ρ_a equals 1 if there is "perfect agreement" among rankings, in the sense previously described, because then each ρ_{im} equals 1. If there is disagreement among rankings ρ_α will be smaller than 1 and may even assume negative values. However, it is not possible for ρ_a to be as small as -1 except in the special case where there are only two blocks, $b = 2$.

The two preceding equations may be combined and simplified as follows. Substitution of (14) into (15) gives

$$\rho_a = \frac{1}{b(b-1)} \left(\sum_{i=1}^{b} \sum_{m=1}^{b} \frac{\sum_{j=1}^{k} \{R(X_{ij}) - [(k+1)/2]\}\{R(X_{mj}) - [(k+1)/2]\}}{k(k+1)(k-1)/12} - b \right)$$

$$= \frac{12}{b(b-1)k(k+1)(k-1)} \sum_{i=1}^{b} \sum_{m=1}^{b} \sum_{j=1}^{k} \left[R(X_{ij}) - \frac{k+1}{2} \right]$$

$$(16) \quad \times \left[R(X_{mj}) - \frac{k+1}{2} \right] - \frac{1}{b-1}$$

This expression is summed first over i, and then over m. Also the fact that

$$R_j = \sum_{i=1}^{b} R(X_{ij})$$

is used to simplify (16), as follows. Summation over i leaves

$$(17) \quad \rho_a = \frac{12}{b(b-1)k(k+1)(k-1)} \sum_{m=1}^{b} \sum_{j=1}^{k} \left[R_j - \frac{b(k+1)}{2} \right]$$

$$\times \left[R(X_{mj}) - \frac{k+1}{2} \right] - \frac{1}{b-1}$$

and summation over m leaves

$$(18) \quad \rho_a = \frac{12}{b(b-1)k(k+1)(k-1)} \sum_{j=1}^{k} \left[R_j - \frac{b(k+1)}{2} \right]^2 - \frac{1}{b-1}$$

A comparison of the above equation for ρ_a with the definition of Friedman's test statistic T, Equation (2), reveals the relationship

$$(19) \qquad\qquad \rho_a = \frac{T}{(b-1)(k-1)} - \frac{1}{b-1}$$

Thus the quantiles for the average Spearman's ρ may be easily obtained from the quantiles of the Friedman test statistic.

The above relationship between the Friedman test statistic and Spearman's ρ illustrates that the Friedman test may be used as a test for linear dependence in the two-sample case where Spearman's ρ was applicable. Spearman's ρ has the advantage of being tabulated for small samples although the exact distribution of Friedman's test statistic could be easily obtained from the distribution of Spearman's ρ. Both tests would be equivalent, and therefore both would have an A.R.E. of .912 when compared with the usual parametric test using Pearson's r as a test statistic in the situation where Pearson's r is appropriate.

AN EXTENSION TO THE CASE OF SEVERAL OBSERVATIONS PER EXPERIMENTAL UNIT. If there are several (m) observations for each treatment in each block, instead of only one observation per experimental unit as before, then the null hypothesis of no differences among treatments may be tested by slightly modifying Friedman's procedure. The observations within each block are ranked as before, with the exception that the ranks range from 1 to mk. The sum of ranks R_j is defined, as before, as the sum of ranks assigned to all observations involving treatment j. Let the observations in block i using treatment j be denoted by $X_{ij1}, X_{ij2}, \ldots, X_{ijm}$. The mean of R_j now becomes

$$E(R_j) = \sum_{i=1}^{b} \sum_{n=1}^{m} E[R(X_{ijn})] = \sum_{i=1}^{b} \sum_{n=1}^{m} \frac{mk+1}{2}$$

(20)
$$= \sum_{i=1}^{b} \frac{m(mk+1)}{2} = \frac{bm(mk+1)}{2}$$

The variance of R_j is again found with the aid of Theorem 1.5.5, in which n is replaced by m and N is replaced by mk;

$$\text{Var}(R_j) = \sum_{i=1}^{b} \text{Var}\left[\sum_{n=1}^{m} R(X_{ijn})\right] = \sum_{i=1}^{b} \frac{m(mk+1)(mk-m)}{12}$$

(21)
$$= \frac{bm^2(mk+1)(k-1)}{12}$$

The same Friedman test statistic

$$T = \sum_{j=1}^{k} \frac{(k-1)}{k} \frac{[R_j - E(R_j)]^2}{\text{Var}(R_j)}$$

as given in (10) is used here, except the mean and variance of R_j are given above in (20) and (21). Thus the Friedman test statistic for use with multiple observations in each experimental unit becomes, after simplification,

(22)
$$T_m = \frac{12}{bkm^2(mk+1)} \sum_{j=1}^{k} \left[R_j - \frac{bm(mk+1)}{2}\right]^2$$

whose distribution may be approximated by the chi-square distribution with $k-1$ degrees of freedom. For one observation per unit $(m=1)$ the T_m above reduces to the usual Friedman test statistic.

FURTHER EXTENSIONS OF THE FRIEDMAN TEST. An extension of the Friedman test to test for what is commonly referred to as "interaction" is given by Wilcoxon (1949). A different extension to the many-factor experiment, in the same way that the median test was extended to the many-factor experiment in Example 4.3.2, appears easily obtainable, although such an extension apparently does not appear in the literature. The analysis would involve reranking for each different factor to be tested, which becomes tedious. In general the nonparametric procedures for analyzing complex experimental designs are awkward and tedious. Until better methods are made available in those areas, the experimenter is almost forced to use parametric procedures with their often unrealistic assumptions.

Some recent references on rank sum tests for two-way layouts include Page (1963) and Hollander (1967b) if the alternative hypothesis specifies an ordering of treatment effects, and Dunn (1964) and McDonald and Thompson (1967) for multiple comparisons. Other methods of analysis are suggested by Doksum (1967), Puri and Sen (1967), and Sen (1968b). Asymptotic efficiency is studied by Mehra and Sarangi (1967) and Sen (1967a). A multivariate extension is given by Gerig (1969). Koch (1970) discusses a split-plot variation with several observations per cell.

PROBLEMS

1. Seven judges were asked to rank the five finalists in a local beauty pageant. The ranks went from 1 for best to 5 for worst. The results were as follows

Girl:	A	B	C	D	E
Judge: 1	5	2	1	3	4
2	5	1	2	3	4
3	3	1	2	4	5
4	2	3	4	1	5
5	3	1	2	5	4
6	4	1	2	3	5
7	4	2	3	1	5

May the null hypothesis of random assignment of ranks be rejected?

2. Twelve randomly selected students are involved in a learning experiment. Four lists of words are made up by the experimenter. Each list contains twenty pairs of words, but different methods of pairing are used on the four lists. Each student is handed a list, given five minutes to study it, and is then examined on his ability to remember the words. This procedure is repeated for all four lists for each student, the order of the lists being rotated from one student to the next. The examination scores are as follows (20 is perfect).

Student:	1	2	3	4	5	6	7	8	9	10	11	12
List 1:	18	7	13	15	12	11	15	10	14	9	8	10
List 2:	14	6	14	10	11	9	16	8	12	9	6	11
List 3:	16	5	16	12	12	9	10	11	13	9	9	13
List 4:	20	10	17	14	18	16	14	16	15	10	14	16

Are some lists easier to learn than other lists? Do some students learn the words better than other students?

3. Show that for $k = 2$ the Friedman test is equivalent to the two-tailed sign test (large sample approximation) of Section 3.4.

4. Show that Kendall's W assumes only values between 0 and 1 inclusive.

5. Use the result in Problem 4 to find the range of values that may be assumed by Spearman's average ρ.

6. If there are only two blocks ($b = 2$) then ρ_a is merely ρ computed between the two samples. The large sample approximation at the bottom of Table 10 is based on the fact that for large n (or k) the distribution of ρ may be approximated by a normal distribution with mean 0 and variance $1/(k - 1)$. This fact and the relationship given by Equation (19) imply that Friedman's T is approximately normally distributed with mean $(k - 1)$ and variance $(k - 1)$. Yet the distribution of T is approximated using the chi-square distribution with $k - 1$ degrees of freedom, which has mean $(k - 1)$ and variance $2(k - 1)$. What accounts for the discrepancy between the two distributions used, and between the two variances? Which appears to be the better approximate distribution to use in the above situation ($b = 2$, large k)?

7. Develop a test for the following situation. Consider b customers (blocks) selected at random. Each customer is shown k brands of a particular product, and asked to rank them by assigning the rank 1 to the brand they prefer, the rank 2 to their second preference, and the rank 3 to all of the other brands. Test the null hypothesis of no difference in preference among the brands against the alternative that some brands are preferred over others.

Modify the Friedman test to fit this situation by computing the mean and the variance of R_j for use in Equation (10). Assume that the usual asymptotic approximation is still valid here.

5.8. THE BALANCED INCOMPLETE BLOCK DESIGN

In the randomized complete block design described at the beginning of the previous section every treatment is applied in every block. However, it is sometimes impractical or impossible for all of the treatments to be applied to each block, especially when the number of treatments is large and the block size is limited. For example, if twenty different foods are to be tasted, each judge (block) may find it quite difficult to rank accurately all twenty foods in order of preference. But if each judge tastes only five of the foods, and then four times as many judges are used (or each judge is used four times), the judging may be easier and more accurate. Those experimental designs in which not all treatments are applied to each block are called incomplete block designs. Furthermore, if the design is balanced so that:

1. Every block contains k experimental units.
2. Every treatment appears in r blocks,

3. Every treatment appears with every other treatment an equal number of times.

The design is then called a *balanced* incomplete block design.

Durbin (1951) presented a rank test which may be used to test the null hypothesis of no differences among treatments in a balanced incomplete block design. Parametric methods of analyzing data obtained using a balanced incomplete block design exist and are based on certain normality assumptions which will not be explained here. The Durbin test may be preferred over the parametric test if the normality assumptions are not met, if an easier method of analysis is desired, or if the observations consist merely of ranks. The Durbin test reduces to the Friedman test if the number of treatments equals the number of experimental units per block. If the third condition given above is not completely satisfied, the Durbin test is still valid in most situations.

The Durbin Test

DATA. We shall use the following notation:

t = The number of treatments to be examined.
k = The number of experimental units per block ($k < t$).
b = The total number of blocks.
r = The number of times each treatment appears ($r < b$).
λ = The number of blocks in which the ith treatment and the jth treatment appear together. (λ is the same for all pairs of treatments.)

The data are arrayed in a balanced incomplete block design, as defined above. Let X_{ij} represent the result of treatment j in the ith block, if treatment j appears in the ith block.

Rank the X_{ij} within each block by assigning the rank 1 to the smallest observation in block i, the rank 2 to the second smallest, and so on to the rank k which is assigned to the largest observation in block i, there being only k observations within each block. Let $R(X_{ij})$ denote the rank of X_{ij} where X_{ij} exists.

Compute the sum of the ranks assigned to the r observed values under the jth treatment and denote this sum by R_j. Then R_j may be written as

$$(1) \qquad\qquad R_j = \sum_{i=1}^{b} R(X_{ij})$$

where only r values of $R(X_{ij})$ exist under treatment j, and therefore only r ranks are added to obtain R_j.

If the observations are nonnumeric but such that they are amenable to ordering and ranking according to some criterion of interest, then the ranking of each observation is noted and the values R_j for $j = 1, 2, \ldots, t$ are computed as described above.

If the ranks may be assigned in several different ways because of several observations being equal to each other, we recommend assigning the average of the disputed ranks to each of the tied observations. This procedure changes the null distribution of the test statistic, but the effect is negligible if the number of ties is not excessive.

ASSUMPTIONS

1. The blocks are mutually independent of each other.
2. Within each block the observations may be arranged in increasing order according to some criterion of interest. (A moderate number of ties may be tolerated.)

HYPOTHESES

H_0: Each ranking of the random variables within each block is equally likely (i.e., the treatments have identical effects).

H_1: At least one treatment tends to yield larger observed values than at least one other treatment.

TEST STATISTIC. The Durbin test statistic is defined as

$$(2) \qquad T = \frac{12(t-1)}{rt(k-1)(k+1)} \sum_{j=1}^{t} \left[R_j - \frac{r(k+1)}{2} \right]^2$$

and may be written in the equivalent machine form

$$(3) \qquad T = \frac{12(t-1)}{rt(k-1)(k+1)} \sum_{j=1}^{t} R_j^2 - 3\frac{r(t-1)(k+1)}{k-1}$$

DECISION RULE. Reject the null hypothesis at the approximate level of significance α if the Durbin test statistic T exceeds the $(1 - \alpha)$ quantile of a chi-square random variable with $t - 1$ degrees of freedom, obtained from Table 2.

Example 1. Suppose an ice cream manufacturer wants to test the taste preferences of several people for his seven varieties of ice cream. He asks each person to taste three varieties and to rank them 1, 2, and 3, with the rank 1 being assigned to his favorite variety. In order to design the experiment so that each variety is compared with every other variety an equal number of times, a Youden square layout given by Federer (1963) is used. Seven people are each given three varieties to taste, and the resulting ranks are as follows.

Variety:	1	2	3	4	5	6	7
Person: 1	2	3		1			
2		3	1		2		
3			2	1		3	
4				1	2		3
5	3				1	2	
6		3				1	2
7	3		1				2
$R_j = $	8	9	4	3	5	6	7

In this experiment,

$t = 7$ = total number of varieties
$k = 3$ = number of varieties compared at one time
$b = 7$ = number of people (blocks)
$r = 3$ = number of times each variety is tasted.
$\lambda = 1$ = number of times each variety is compared with each other variety.

Therefore the design is a balanced incomplete block design, and the Durbin test may be used to test the null hypothesis that no variety of ice cream tends to be preferred over any other variety of ice cream.

The critical region of approximate size $\alpha = .05$ corresponds to all values of T greater than 12.59, which is the .95 quantile of a chi-square random variable with $t - 1 = 6$ degrees of freedom, obtained from Table 2. The value of the Durbin test statistic is found as follows.

$$T = \frac{12(t - 1)}{rt(k - 1)(k + 1)} \sum_{j=1}^{t} \left[R_j - \frac{r(k + 1)}{2} \right]^2$$

$$= \frac{(12)(6)}{(3)(7)(2)(4)} [(8 - 6)^2 + (9 - 6)^2 + \ldots + (7 - 6)^2]$$

$$= 12$$

which is not in the critical region, so H_0 is accepted. However the critical level is quite small, and is estimated by interpolation to be about .065.

Theory.   The theoretical development of the Durbin test is very similar to that of the Friedman test. That is, the exact distribution of the Durbin test statistic is found under the assumption that each arrangement of the k ranks within a block is equally likely, because of no differences between treatments. There are $k!$ equally likely ways of arranging the ranks within each block, and there are b blocks. Therefore each arrangement of ranks over the entire array of b blocks is equally likely, and has probability $1/(k!)^b$ associated with it, because there are $(k!)^b$ different arrays possible. The Durbin test statistic is calculated for each array, and then the distribution function is determined, just as it was for the Friedman test statistic in the previous section.

The exact distribution is not practical to find in most cases, and so the distribution of the Durbin test statistic is approximated by the chi-square distribution with $t - 1$ degrees of freedom, if the number of repetitions r of each treatment is large. The justification for this approximation is as follows.

If the number r of repetitions of each treatment is large, then the sum of the ranks, R_j, under the jth treatment is approximately normal, according to the Central Limit Theorem. Therefore the random variable

$$\frac{R_j - E(R_j)}{\sqrt{\text{Var}(R_j)}}$$

has approximately a standard normal distribution. As in the previous section, if the R_j were independent the statistic

$$(4) \qquad T' = \sum_{j=1}^{t} \frac{[R_j - E(R_j)]^2}{\text{Var}(R_j)}$$

could be considered as the sum of t independent, approximately chi-square, random variables and the distribution of T' then could be approximated with a chi-square distribution with t degrees of freedom. But the R_j are not independent. Their sum is fixed as

$$(5) \qquad \sum_{j=1}^{t} R_j = \frac{bk(k+1)}{2}$$

so that the knowledge of $t - 1$ of the R_j enables us to state the value of the remaining R_j. Durbin (1951) shows that multiplication of T' by $(t - 1)/t$ results in a statistic which is approximately chi-square with $t - 1$ degrees of freedom, with the form

$$(6) \qquad T = \frac{t-1}{t} T' = \frac{t-1}{t} \sum_{j=1}^{t} \frac{[R_j - E(R_j)]^2}{\text{Var}(R_j)}$$

It only remains to find the mean and variance of R_j in order to transform (6) into the usual form given by (2).

The sum of ranks R_j is the sum of independent random variables $R(X_{ij})$.

$$(7) \qquad R_j = \sum_{i=1}^{b} R(X_{ij})$$

Each $R(X_{ij})$, where it exists, is a randomly selected integer from 1 to k. Therefore the mean and variance of $R(X_{ij})$ are given by Theorem 1.5.5 as

$$(8) \qquad E[R(X_{ij})] = \frac{k+1}{2}$$

and

$$(9) \qquad \text{Var}[R(X_{ij})] = \frac{(k+1)(k-1)}{12}$$

Then the mean and variance of R_j are easily found to be

(10) $$E(R_j) = \sum_{i=1}^{b} E[R(X_{ij})] = \frac{r(k+1)}{2}$$

and

(11) $$\text{Var }(R_j) = \sum_{i=1}^{b} \text{Var }[R(X_{ij})] = \frac{r(k+1)(k-1)}{12}$$

The mean and variance of R_j given above are substituted into the Durbin test statistic given by equation (6) to obtain

(12)
$$T = \frac{t-1}{t} \sum_{j=1}^{t} \frac{[R_j - r(k+1)/2]^2}{r(k+1)(k-1)/12}$$
$$= \frac{12(t-1)}{rt(k+1)(k-1)} \sum_{j=1}^{t} \left[R_j - \frac{r(k+1)}{2} \right]^2$$

which is in the same form as given in the explanation of the Durbin test above.

The chi-square approximation is based on the assumption that the number r of repetitions of each treatment is reasonably large. In practice the approximation is used even if r is as small as 3 or 2, out of sheer necessity. The stated α level is probably not very accurate in those circumstances.

The Durbin test has been generalized to the case where some experimental units may contain several observations, by Benard and van Elteren (1953). Noether (1967a) also discusses the Durbin test and its generalizations, and shows that the A.R.E. of the Durbin test relative to its parametric counterpart is the same as that of the Friedman test relative to its parametric counterpart. See the preceding section for details. The case of paired comparisons ($k = 2$) is discussed by Puri and Sen (1969b).

PROBLEMS

1. Seven types of automobile tires are being tested for durability. It is felt that the best test is to see how the tires perform under actual driving conditions. However, only four tires may be compared at a time because only four-wheeled vehicles are available for testing. Therefore the experiment is designed using a balanced incomplete block design.

   Each of seven drivers is given four tires which are placed on his car in a random order and rotated regularly during the experiment. The tires are replaced when necessary, and ranks are assigned to the original tires according to the order of replacement.

Tire type	1	2	3	4	5	6	7
Driver 1			3		1	4	2
2	1			3		4	2
3	2	1			3		4
4	1	2	4			3	
5		1	4	3			2
6	2			4	1	3	
7		1		2	3	4	

Do the results indicate a significant difference in durability? (First examine the experiment to be sure it follows a balanced incomplete block design.)

2. Show that kb equals rt. (*Hint:* Count the number of observations in two different ways.)

3. Show that λ equals $r(k-1)/(t-1)$. (*Hint:* First note that any particular treatment occurs in r blocks. Then count the number of units in which the treatment does not appear in those r blocks, and count them two different ways.)

5.9. TESTS WITH EFFICIENCY OF ONE OR MORE

The tests described in this section all share one property in common. The A.R.E. of each of these tests is 1 when compared with the usual parametric test in situations where the parametric test is appropriate. If the normality assumptions underlying the parametric test are not satisfied, then under certain conditions which are easily met the A.R.E. is always greater than 1, and may be as high as infinity. It seems like a rather strong statement to say that the tests of this section are always at least as good as the usual parametric tests, such as the t test and the F test, as measured by their asymptotic relative efficiencies. However it is true. Remember, though, that A.R.E. is only one way to compare tests, although admittedly A.R.E. is probably the most universally accepted method of comparison. Just plain relative efficiency, without the word "asymptotic," is a method of comparison also. It compares the sample sizes required for two tests to have the same power under identical conditions, where the sample sizes are finite. On the basis of relative efficiencies the tests in this section may be better or worse than their usual parametric counterparts, depending on the circumstances. It is not possible to examine all of the possible circumstances, and so that is why the A.R.E. is usually used to compare tests.

Unlike the previous sections of this chapter, no new experimental situations are introduced in this section. We have already introduced nonparametric methods, based on ranks, of handling the correlation situation in Section 5.5, the one-way layout in Section 5.6, and the randomized complete block design in Section 5.7. Those methods are widely accepted, reasonably powerful, and are not too difficult to administer. By comparison the methods of this section are usually slightly more powerful, but they are also slightly more difficult to administer. The assumptions behind these tests are practically identical to the assumptions underlying the earlier tests. Indeed, these tests are basically rank tests with a little dressing to improve the A.R.E. The user may decide whether to use these tests or the previously introduced tests, there is no solid statistical basis for preferring some to others.

The first tests we shall describe were introduced by Bell and Doksum (1965). These tests use random normal deviates instead of ranks as used in the previous tests. By random normal deviates we mean numbers which appear to be observations on independent standard normal random variables. In practice these numbers are usually obtained from special tables, such as one large table furnished by the Rand Corporation (1955), or by a computer using special programs. For our purposes we shall rely on a short table of random normal deviates given in the appendix as Table 13.

The principle behind the Bell and Doksum tests is to use order statistics from normally distributed random samples instead of using ranks. That is, instead of using the rank i, say, where ranks ranged from 1 to n, we now use the ith smallest observation from n random normal deviates. The arithmetic of the tests is performed on the order statistics in the same way that it was performed on the ranks in the rank tests, or for that matter in the same way that it is performed on the original observations in the usual parametric tests. What will appear to be differences between these tests and their corresponding rank tests are actually only simplifications in arithmetic.

The principle of using random normal deviates is not new with Bell and Doksum, but has been traced by Doksum back through an article by Durbin (1961), the last problem in Fraser (1957), to a mention of the technique by Ehrenberg (1951). The theory underlying these tests by Bell and Doksum, and all of the tests in this section, is beyond the scope of this book and is therefore omitted. The first procedure introduced is the Bell-Doksum modification of Spearman's ρ and Pearson's r.

The Bell-Doksum Measure of Correlation

DATA. The data may consist of a bivariate random sample of size n, (X_1, Y_1), $(X_2, Y_2), \ldots, (X_n, Y_n)$. Let $R(X_i)$ and $R(Y_i)$ be the rank of the X_i and the Y_i, as defined in the description of Spearman's ρ in Section 5.5. After ranks have been assigned, draw two groups of random normal deviates, each of size n, from a table such as Table 13. Arrange each of the groups from smallest to largest, and let $Z_1(i)$ represent the ith smallest value from group one, and similarly $Z_2(i)$ for the second group, for each i from 1 to n. These are the order statistics of rank i, as defined in

Definition 2.1.4. Assign the value $Z_1(i)$ to the X, in the original data, that has rank i, for each i from 1 to n. Similarly, assign the value $Z_2(i)$ to the Y that has rank i, for each i. In other words, assign $Z_1[R(X_j)]$ to X_j and $Z_2[R(Y_j)]$ to Y_j for each j.

As with Spearman's ρ, the data may also be nonnumeric, as long as it may be ranked. Ties may be handled in a manner analogous to Spearman's ρ. That is, in case of ties, assign to each tied X or Y the average value of the Z's that would have been assigned to the tied values had the tied values actually been slightly unequal to each other.

MEASURE OF CORRELATION. As a measure of correlation use

$$(1) \qquad T_1 = \frac{1}{n} \sum_{i=1}^{n} Z_1[R(X_i)] \cdot Z_2[R(Y_i)]$$

In general, T_1 is not bounded by -1 and $+1$ as the other measures of correlation have been. However, if X and Y are uncorrelated, and such is the case when they are independent, then T_1 will tend to be close to zero. When X and Y are perfectly correlated, T_1 will tend to be close to $+1$ or -1 depending on whether the correlation is positive or negative.

HYPOTHESIS TEST. The same hypotheses given under Spearman's ρ in Section 5.5 are applicable here. The two-tailed critical region of size α corresponds to values of T_1 less than $y_{\alpha/2}$ or greater than $y_{1-\alpha/2}$, and the one-tailed critical region of size α corresponds to values of T_1 greater than $y_{1-\alpha}$ in the test for positive correlation, or less than y_α in the test for negative correlation. The quantile y_p may be approximated by $x_p/\sqrt{n}$, where x_p is the p quantile of a standard normal random variable obtained from Table 1. The exact distribution of T_1 is the same as the difference of two chi-square random variables, each with n degrees of freedom, and each divided by $2n$, but apparently no tables exist for the exact distribution.

Example 1. Nine pairs of observations on the bivariate random variable (X, Y) are obtained, as listed in the first column below. Their ranks are given in the second column. Then two groups of nine random normal deviates are obtained from Table 13, by arbitrarily starting with row 6 and reading across. These are listed in the third and fourth columns. The numbers in the third column are ordered, and assigned to the X_i of the same rank as illustrated in the left half of column five. The numbers in the fourth column are assigned to the Y_i's in a similar manner as shown in the right half of column five.

	X_j, Y_j	$[R(X_j), R(Y_j)]$	Z_1	Z_2	$\{Z_1[R(X_j)], Z_2[R(Y_j)]\}$
$j = 1$	(2.6, 17)	(7, 7)	$-.18$	$-.73$	(.76, .52)
$j = 2$	(1.4, 11)	(2, 2)	.95	.52	$(-1.10, -.73)$
$j = 3$	(1.3, 13)	(1, 4)	-1.10	-1.37	$(-2.50, -.32)$
$j = 4$	(1.8, 10)	(3, 1)	.67	1.64	$(-1.02, -1.37)$
$j = 5$	(3.2, 20)	(9, 8)	.76	.53	(1.61, .53)
$j = 6$	(2.7, 21)	(8, 9)	.21	$-.60$	(.95, 1.64)
$j = 7$	(2.5, 16)	(6, 6)	1.61	$-.22$	(.67, .41)
$j = 8$	(1.9, 14)	(4, 5)	-1.02	$-.32$	$(-.18, -.22)$
$j = 9$	(2.0, 12)	(5, 3)	-2.50	.41	$(.21, -.60)$

The Bell-Doksum measure of correlation

$$T_1 = \tfrac{1}{9} \sum_{j=1}^{9} \{Z_1[R(X_j)]\}\{Z_2[R(Y_j)]\}$$

$$= \tfrac{1}{9}[(.76)(.52) + (-1.10)(-.73) + \ldots + (.21)(-.60)]$$

$$= \tfrac{1}{9}(5.995)$$

$$= .666$$

A two-tailed critical region of size $\alpha = .05$ corresponds approximately to values of T_1 less than $(-1.96)/3 = -.653$ or greater than $+.653$. Therefore the null hypothesis of no correlation between X and Y may be rejected at the .05 level of significance.

The second random normal deviates procedure we shall introduce is the Bell-Doksum modification of the Kruskal-Wallis analysis of variance using ranks, which was introduced in Section 5.6 for the completely randomized design.

The Bell-Doksum Test for Several Independent Samples

DATA. The data consist of k random samples of possibly unequal sample sizes. Denote the ith sample, of size n_i, by $X_{i1}, X_{i2}, \ldots, X_{in_i}$. Let N denote the total number of observations. Rank all N values from rank 1 to rank N, as was explained in the Kruskal-Wallis test. Let $R(X_{ij})$ denote the rank of X_{ij}.

Then draw a group of N numbers from a table of random normal deviates, such as Table 13. Assign the rth smallest of these numbers to the particular X_{ij} which has rank r in the original data. That is, if $Z(r)$ represents the rth smallest random normal deviate, then $Z[R(X_{ij})]$ is assigned to X_{ij}.

Compute the average Z value for each of the k samples

$$(2) \qquad Z_i = \frac{1}{n_i} \sum_{i=1}^{N} Z[R(X_{ij})] \qquad \text{for } i = 1, 2, \ldots, k$$

and the overall average Z value

$$(3) \qquad \bar{Z} = \frac{1}{N} \sum_{r=1}^{N} Z(r)$$

If there are ties among the X_{ij}'s, then assign to each tied value the average of the Z values that would have been assigned had there been no ties. This will affect the distribution of the test statistic, but if ties are not extensive the effect should be minor.

ASSUMPTIONS. The assumptions here are the same as in the Kruskal-Wallis test, with one addition. We also assume that the group of N random normal deviates resembles in all respects a random sample of size N from the standard normal distribution.

HYPOTHESES

H_0: All of the k population distribution functions are identical.
H_1: At least one of the populations tends to yield larger observations than at least one of the other populations.
As in the Kruskal-Wallis test, H_1 is sometimes stated as follows.

H_1: The k populations do not all have identical means.

TEST STATISTIC. The test statistic T_2 is defined as

$$(4) \qquad T_2 = \sum_{i=1}^{k} n_i (Z_i - \bar{Z})^2$$

where Z_i and $\bar{Z}$ are given by (2) and (3).

DECISION RULE. Reject H_0 at the level α if T_2 exceeds the $1 - \alpha$ quantile of a chi-square random variable with $k - 1$ degrees of freedom, given in Table 2. Note that this distribution is the exact distribution.

Example 2.   The same example that was used to illustrate the Kruskal-Wallis test in Section 5.6 and the median test in Section 4.3 will also be used here, for purposes of illustrating the Bell-Doksum test and for comparisons.

Four methods of growing corn resulted in the following observations, along with their ranks.

Method	1		2		3		4	
	Obs.	*Rank*	*Obs.*	*Rank*	*Obs.*	*Rank*	*Obs.*	*Rank*
	83	11	91	23	101	34	78	2
	91	23	90	19.5	100	33	82	9
	94	28.5	81	6.5	91	23	81	6.5
	89	17	83	11	93	27	77	1
	89	17	84	13.5	96	31.5	79	3
	96	31.5	83	11	95	30	81	6.5
	91	23	88	15	94	28.5	80	4
	92	26	91	23			81	6.5
	90	19.5	89	17				
			84	13.5				

Because there are $N = 34$ observations in all, 34 numbers were selected from Table 13. We arbitrarily decided to start with the top number in the third column in Table 13 and work down. The first number is $-.08$, the second number is $-.71$, and so on. After using all of the numbers in column three we started down column four, until we had 34 numbers. These numbers are ordered, and the ordered numbers, with their ranks, are given below.

$Z(r)$	r	$Z(r)$	r	$Z(r)$	r	$Z(r)$	r
-1.42	1	$-.62$	10	.52	19	1.14	28
-1.37	2	$-.44$	11	.52	20	1.16	29
-1.31	3	$-.37$	12	.67	21	1.33	30
-1.25	4	$-.28$	13	.74	22	1.49	31
-1.10	5	$-.18$	14	.77	23	1.49	32
$-$ .87	6	$-.09$	15	.93	24	1.61	33
$-$ .77	7	$-.08$	16	1.00	25	2.76	34
$-$ .71	8	$-.05$	17	1.08	26		
$-$ .67	9	.43	18	1.14	27		

Before these numbers can be assigned to the original X_{ij} values, an adjustment must be made to account for the ties among the X_{ij}'s. The Z values which are to be assigned to each group of tied observations are first averaged, and are then assigned.

Ranks of Tied Observations	Average Rank	Average of the Z Numbers
5, 6, 7, 8	6.5	$-.86$
10, 11, 12	11	$-.48$
13, 14	13.5	$-.23$
16, 17, 18	17	$-.10$
19, 20	19.5	.52
21, 22, 23, 24, 25	23	.82
28, 29	28.5	1.15
31, 32	31.5	1.49

Now the random normal deviates, or their averages in the case of ties, may be assigned to the original observations.

Method	1			2			3			4		
Obs.	Rank	Z	Obs.	Rank	Z	Obs.	Rank	Z	Obs.	Rank	Z
83	11	$-.48$	91	23	.82	101	34	2.76	78	2	-1.37
91	23	.82	90	19.5	.52	100	33	1.61	82	9	$-$ .67
94	28.5	1.15	81	6.5	$-.86$	91	23	.82	81	6.5	$-$ .86
89	17	$-.10$	83	11	$-.48$	93	27	1.14	77	1	-1.42
89	17	$-.10$	84	13.5	$-.23$	96	31.5	1.49	79	3	-1.31
96	31.5	1.49	83	11	$-.48$	95	30	1.33	81	6.5	$-$ .86
91	23	.82	88	15	$-.09$	94	28.5	1.15	80	4	-1.25
92	26	1.08	91	23	.82		Total $= \overline{10.30}$		81	6.5	$-$ .86
90	19.5	.52	89	17	$-.10$		$n_3 = 7$			Total $= \overline{-8.60}$	
	Total $= \overline{5.20}$		84	13.5	$-.23$		$Z_3 = 1.471$			$n_4 = 8$	
	$n_1 = 9$			Total $= \overline{-.31}$						$Z_4 = -1.075$	
	$Z_1 =$ .578			$n_2 = 10$							
				$Z_2 = -.031$							

The average Z under each method is computed using Equation (2) and is given above. The overall average $\bar{\bar{Z}}$ is computed from (3) and is

$$\bar{\bar{Z}} = \frac{1}{34} (6.59) = .194$$

The test statistic T_2, defined by (4), may now be computed.

$$T_2 = \sum_{i=1}^{4} n_i (\bar{Z}_i - \bar{\bar{Z}})^2$$

$$= 9(.384)^2 + 10(-.225)^2 + 7(1.277)^2 + 8(-1.269)^2$$

$$= 26.13$$

A comparison of $T_2 = 26.13$ with 7.815, the .95 quantile of a chi-square random variable with $4 - 1 = 3$ degrees of freedom, shows that T_2 is larger, and so H_0 is easily rejected at the .05 level.

The Bell-Doksum test statistic, the Kruskal-Wallis test statistic, and the median test statistic all use the chi-square distribution with 3 degrees of freedom. Therefore the test statistics may be compared directly with each other to see which test appears to be the most powerful for this set of data. In Example 4.3.1 the median test statistic equaled 17.6 for these data. In Example 5.6.1 the Kruskal-Wallis test statistic was shown to be 25.46 for these data, a reflection of the greater power of the Kruskal-Wallis test. The Bell-Doksum test statistic, given above as 26.13, is larger than both of the others, which suggests that the Bell-Doksum test may have slightly more power than the Kruskal-Wallis test, at least for this set of data.

The third procedure involving random normal deviates is the Bell-Doksum test for several related samples, which is an analogue of the Friedman test for several related samples, presented in Section 5.7.

The Bell-Doksum Test for Several Related Samples

DATA. The data consist of b blocks $(X_{i1}, X_{i2}, \ldots, X_{ik})$, $i = 1, 2, \ldots, b$, with k observations in each block. The ranks 1 to k are assigned to the observations in each block, as was described in the Friedman test of Section 5.7. Let $R(X_{ij})$ be the rank, from 1 to k, assigned to X_{ij} within block (row) i.

Then draw b groups of k numbers each from a table of random normal deviates, such as Table 13. Let $Z_i(r)$ be the rth smallest number in the ith group of numbers, $r = 1, 2, \ldots, k$, and $i = 1, 2, \ldots, b$. Assign $Z_i(r)$ to the X_{ij} of rank r in block i, for each r and i, and denote that Z value by $Z_i[R(X_{ij})]$. Compute the average Z for each treatment (column);

$$(5) \qquad Z_j = \frac{1}{b} \sum_{i=1}^{b} Z_i[R(X_{ij})] \qquad \text{for } j = 1, 2, \ldots, k$$

and compute the average of all Z values

$$(6) \qquad \bar{Z} = \frac{1}{bk} \sum_{i=1}^{b} \sum_{r=1}^{k} Z_i(r)$$

As a computation check, $\bar{Z}$ in (6) should equal the average of the kZ_j's computed in (5).

In case of ties, we recommend averaging the Z's that are to be assigned to each group of tied observations, and then assigning that average Z value to each observation in the group of tied observations. This is analogous to the average rank method used in the Friedman test.

ASSUMPTIONS. The assumptions behind this Bell-Doksum test are the same as for the Friedman test, except for an additional assumption needed for this test. We assume that the random normal deviates used in the test resemble in all respects observations on independent random variables with the standard normal distribution.

HYPOTHESES

H_0: The random variables within each block are identically distributed (i.e., there are no treatment effects).

H_1: At least one of the treatments tends to yield larger observed values than at least one other treatment.

TEST STATISTIC. The test statistic is the same here as for the previous Bell-Doksum test, except for the difference in notation

$$(7) \qquad T_3 = b \sum_{j=1}^{k} (Z_j - \bar{Z})^2$$

where Z_j and $\bar{Z}$ are given by (5) and (6).

DECISION RULE. Reject the null hypothesis at the level α if T_3 exceeds the $1 - \alpha$ quantile of a chi-square random variable with $k - 1$ degrees of freedom, obtained from Table 2. Note that this is the exact distribution of T_3.

Example 3. For comparison purposes the Bell-Doksum test for related samples will be applied to the same data used in Example 5.7.1 to illustrate the Friedman test.

Twelve homeowners were asked to rank four types of grass from rank 1 for least preferred to rank 4 for the favorite. The null hypothesis was that there is no difference in preference for the various grass types. Each homeowner produced his own ranking, with the following results.

The ranks assigned to the grasses by the homeowners

Grass	1	2	3	4
Homeowner 1	4	3	2	1
2	4	2	3	1
3	3	1	2	4
4	3	1	2	4
5	4	2	1	3
6	3	1	2	4
7	1	3	2	4
8	2	4	1	3
9	3	1	2	4
10	4	1	3	2
11	4	2	3	1
12	3	1	2	4

Now 48 random normal deviates are drawn from Table 13, in 12 groups of 4 each. We arbitrarily decide to select the numbers in rows 1 through 12, columns 5 through 8; that is, we select a block of numbers in the upper right corner of Table 13. The four numbers from row 1 are assigned to homeowner 1, row 2 to homeowner 2, and so on. For instance, the four numbers from row 1 in Table 13 are

Number	.45	−.33	.70	−.76
Rank	3	2	4	1

These numbers are assigned to the corresponding ranks in the original data.

Grass	1	2	3	4
Homeowner 1	4	3	2	1
Assigned Z's	0.70	0.45	−0.33	−0.76

The procedure is repeated for each row, with the following result.

The random normal deviates assigned to each original observation

Grass	1	2	3	4
Homeowner 1	.70	.45	− .33	− .76
2	− .31	− .84	− .72	−1.00
3	1.40	.57	1.03	1.50
4	1.09	−1.18	− .97	2.56
5	.38	−1.00	−1.65	− .55
6	.76	−1.02	.21	1.61
7	− .60	.53	− .22	1.64
8	− .59	− .35	−1.32	− .53
9	1.01	− .94	.37	2.83
10	1.30	−1.06	.84	.51
11	1.05	− .11	.79	− .25
12	1.65	.78	1.13	1.88
total	7.84	−4.17	− .84	9.44
$Z_j = \dfrac{\text{total}}{12} =$	.653	− .348	− .070	.787

The treatment averages Z_j are also computed above for each treatment. The overall average is

$$\bar{Z} = \frac{1}{48} \sum_{i=1}^{12} \sum_{j=1}^{4} Z_i(r) = \frac{1}{48} \quad (12.27)$$

$$= 0.256$$

The test statistic is computed from Equation (7) as

$$T_3 = b \sum_{j=1}^{k} (Z_j - \bar{Z})^2$$
$$= 12[(.397)^2 + (-.604)^2 + (-.326)^2 + (.531)^2]$$
$$= 10.9$$

This value of T_3 is greater than 7.815, the .95 quantile of a chi-square random variable with $4 - 1 = 3$ degrees of freedom, so the null hypothesis of equal grass preferences is rejected at the .05 level of significance.

The value 10.9 for the Bell-Doksum test statistic may be compared with the value 8.9 for the Friedman test statistic on the same set of data, because both statistics use the chi-square distribution with 3 degrees of freedom for their distributions, even though that distribution is only approximate in the Friedman test. The larger value of the Bell-Doksum test statistic indicates that it is slightly more sensitive to the differences present in this particular set of data. However, a different choice of random normal deviates may have resulted in a different value of the Bell-Doksum test statistic and thus a different comparison.

Further modifications of the above tests to fit any experimental design seem to be easily obtainable, but they have not yet appeared in the literature.

The Bell-Doksum tests have the disadvantage that the test statistic depends not only on the ranks of the observations, but also on the particular random normal deviates selected. Two people may use the same test to analyze the same data and get different results. However, this situation is much the same as that of two people running the same experiment and getting different sets of data, an occurrence readily accepted and expected. Still, the Bell-Doksum test introduces additional unwanted variation. There has apparently been no investigation into the amount of variation that several different sets of random normal deviates would introduce into the Bell-Doksum test statistics. Judging from the close agreement of the Bell-Doksum test statistics' values with the rank test statistics' values in the preceding two examples, we may suspect that the unwanted variation is small when the number of observations is reasonably large.

One way of avoiding the unwanted variation of the random normal deviates in the Bell-Doksum tests is to use some of the techniques suggested below. The first technique involves using the mean of $Z(r)$, instead of $Z(r)$

as Bell and Doksum suggested. The order statistic $Z(r)$ varies from sample to sample as we saw above, but the mean of $Z(r)$ depends only the rank r of the order statistic and the sample size n. Instead of using a table of random normal deviates to find $Z(r)$, we consult special tables that give the means of order statistics from normally distributed samples. Tests of this type are called *expected normal scores tests* by Bradley (1968), who presents a good explanation of how to use these tests. Bradley also gives tables of the required means for $n \leq 20$. Other tables for $n \leq 50$ are given by Fisher and Yates (1957) and Pearson and Hartley (1962). If more than 50 items are ranked together these tests may not be used simply because of the lack of tables. Also, special tables are required to see if the test statistic is significant, unless the sample sizes are large enough to justify the use of the asymptotic distributions.

The second technique for avoiding the unwanted variation of the random normal numbers is to use the $r/(n + 1)$ quantile of a standard normal random variable instead of $Z(r)$, the rth order statistic from a sample of size n, as the Bell-Doksum tests use. These quantiles may be obtained from Table 1 or from more extensive tables of the standard normal distribution, such as that of Pearson and Hartley (1962). Tests of this type are called *inverse normal scores tests* or, more simply, *normal scores tests*. The better known of these tests were introduced by van der Waerden (1952/53) for the case of two independent samples and Klotz (1962) as an analogue of Mood's test for equal variances, discussed in Section 5.3. A more complete presentation of normal scores tests may be found in Bradley (1968) and Hájek and Šidák (1967).

Once a rank test has been invented for a particular experimental situation, the other types of tests follow easily. The same model structure, data, hypotheses, and assumptions may be used, but instead of using ranks in the test statistic we use order statistics from a normal distribution if we want a Bell-Doksum test, means of those order statistics if we want an expected normal scores test, or quantiles from the normal distribution if we want an inverse normal scores test. The distribution of the test statistic varies slightly under these different conditions, but as the sample sizes go to infinity the tests become almost equivalent. The A.R.E. is the same for all three variations of the rank tests, and is greater than or equal to 1, relative to the usual parametric t test or F test under the alternative hypothesis of differences only in location parameters. The A.R.E. of these tests relative to the corresponding rank tests may be greater than 1 or less than 1 depending on the particular alternative being considered. That is, even though the rank tests may be worse than the parametric tests at times, as measured by their A.R.E., in other situations the rank tests may be better than the parametric tests and better than the tests presented in this section. A complete discussion of A.R.E.

may be found in Hájek and Šidák (1967) but is far beyond the general level of mathematical competence assumed in this book.

A recent discussion of random normal numbers appears in Marsaglia (1968). Jogdeo (1966) shows that the relative efficiency of the Bell-Doksum tests is less than one for some fixed alternatives. Other related tests are discussed by Bartlett and Govindarajulu (1968). The efficiency of various types of normal scores tests is examined in recent articles by Raghavachari (1965b), Thompson, Govindarajulu, and Doksum (1967), Bhattacharyya (1967), Stone (1968), and Gokhale (1968).

PROBLEMS

1. Work Example 1, using a different set of random normal deviates than was used in the example.
2. Compute the Bell-Doksum measure of correlation for Example 5.5.1, and compare it with Spearman's ρ. Also compare their critical levels.
3. Work Example 2, using a different set of random normal deviates than was used in the example.
4. Compute the Bell-Doksum test statistic for the data in Problem 5.6.1. Also compute the Kruskal-Wallis test statistic if you haven't already done so, and compare the two values.
5. Work Example 3, using a different set of random normal deviates than was used in the example.
6. Compute the Bell-Doksum test statistic for the data in Problem 5.7.1. Also compute the Friedman test statistic if you have not already done so. Compare the two results.

CHAPTER 6

Statistics of the Kolmogorov-Smirnov Type

6.1. TESTS OF GOODNESS OF FIT

We shall begin this chapter with a test for goodness of fit which was introduced by Kolmogorov (1933). This test is perhaps the most useful of the tests in this chapter, partly because it furnishes us with an alternative, designed for ordinal data, to the chi-square test for goodness of fit introduced in Section 4.5, which was designed for nominal type data, and partly because the Kolmogorov test statistic enables us to form a "confidence band" for the unknown distribution function, as we shall explain in this section.

A test for goodness of fit usually involves examining a random sample from some unknown distribution in order to test the null hypothesis that the unknown distribution function is in fact a known, specified function. That is, the null hypothesis specifies some distribution function $F^*(x)$, perhaps graphically as in Figure 1, or perhaps as a mathematical function which may be graphed. A random sample $X_1, X_2, \ldots, X_n$ is then drawn from some population, and is compared with $F^*(x)$ in some way to see if it is reasonable to say that $F^*(x)$ is the true distribution function of the random sample.

One logical way of comparing the random sample with $F^*(x)$ is by means of the empirical distribution function $S(x)$, which was defined by Definition 2.2.1 as the fraction of X_i's which are less than or equal to x for each x, $-\infty < x < +\infty$. We learned in Section 2.2 that the empirical distribution function $S(x)$ is useful as an estimator of $F(x)$, the unknown distribution

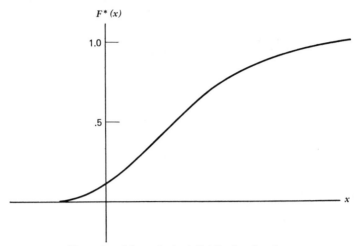

Figure 1. A hypothesized distribution function.

function of the X_i's. So we can compare the empirical distribution function $S(x)$ with the hypothesized distribution function $F^*(x)$ to see if there is good agreement. If there is not good agreement, then we may reject the null hypothesis and conclude that the true but unknown distribution function, $F(x)$, is in fact not given by the function $F^*(x)$ in the null hypothesis.

But what sort of test statistic can we use as a measure of the discrepancy between $S(x)$ and $F^*(x)$? One of the simplest measures imaginable is the largest distance between the two graphs $S(x)$ and $F^*(x)$, measured in a vertical direction. This is the statistic suggested by Kolmogorov (1933). That is, if $F^*(x)$ is given by Figure 1, and a random sample of size five is drawn from the population, then the empirical distribution function $S(x)$

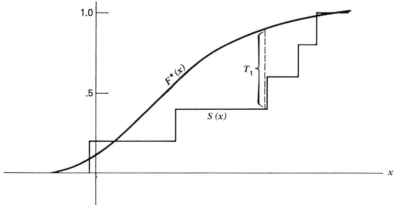

Figure 2. The hypothesized distribution function $F^*(x)$, the empirical distribution function $S(x)$, and Kolmogorov's statistic T_1.

may be drawn on the same graph along with $F^*(x)$, as is shown in Figure 2. If $F^*(x)$ and $S(x)$ are as given, then the maximum vertical distance between the two graphs occurs just before the third step of $S(x)$. This distance is about .5 in Figure 2, and therefore the Kolmogorov statistic T_1 equals .5 in this case. Large values of T_1 as determined by Table 14 lead to rejection of $F^*(x)$ as a reasonable approximation to the unknown true distribution function $F(x)$.

The Kolmogorov test may be preferred over the chi-square test for goodness of fit if the sample size is small, because the Kolmogorov test is exact even for small samples, while the chi-square test assumes that the number of observations is large enough so that the chi-square distribution provides a good approximation as the distribution of the test statistic. There is controversy over which test is the more powerful, but the general feeling seems to be that the Kolmogorov test is probably more powerful than the chi-square test in most situations. For further comparisons see a recent paper by Slakter (1965).

The title of this chapter is "Statistics of the Kolmogorov-Smirnov Type." Statistics that are functions of the vertical distance between $S(x)$ and $F^*(x)$ are considered to be Kolmogorov-type statistics. Statistics which are functions of the vertical distance between two empirical distribution functions are of the Smirnov type. This entire chapter is concerned with statistics that are determined only by the vertical distances between distribution functions, either hypothesized or empirical distribution functions.

The Kolmogorov Goodness of Fit Test

DATA. The data consist of a random sample $X_1, X_2, \ldots, X_n$ of size n associated with some unknown distribution function, denoted by $F(x)$.

ASSUMPTIONS
1. The sample is a random sample.
2. If the hypothesized distribution function, $F^*(x)$ in H_0 below, is continuous the test is exact. Otherwise the test is conservative (Noether, 1967a).

HYPOTHESES. Let $F^*(x)$ be a completely specified hypothesized distribution function.

A. (two-sided test)
$$H_0: F(x) = F^*(x) \text{ for all } x \text{ from } -\infty \text{ to } +\infty$$
$$H_1: F(x) \neq F^*(x) \text{ for at least one value of } x$$

B. (one-sided test)
$$H_0: F(x) \geq F^*(x) \text{ for all } x \text{ from } -\infty \text{ to } +\infty$$
$$H_1: F(x) < F^*(x) \text{ for at least one value of } x$$

C. (one-sided test)

$$H_0: F(x) \leq F^*(x) \text{ for all } x \text{ from } -\infty \text{ to } +\infty$$
$$H_1: F(x) > F^*(x) \text{ for at least one value of } x$$

TEST STATISTIC. Let $S(x)$ be the empirical distribution function based on the random sample $X_1, X_2, \ldots, X_n$. The test statistic is defined differently for the three different sets of hypotheses, A, B, and C.

A. (two-sided test) Let the test statistic T_1 be the greatest (denoted by "sup" for supremum) vertical distance between $S(x)$ and $F^*(x)$. In symbols we say

(1) $$T_1 = \sup_x |F^*(x) - S(x)|$$

which is read "T_1 equals the supremum, over all x, of the absolute value of the difference $F^*(x)$ minus $S(x)$."

B. (one-sided test) Denote this test statistic by T_1^+ and let it equal the greatest vertical distance attained by $F^*(x)$ above $S(x)$. That is,

(2) $$T_1^+ = \sup_x [F^*(x) - S(x)]$$

which is similar to T_1 except that we consider only the greatest difference where the function $F^*(x)$ is above the function $S(x)$.

C. (one-sided test) For this test use the test statistic T_1^-, defined as the greatest vertical distance attained by $S(x)$ above $F^*(x)$. Formally this becomes

(3) $$T_1^- = \sup_x [S(x) - F^*(x)]$$

DECISION RULE. Reject H_0 at the level of significance α if the appropriate test statistic, T_1, T_1^+, or T_1^- exceeds the $1 - \alpha$ quantile $w_{1-\alpha}$ as given by Table 14. Quantiles are provided for use in two-sided tests at $\alpha = .20, .10, .05, .02,$ and $.01$, and for one-sided tests at α values of $.10, .05, .025, .01,$ and $.005$. The tables are exact for $n \leq 20$ in the two-sided test. For the one-sided test, and for $n > 20$ in the two-sided test the tables provide good approximations which are exact in most cases. The approximation for n greater than 40 is based on the asymptotic distribution of the test statistics, and is not very accurate until n becomes large.

Example 1. A random sample of size 10 is obtained: $X_1 = .621$, $X_2 = .503$, $X_3 = .203$, $X_4 = .477$, $X_5 = .710$, $X_6 = .581$, $X_7 = .329$, $X_8 = .480$, $X_9 = .554$, $X_{10} = .382$. The null hypothesis is that the distribution function is the uniform distribution function whose graph is given in Figure 3. The mathematical expression for the hypothesized distribution function is

$$F^*(x) = 0 \quad \text{if} \quad x < 0$$
(4) $$ = x \quad \text{if} \quad 0 \leq x < 1$$
$$ = 1 \quad \text{if} \quad 1 \leq x$$

Formally, the hypotheses are given by

$$H_0: F(x) = F^*(x) \text{ for all } x$$
$$H_1: F(x) \neq F^*(x) \text{ for at least one } x$$

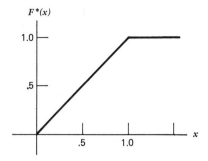

Figure 3. The hypothesized distribution function.

where $F(x)$ is the unknown distribution function common to the X_i's, and where $F^*(x)$ is given by (4) above.

The two-sided Kolmogorov test for goodness of fit is used. The critical region of size $\alpha = .05$ corresponds to values of T_1 greater than the .95 quantile .409, obtained from Table 14 for $n = 10$. The value of T_1 is obtained by graphing the empirical distribution function $S(x)$ on top of the hypothesized distribution function $F^*(x)$, as shown in Figure 4. The largest vertical distance separating the two graphs in Figure 4 is .290, which occurs at $x = .710$ because $S(.710) = 1.000$ and $F^*(.710) = .710$. In other words,

$$T_1 = \sup_x |F^*(x) - S(x)|$$
$$= |F^*(.710) - S(.710)|$$
$$= .290$$

Since $T_1 = .290$ is less than .409, the null hypothesis is accepted. The critical level $\hat{\alpha}$ is seen, from Table 14, to be somewhat larger than .20.

If we had wished to test the null hypothesis

$$H_0: F(x) \geq F^*(x) \text{ for all } x$$

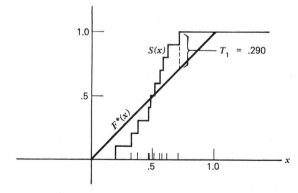

Figure 4. Graphs of $F^*(x)$ and $S(x)$, with T_1.

against the one-sided alternative

$$H_1 : F(x) < F^*(x) \text{ for some } x$$

then the test statistic T_1^+ would have been used. The decision rule is to reject H_0 at $\alpha = .05$ if T_1^+ exceeds the .95 quantile for a one-sided test, .369, as given by Table 14 for $n = 10$. The value for T_1^+ in this case is computed just to the left of the second jump of $S(x)$;

$$T_1^+ = \sup_x \, [F^*(x) - S(x)]$$

$$= F^*(.3289) - S(.3289)$$

$$= .3289 - .100$$

$$= .2289$$

To be more precise, we should say that T_1^+ equals $.228999 \ldots$, which is rounded off to .229. The end result is the same.

A one-sided test in the other direction would have resulted in

$$T_1^- = \sup_X \, [S(x) - F^*(x)]$$

$$= S(.710) - F^*(.710)$$

$$= 1.000 - .710$$

$$= .290$$

The two-sided test is the appropriate test for this situation. The one-sided tests were presented merely to show how their test statistics are evaluated. In general, of course, the two-sided test statistic T_1 always equals the larger of the two one-sided test statistics T_1^+ and T_1^-.

COMMENT. One of the most valuable features of the Kolmogorov two-sided test statistic is that its $1 - \alpha$ quantile $w_{1-\alpha}$ may be used to form a confidence band for the true unknown distribution function. Recall that in finding a confidence interval for some unknown parameter, we first drew a random sample, and then from that sample we computed an upper value U and a lower value L which contained the unknown parameter between them with a certain probability $1 - \alpha$, called the confidence coefficient. It would be convenient if we could do the same thing to obtain a "confidence band" within which the entire unknown distribution function would lie, with probability $1 - \alpha$. Then we could draw a random sample for some population whose distribution function is completely unknown, and we could place some bounds on a graph and make the statement that the unknown distribution function lies entirely within those bounds, with some probability $1 - \alpha$ that the statement is correct.

Confidence Band for the Population Distribution Function

DATA. The data consist of a random sample $X_1, X_2, \ldots, X_n$ of size n associated with some unknown distribution function, denoted by $F(x)$.

ASSUMPTIONS
1. The sample is a random sample.
2. For the confidence coefficient to be exact, the random variables should be continuous. If the random variables are discrete, the confidence band is conservative; that is, the true but unknown confidence coefficient is greater than the stated one.

METHOD. Draw a graph of the empirical distribution function $S(x)$, based on the random sample. To form a confidence band with a confidence coefficient $1 - \alpha$, find the $1 - \alpha$ quantile of the Kolmogorov test statistic from Table 14, for the two-sided test (if a two-sided confidence band is desired) and for the appropriate sample size n. Let $w_{1-\alpha}$ denote this quantile. Draw a graph above $S(x)$ a distance $w_{1-\alpha}$ and call this graph $U(x)$. Draw a second graph a distance $w_{1-\alpha}$ below $S(x)$ and call this second graph $L(x)$. Then $U(x)$ and $L(x)$ form the upper and lower boundaries, respectively, of a $1 - \alpha$ confidence band which contains the unknown $F(x)$ completely within its boundaries.

There is no reason for $U(x)$ to be drawn above 1.0, even though $S(x) + w_{1-\alpha}$ might exceed 1.0, because we know that no distribution function ever exceeds 1.0. For the same reason $L(x)$ should not extend below the horizontal axis. The formal mathematical definitions of $U(x)$ and $L(x)$ are as follows:

(5)
$$U(x) = S(x) + w_{1-\alpha} \quad \text{if} \quad S(x) + w_{1-\alpha} \leq 1$$
$$U(x) = 1.0 \quad \text{if} \quad S(x) + w_{1-\alpha} > 1$$

(6)
$$L(x) = S(x) - w_{1-\alpha} \quad \text{if} \quad S(x) - w_{1-\alpha} \geq 0$$
$$L(x) = 0 \quad \text{if} \quad S(x) - w_{1-\alpha} < 0$$

The resulting probability statement is

(7)
$$P[L(x) \leq F(x) \leq U(x), \text{ for all } x] \geq 1 - \alpha$$

where the last inequality applies only when the random variables are discrete.

Example 2. Suppose we wish to form a 90% confidence band for an unknown distribution function $F(x)$. A random sample of size 20 was obtained from the population with that distribution function. The results were ordered from smallest to largest for convenience.

| 16.7 | 17.4 | 18.1 | 18.2 | 18.8 | 19.3 | 22.4 | 22.4 | 24.0 | 24.7 |
| 25.9 | 27.0 | 35.1 | 35.8 | 36.5 | 37.6 | 39.8 | 42.1 | 43.2 | 46.2 |

The .90 quantile is found from Table 14 to equal $w_{.90} = .265$ for $n = 20$. The confidence band is $S(x) \pm .265$ as long as the band is between 0 and 1.

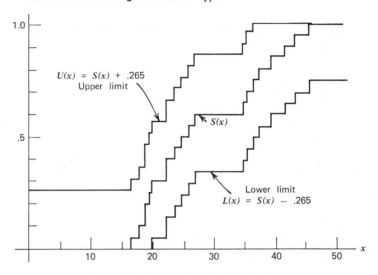

Figure 5. A confidence band for $F(x)$.

Figure 5 shows $S(x)$, $U(x)$, and $L(x)$. The statement "$F(x)$ lies entirely between $U(x)$ and $L(x)$" is true with probability .90.

Theory.   The derivation of the distribution of the Kolmogorov statistic is complicated and is not presented here. We shall only mention some basic papers on the subject.

The asymptotic distribution of the two-sided statistic T_1 was found by Kolmogorov (1933) and was tabulated by Smirnov (1948). The asymptotic distributions of the one-sided statistics T_1^+ and T_1^- were obtained by Smirnov (1939). The exact distribution of the test statistics for finite (small) sample sizes was studied by Wald and Wolfowitz (1939), and tabulated by Massey (1950a). The distribution function of T_1^+ for finite sample sizes was derived by Birnbaum and Tingey (1951), and comparisons were made between the exact quantiles obtained from their distribution function and the asymptotic quantiles given by Smirnov (1939, 1948). It was found that use of the asymptotic quantiles leads to a conservative test.

The two-sided Kolmogorov test has the desirable property of being consistent against *all* differences between $F(x)$ and $F^*(x)$, the true and hypothesized distribution functions. However, it is biased for finite sample sizes (Massey, 1950b). A lower bound for the power of the two-sided test is given by Massey (1950b). The greatest lower bound for the power, under a certain class of alternative hypotheses was obtained by

Birnbaum (1953), and another greatest lower bound for the power, under a different class of alternative hypotheses, was obtained by Lee (1966).

Lee (1966) also compared the exact power of the Kolmogorov test with a standard parametric test. Some of his findings will now be presented. A random sample of size five was considered drawn from a population with the normal distribution with mean μ_1 and variance σ^2. The null hypothesis is that the distribution is normal with mean μ_0, not μ_1, but with the same variance. The power of the Kolmogorov test was obtained for several differences between μ_0 and μ_1, relative to the size of the standard deviation σ, and compared to the "normal test," which is the most powerful parametric test that exists for that situation. Even under these most unfavorable conditions, the power of the Kolmogorov test is not much worse than the normal test.

Power when $\alpha = .10$			$\alpha = .05$		$\alpha = .01$	
$\mu_0 - \mu_1$	*Kolm.* *Test*	*Normal* *Test*	*Kolm.* *Test*	*Normal* *Test*	*Kolm.* *Test*	*Normal* *Test*
σ						
.5	31.65%	43.48%	25.09%	29.91%	8.66%	11.35%
1.0	74.95	82.99	61.57	72.27	33.16	46.41
1.5	94.93	98.09	89.50	95.58	68.19	84.80
2.0	99.50	99.93	97.89	99.77	90.53	98.40

Other power comparisons were made by van der Waerden (1953). It should be noted that if a deviation from the hypothesized variance existed, rather than a deviation from the hypothesized mean as above, then the normal test is powerless to detect the difference, and the Kolmogorov test would be more powerful than the normal test. More recent power studies have been made by Suzuki (1968), who includes a more extensive set of references, and Shapiro, Wilk, and Chen (1968), who present some more powerful tests of goodness of fit when the null hypothesis specifies a normal distribution. The latter article is of interest because it also discusses some goodness of fit tests which may be used to test the composite hypothesis of normality, where the mean μ and variance σ^2 are not specified, as contrasted with the simple hypothesis required for the Kolmogorov test, with μ and σ^2 specified. The Kolmogorov test has recently been modified by Lilliefors (1967) so that it may be used to test the composite hypothesis of normality, without specifying μ and σ^2. This very useful modification is described as follows.

The Lilliefors Test

DATA. The data consist of a random sample $X_1, X_2, \ldots, X_n$ of size n associated with some unknown distribution function, denoted by $F(x)$. Compute the sample mean

$$(8) \qquad \bar{X} = \frac{1}{n} \sum_{i=1}^{n} X_i$$

for use as an estimate of μ, and compute

$$(9) \qquad s = \sqrt{\frac{1}{n-1} \sum_{i=1}^{n} (X_i - \bar{X})^2}$$

as an estimate of σ. Then compute the "normalized" sample values Z_i, defined by

$$(10) \qquad Z_i = \frac{X_i - \bar{X}}{s} \qquad i = 1, 2, \ldots, n$$

The test statistic is computed from the Z_i's instead of from the original random sample.

ASSUMPTIONS
1. The sample is a random sample.

HYPOTHESES
H_0: The random sample has the normal distribution, with unspecified mean and variance.
H_1: The distribution function of the X_i's is nonnormal.

TEST STATISTIC. Ordinarily the test statistic is the usual two-sided Kolmogorov test statistic, defined as the maximum vertical distance between the empirical distribution function of the X_i's and the normal distribution function with mean $\bar{X}$ and standard deviation s, as given by (8) and (9). However, the following method of computing the test statistic is slightly easier, and is equivalent to the method indicated above.

Draw a graph of the standard normal distribution function, and call it $F^*(x)$. Table 1 may be of assistance. Also draw a graph of the empirical distribution function of the normalized sample, the Z_i's defined by Equation (10), using the same set of coordinates as was used above for $F^*(x)$. Find the maximum vertical distance between the two graphs, $F^*(x)$ and the empirical distribution function which we shall call $S(x)$. This distance is the test statistic. That is, the Lilliefors test statistic T_2 is defined by

$$(11) \qquad T_2 = \sup_x \left| F^*(x) - S(x) \right|$$

The difference between T_2 and the Kolmogorov test statistic T_1 is that the empirical distribution function $S(x)$ in (11) was obtained from the normalized sample, while $S(x)$ in the Kolmogorov test was based on the original unadjusted observations.

DECISION RULE. Reject H_0 at the approximate level of significance α if T_2 exceeds the $1 - \alpha$ quantile as given in Table 15.

Example 3. The same data used to illustrate the chi-square test for normality in Example 4.5.3 will be used to illustrate the Lilliefors test.

Fifty two-digit numbers were drawn at random from a telephone book. Although the random variable sampled is clearly discrete, we may still justify testing for normality, if we realize that acceptance of the null hypothesis of normality does not imply that the random variable has the normal distribution and is therefore continuous, but merely indicates that the difference between the normal distribution function and the true distribution function is sufficiently insignificant so as to remain undetected.

The numbers X_i are arranged from smallest to largest, and converted to Z_i by subtracting $\overline{X} = 55.04$, and dividing by $s = 19.00$, as computed from Equations (8) and (9).

X_i	Z_i	X_i	Z_i	X_i	Z_i	X_i	Z_i	X_i	Z_i
23	-1.69	36	-1.00	54	$-.05$	61	.31	73	.95
23	-1.69	37	$-.95$	54	$-.05$	61	.31	73	.95
24	-1.63	40	$-.79$	56	.05	62	.37	74	1.00
27	-1.48	42	$-.69$	57	.10	63	.42	75	1.05
29	-1.37	43	$-.63$	57	.10	64	.47	77	1.16
31	-1.27	43	$-.63$	58	.16	65	.52	81	1.37
32	-1.21	44	$-.58$	58	.16	66	.58	87	1.68
33	-1.16	45	$-.53$	58	.16	68	.68	89	1.79
33	-1.16	48	$-.37$	58	.16	68	.68	93	2.00
35	-1.05	48	$-.37$	59	.21	70	.79	97	2.21

The null hypothesis of normality is tested with the Lilliefors test statistic

$$T_2 = \sup_x |F^*(x) - S(x)|$$

where $F^*(x)$ is the standard normal distribution function, and where $S(x)$ is the empirical distribution function of the Z_i's. Figure 6 presents the graphs of $F^*(x)$ and $S(x)$. The maximum vertical distance between $F^*(x)$ and $S(x)$ is seen from Figure 6 to occur just to the left of $x = -.05$, where $S(x)$ equals .40, $F^*(x)$ equals .48, and so T_2 equals .08. The vertical distance between the two curves equals .08 at other points too, such as at $x = +.05$ and $x = .10$. But at no point does the distance separating the two curves exceed .08.

The Lilliefors test calls for rejection of H_0 at $\alpha = .05$ if T_2 exceeds its .95 quantile, which is given by Table 15 as

$$w_{.95} = \frac{.886}{\sqrt{n}} = \frac{.886}{\sqrt{50}} = .125$$

Because T_2 equals .08, and is less than .125, the null hypothesis is accepted. In fact, the null hypothesis would still be accepted at $\alpha = .20$, because the .80

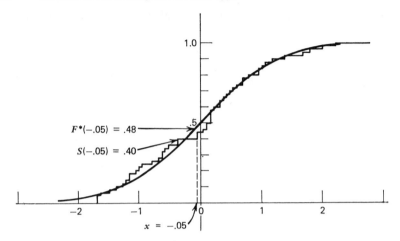

Figure 6. Graphs of $F^*(x)$ and $S(x)$, showing the maximum distance between them.

quantile is found to equal .104. Because Table 15 does not present smaller quantiles, we conclude that the critical level is some value greater than .20. Recall that the chi-square test resulted in about the same conclusion.

Acceptance of the null hypothesis does not mean that the parent population is normal. But it does mean that the normal distribution does not appear to be an unreasonable approximation to the true unknown distribution, and therefore either nonparametric methods or the parametric statistical procedures which assume a normal parent distribution may be appropriate for further testing with these data.

Theory.   One of the principal reasons for presenting the Lilliefors test is to show how the quantiles in Table 15 were obtained. The problem of finding the distribution of T_2, so that the Kolmogorov test could be used to test the composite hypothesis of normality with unspecified mean and variance, had been too difficult to solve analytically. Therefore, Lilliefors used a high speed computer and random numbers to obtain an approximate solution. This same technique, described below, may be used to obtain an approximate solution to almost any problem in statistical inference.

Recall that to develop a statistical hypothesis test one must first invent a test statistic that acts as a reasonably sensitive indicator, indicating whether the null hypothesis is true or false. The statistic T_2 satisfies this requirement. Then one must select a certain region of values, corresponding to the critical region, which is unlikely to occur if H_0 is true, but which is more likely to occur if H_0 is false. Large values of T_2 meet this requirement. Then the difficulty comes in trying to find α, the probability of getting a point in the critical region when H_0 is true.

To do this Lilliefors generated random normal deviates on a high speed computer. Random normal deviates, as we mentioned in Section 5.9, are numbers which appear to be observations on independent standard normal random variables, such as the numbers in Table 13. These numbers were grouped into samples of size n, for various values of n. For illustration let us say n equals 8. A simulated sample of size 8 from a standard normal distribution, so that H_0 is true, was obtained from the computer. The sample mean $\bar{X}$ was computed and subtracted, and the result was divided by s as computed by Equation (9) for that sample, to obtain the Z_i values. The empirical distribution function based on those Z_i's was compared with the standard normal distribution function and the maximum vertical difference T_2 was written down. The process was repeated with another set of eight computer generated numbers to obtain another observed value of T_2. In all, over 1000 samples of size 8 were obtained, and over 1000 values of T_2 were computed under the condition that H_0 is true. The empirical distribution function based on those 1000 or more values of T_2 was then used as an approximation to the true but unknown distribution function of T_2. From that empirical distribution function the selected quantiles given in Table 15 for $n = 8$ were obtained. Now a critical region of approximate size α may be specified.

The same procedure was repeated for other sample sizes ranging from $n = 4$ to $n = 30$. To obtain the approximation suggested by Lilliefors for n greater than 30, samples of size 40 were obtained, the quantiles were determined in the manner described above, and then the quantiles were multiplied by $\sqrt{40}$ and given in the table. This procedure is based on the unproved conjecture that T_2 approaches its asymptotic distribution in much the same way that the Kolmogorov statistic is known to approach its limiting asymptotic distribution, as a function of $\sqrt{n}$. The conjecture seems reasonable in view of the approximate quantiles obtained by Lilliefors in Table 15 for various values of n.

Lilliefors (1967) also compared the power of his test with the power of the chi-square test in several nonnormal situations, and found his test to be more powerful in the situations reported.

The Lilliefors test is designed specifically to test only the composite null hypothesis of normality. While this is probably the null hypothesis most frequently tested by goodness of fit tests, general all-purpose goodness of fit tests such as the Kolmogorov test are still indispensable because of their versatility.

Another all-purpose goodness of fit test was suggested by Cramér (1928), von Mises (1931), and Smirnov (1936). It is called the Cramér-von Mises one sample, or goodness of fit, test. Like the Kolmogorov test, the Cramér-von Mises test is suitable only for simple null hypotheses. Also like the Kolmogorov test the Cramér-von Mises test is a function of the vertical distance between $F^*(x)$ and $S(x)$, the hypothesized and the empirical distribution functions. However, unlike the Kolmogorov test, the Cramér-von Mises test not only considers the largest difference, but also considers n differences between the two curves. Intuitively it appears that the Cramér-von Mises test statistic makes more complete use of the data and therefore should be more effective than the Kolmogorov test statistic, but the facts fail either to prove or disprove such intuition. We are presenting the Cramér-von Mises test because it is an alternative to the two-sided Kolmogorov test, and because it may be intuitively more appealing than the Kolmogorov test to some people.

The Cramér-von Mises Goodness of Fit Test

DATA. The data consist of a random sample $X_1, X_2, \ldots, X_n$ of size n associated with some unknown distribution function $F(x)$. Denote the ordered random sample by $X^{(1)} \leq X^{(2)} \leq \ldots \leq X^{(n)}$ where $X^{(i)}$ is the order statistic of rank i.

ASSUMPTIONS
1. The sample is a random sample.

HYPOTHESES. Let $F^*(x)$ be a completely specified hypothesized distribution function. This test requires that $F^*(x)$ be continuous.

$H_0: F(x) = F^*(x)$ for all x from $-\infty$ to $+\infty$.

$H_1: F(x) \neq F^*(x)$ for at least one value of x.

TEST STATISTIC. Let $F^*(X^{(i)})$ be the value of the hypothesized distribution function $F^*(x)$ at the ith smallest number observed in the random sample, for $i = 1, 2, \ldots, n$. The test statistic is a function of those values of $F^*(X^{(i)})$.

$$(12) \qquad T_3 = \frac{1}{12n} + \sum_{i=1}^{n} \left[F^*(X^{(i)}) - \frac{2i-1}{2n} \right]^2$$

That is, the ordinate of $F^*(x)$ is found at each $X^{(i)}$ value in the random sample, and from this is subtracted the quantity $(2i-1)/2n$ which is the average of $S(x)$ just before and just after the jump that occurs at $X^{(i)}$, that is, the average of $(i-1)/n$ and i/n. That difference is squared so that positive differences don't cancel the negative differences, and the results are added together.

DECISION RULE. The quantiles of T_3 are approximated by using the asymptotic distribution function of T_3 as given by Anderson and Darling (1952). The approximate quantiles w_p of T_3 are as follows:

$$w_{.10} = .046 \qquad w_{.50} = .119 \qquad w_{.90} = .347$$
$$w_{.20} = .062 \qquad w_{.60} = .147 \qquad w_{.95} = .461$$
$$w_{.30} = .079 \qquad w_{.70} = .184 \qquad w_{.99} = .743$$
$$w_{.40} = .097 \qquad w_{.80} = .241 \qquad w_{.999} = 1.168$$

Reject H_0 at the approximate level α if T_3 exceeds $w_{1-\alpha}$ as given above, and conclude that the true distribution function differs from the hypothesized distribution function. Otherwise accept H_0. Exact quantiles for finite sample sizes are given by Stephens and Maag (1968).

Example 4. For the same data and hypotheses tested by the two sided Kolmogorov test in Example 1 the following table may be formed.

i	$X^{(i)}$	$F^*(X^{(i)})$	$F^*(X^{(i)}) - \dfrac{2i-1}{2n}$
1	.203	.203	.153
2	.329	.329	.179
3	.382	.382	.132
4	.477	.477	.127
5	.480	.480	.030
6	.503	.503	−.047
7	.554	.554	−.096
8	.581	.581	−.169
9	.621	.621	−.229
10	.710	.710	−.240

The null hypothesis that $F(x)$ is the uniform distribution function will be rejected at $\alpha = .05$ if T_3 exceeds .461. In this case the use of Equation (12) furnishes

$$T_3 = \tfrac{1}{120} + .240 = .248$$

which is less than .461, so H_0 is accepted. The critical level is estimated at about $\hat{\alpha} \cong .19$. Although H_0 is accepted, the Cramér-von Mises test came closer to rejection of H_0 than the Kolmogorov test did in Example 1.

Theory.   The asymptotic distribution of the Cramér-von Mises test statistic is the result of work done by Anderson and Darling (1952) in a paper that is considered a definitive work in the area of Kolmogorov-Smirnov type statistics. The theory is well beyond the scope of this book and is therefore omitted.

Recent studies on the distributions of the Kolmogorov and Cramér-von Mises test statistics are by Stephens (1964, 1965a), Tiku (1965), Suzuki (1967), Cronholm (1968), and Noé and Vandewiele (1968). Exact tables of the latter statistic for finite sample sizes are given by Stephens and Maag (1968). Effects of discrete random variables on the two tests are discussed by Walsh (1960, 1963). Bias and power of the Cramér-von Mises test are examined by Thompson (1966) and Weiss (1966). Relative efficiency of the Kolmogorov test is studied in Gelzer and Pyke (1965), Quade (1965), and Abrahamson (1967).

Goodness of fit for a sample density is discussed by Woodroofe (1967), for a circle by Stephens (1969), and in general by Riedwyl (1967). A different type of confidence interval for the distribution function is introduced by Durbin (1968).

A new test by Lilliefors (1969) tests the composite null hypothesis of an exponential parent distribution. A general discussion of simulation and presimulation days is found in Teichroew (1965).

PROBLEMS

1. Give a 90% confidence band, either graphically or in tabled form, for the distribution function from which the random sample 6.3, 4.2, 4.7, 6.0, 5.7 was obtained.

2. For the data in Problem 1, test the hypothesis that $F(x) = F^*(x)$ for all x, where

$$F^*(x) = 0 \qquad \text{for} \quad x < 4$$

$$= \frac{x - 4}{4} \qquad \text{for} \quad 4 \le x < 8$$

$$= 1 \qquad \text{for} \quad 8 \le x$$

using

(a) the Kolmogorov test
(b) the Cramér-von Mises test

3. For the data in Problem 1, test the null hypothesis of normality.

4. In order to show that the confidence band, given as Equation (7), is valid, show that if $w_{1-\alpha}$ is a $1 - \alpha$ quantile of the statistic sup $|F(x) - S(x)|$, then (7) holds. That is, show that if

$$P[\sup_x |F(x) - S(x)| \le w_{1-\alpha}] \ge 1 - \alpha$$

is true, then it follows that

$$P[S(x) - w_{1-\alpha} \le F(x) \le S(x) + w_{1-\alpha}, \text{ for all } x] \ge 1 - \alpha$$

is also true.

6.2. TESTS FOR TWO INDEPENDENT SAMPLES

The tests presented in this section are useful in situations where two samples are drawn, one from each of two possibly different populations, and the experimenter wishes to determine whether the two distribution functions associated with the two populations are identical or not. While other tests may also be appropriate, such as the median test, the Mann-Whitney test, or the parametric t test, these other tests are sensitive to differences between the two means or medians, but may not detect differences of other types, such as differences in variances, etc. One of the advantages of the two two-sided tests presented in this section is that both tests are consistent against all types of differences that may exist between the two distribution functions.

The first test presented is the Smirnov test (Smirnov, 1939). It is a two-sample version of the Kolmogorov test presented in the previous section, and is sometimes called the Kolmogorov-Smirnov two-sample test while the Kolmogorov test is sometimes called the Kolmogorov-Smirnov one-sample test. The Smirnov test is presented in the one-sided and two-sided versions. Another two-sided test is also presented, the Cramér-von Mises test for two samples, which resembles in form the Cramér-von Mises test for goodness of fit given in the previous section. It is slightly more difficult to compute than the Smirnov test, but it appeals to some people because it appears to make more effective use of the data. Actually there is probably little difference in power between the two tests.

The Smirnov Test

DATA. The data consist of two independent random samples, one of size n, $X_1, X_2, \ldots, X_n$, and the other of size m, $Y_1, Y_2, \ldots, Y_m$. Let $F(x)$ and $G(x)$ represent their respective, unknown, distribution functions.

ASSUMPTIONS
1. The samples are random samples.
2. The two samples are mutually independent.
3. The measurement scale is at least ordinal.
4. For this test to be exact the random variables are assumed to be continuous. If the random variables are discrete, the test is still valid but becomes conservative (Noether, 1967a).

HYPOTHESES
A. (two-sided test)

$$H_0: F(x) = G(x) \text{ for all } x \text{ from } -\infty \text{ to } +\infty.$$

$$H_1: F(x) \neq G(x) \text{ for at least one value of } x.$$

B. (one-sided test)

$$H_0: F(x) \leq G(x) \text{ for all } x \text{ from } -\infty \text{ to } +\infty.$$

$$H_1: F(x) > G(x) \text{ for at least one value of } x.$$

This alternative hypothesis is sometimes stated as, "The X's tend to be *smaller* than the Y's," if the X's and Y's differ only by a location parameter (means or medians).

C. (one-sided test)

$$H_0: F(x) \geq G(x) \text{ for all } x \text{ from } -\infty \text{ to } +\infty.$$

$$H_1: F(x) < G(x) \text{ for at least one value of } x.$$

This is the one-sided test to use if it is suspected that the X's might be shifted to the right (i.e., larger) of the Y's.

TEST STATISTIC. Let $S_1(x)$ be the empirical distribution function based on the random sample $X_1, X_2, \ldots, X_n$, and let $S_2(x)$ be the empirical distribution function based on the other random sample $Y_1, Y_2, \ldots, Y_m$. The test statistic is defined differently for the three different sets of hypotheses.

A. (two-sided test) Define the test statistic T_1 as the greatest vertical distance between the two empirical distribution functions.

(1)
$$T_1 = \sup_x |S_1(x) - S_2(x)|$$

B. (one-sided test) Denote the test statistic by T_1^+, and let it equal the greatest vertical distance attained by $S_1(x)$ above $S_2(x)$.

(2)
$$T_1^+ = \sup_x [S_1(x) - S_2(x)]$$

C. (one-sided test) For the one-sided hypotheses in C above, let the test statistic, denoted by T_1^-, be the greatest vertical distance attained by $S_2(x)$ above $S_1(x)$.

(3)
$$T_1^- = \sup_x [S_2(x) - S_1(x)]$$

DECISION RULE. Reject H_0 at the level of significance α if the appropriate test statistic T_1, T_1^+, or T_1^- as the case may be, exceeds its $1 - \alpha$ quantile as given by Table 16 if $n = m$, and by Table 17 if $n \neq m$. For the one-sided tests those tables give the .90, .95, .975, .99, and .995 quantiles. For the two-sided test the .80, .90, .95, .98, and .99 quantiles are furnished. The large sample approximations given at the end of the tables may be used for the sample sizes not covered by the tables.

Example 1.   A random sample of size 9, $X_1, \ldots, X_9$ is obtained from one population, and a random sample of size 15, $Y_1, \ldots, Y_{15}$ is obtained from a second population. The null hypothesis is that the two populations have identical distribution functions. If the respective distribution functions are denoted by $F(x)$ and $G(x)$, then the null hypothesis may be written as

$$H_0: F(x) = G(x) \text{ for all } x \text{ from } -\infty \text{ to } +\infty$$

The alternative hypothesis may be stated as

$$H_1: F(x) \neq G(x) \text{ for at least one value of } x$$

The two samples are ordered from smallest to largest for convenience, and their values, along with other pertinent information about their empirical distribution functions, are given below.

X_i	Y_i	$S_1(x) - S_2(x)$	X_i	Y_i	$S_1(x) - S_2(x)$
	5.2	$0 - 1/15 = -1/15$	9.8		$5/9 - 8/15 = 1/45$
	5.7	$0 - 2/15 = -2/15$	9.9		$6/9 - 8/15 = 2/15$
	5.9	$0 - 3/15 = -1/5$	10.1		$7/9 - 8/15 = 11/45$
	6.5	$0 - 4/15 = -4/15$	10.6		$8/9 - 8/15 = 16/45$
	6.8	$0 - 5/15 = -1/3$		10.8	$8/9 - 9/15 = 13/45$
7.6		$1/9 - 5/15 = -2/9$		11.2	$1 - 9/15 = 2/5$
	8.2	$1/9 - 6/15 = -13/45$		11.3	$1 - 10/15 = 1/3$
8.4		$2/9 - 6/15 = -8/45$		11.5	$1 - 11/15 = 4/15$
8.6		$3/9 - 6/15 = -1/15$		12.3	$1 - 12/15 = 1/5$
8.7		$4/9 - 6/15 = 2/45$		12.5	$1 - 13/15 = 2/15$
	9.1	$4/9 - 7/15 = -1/45$		13.4	$1 - 14/15 = 1/15$
9.3		$5/9 - 7/15 = 4/45$		14.6	$1 - 1 = 0$

The test statistic for the two-sided test is given by Equation (1) as

$$T_1 = \sup_x |S_1(x) - S_2(x)|$$

$$= 2/5 = .400$$

the largest absolute difference between $S_1(x)$ and $S_2(x)$, which happens to occur between $x = 11.2$ and $x = 11.3$. The value of .400 for T_1 could also have been determined graphically, by drawing graphs of $S_1(x)$ and $S_2(x)$ on the same coordinate axes. From the graphs one can easily see that the difference $S_1(x) - S_2(x)$ changes only at those observed values $x = X_i$ or $x = Y_j$, and that is why it is sufficient to compute $S_1(x) - S_2(x)$ only at the observed sample values, as was done above.

From Table 17 we see that the .95 quantile of T_1, for the two-sided test and for $n = 9 = N_1$ and $m = 15 = N_2$, is given as $w_{.95} = 8/15$. For these data T_1 equals 2/5 or 6/15. Therefore H_0 is accepted at the .05 level. From the table, the critical level $\hat{\alpha}$ may be estimated as slightly larger than .20.

For the sake of comparison, the approximate .95 quantile based on the asymptotic distribution is found to be

$$w_{.95} \cong 1.36 \sqrt{\frac{m+n}{mn}} = .573$$

which is slightly larger than the exact value of $8/15 = .533$. This illustrates the tendency of the asymptotic approximation to furnish a conservative test.

It should be noted that many of the calculations performed in this example could have been eliminated, because by either an inspection of the data or a preliminary sketch of $S_1(x)$ and $S_2(x)$ many of the values of X_i and Y_j may be seen to be unlikely candidates for yielding the maximum value of $|S_1(x) - S_2(x)|$, and therefore may be ignored in favor of the more likely values of X_i and Y_j.

If a one-sided test had been appropriate instead of the two-sided test, the statistics

$$T_1^+ = \sup_x \, [S_1(x) - S_2(x)] = 2/5 = .400$$

for the set B of hypotheses, and

$$T_1^- = \sup_x \, [S_2(x) - S_1(x)] = 1/3 = .333$$

for the set C of hypotheses are easily determined from the above table of data. The critical levels for both of the one sided tests are seen from Table 17 to be greater than .10.

Theory.   Although it may not be apparent at first, the statistics T_1, T_1^+ and T_1^- depend only on the order of the X's and Y's in the ordered combined sample of X's and Y's, and do not require actual knowledge of the numerical values of the observations. To illustrate this, suppose there are 3 X's and 2 Y's. There are $\binom{5}{2} = 10$ different ordered arrangements of the combined sample. These arrangements are given below, along with the values of T_1, T_1^+ and T_1^- for each ordered arrangement.

Arrangement	T_1	T_1^+	T_1^-	Arrangement	T_1	T_1^+	T_1^-
$X < X < X < Y < Y$	1	1	0	$X < Y < X < Y < X$	$\frac{1}{3}$	$\frac{1}{3}$	$\frac{1}{3}$
$X < X < Y < X < Y$	$\frac{2}{3}$	$\frac{2}{3}$	0	$Y < X < X < Y < X$	$\frac{1}{2}$	$\frac{1}{6}$	$\frac{1}{2}$
$X < Y < X < X < Y$	$\frac{1}{2}$	$\frac{1}{2}$	$\frac{1}{6}$	$X < Y < Y < X < X$	$\frac{2}{3}$	$\frac{1}{3}$	$\frac{2}{3}$
$Y < X < X < X < Y$	$\frac{1}{2}$	$\frac{1}{2}$	$\frac{1}{2}$	$Y < X < Y < X < X$	$\frac{2}{3}$	0	$\frac{2}{3}$
$X < X < Y < Y < X$	$\frac{2}{3}$	$\frac{2}{3}$	$\frac{1}{3}$	$Y < Y < X < X < X$	1	0	1

If the null hypothesis in the two-sided test is true, then the two distribution functions are equal and each ordered arrangement is equally likely, under the assumption of continuous random variables. This same point was discussed more thoroughly in connection with the Mann-Whitney test in Section 5.3. Therefore in the two-sided test the probability associated with each ordered arrangement is given by

(4) $$\text{probability} = \frac{1}{\binom{m+n}{n}} = \frac{1}{\binom{5}{3}} = \frac{1}{10}$$

and from this the following probability distributions can be deduced.

$$P(T_1 = \tfrac{1}{3}) = \tfrac{1}{10} \qquad P(T_1^+ = 0) = \tfrac{1}{5} \qquad P(T_1^- = 0) = \tfrac{1}{5}$$

$$P(T_1 = \tfrac{1}{2}) = \tfrac{3}{10} \qquad P(T_1^+ = \tfrac{1}{6}) = \tfrac{1}{10} \qquad P(T_1^- = \tfrac{1}{6}) = \tfrac{1}{10}$$

$$P(T_1 = \tfrac{2}{3}) = \tfrac{2}{5} \qquad P(T_1^+ = \tfrac{1}{3}) = \tfrac{1}{5} \qquad P(T_1^- = \tfrac{1}{3}) = \tfrac{1}{5}$$

$$P(T_1 = 1) = \tfrac{1}{5} \qquad P(T_1^+ = \tfrac{1}{2}) = \tfrac{1}{5} \qquad P(T_1^- = \tfrac{1}{2}) = \tfrac{1}{5}$$

$$P(T_1^+ = \tfrac{2}{3}) = \tfrac{1}{5} \qquad P(T_1^- = \tfrac{2}{3}) = \tfrac{1}{5}$$

$$P(T_1^+ = 1) = \tfrac{1}{10} \qquad P(T_1^- = 1) = \tfrac{1}{10}$$

It is no coincidence that the distributions of T_1^+ and T_1^- are identical with each other for $n = 3$ and $m = 2$. They are identical with each other for all choices of n and m. However, the space-saving technique used in Tables 16 and 17 of stating that the $1 - \alpha$ quantile of T_1 in the two-sided test equals the $1 - \alpha/2$ quantile of T_1^+ in the one-sided test is a valid technique only if α is small. Notice for example in the above illustration that $P(T_1 \geq 1)$ equals twice $P(T_1^+ \geq 1)$, and $P(T_1 \geq 2/3)$ equals twice $P(T_1^+ \geq 2/3)$, but $P(T_1 \geq 1/2)$ does not equal twice $P(T_1^+ \geq 1/2)$.

The null distribution (that is, the distribution when H_0 is true) in the one-sided tests is also found in the manner described above, because under the one-sided null hypotheses the size of the critical region is a maximum when $F(x)$ is identical with $G(x)$. If the two samples are of equal size, it is not necessary to use the above method to find the upper quantiles, because the distribution functions for T_1, T_1^+, and T_1^- were derived as a function of the sample size n by Gnedenko and Korolyuk (1951). The derivation of these distribution functions is interesting, and it is within the presumed mathematical grasp of the reader, but its length precludes its presentation here. The reader is referred to Fisz (1963) for a readable presentation of the derivation.

For samples of unequal size the method of finding quantiles is essentially as illustrated above. However, many refinements using path counting methods have simplified the bookkeeping enough so that extensive tables exist, although not in published form. See Steck (1969) for references on these tables, as well as for a general discussion of the Smirnov test. Kim (1969) gives some closer approximations to the exact quantiles when exact tables are not available.

A modification of the Smirnov test was suggested by Tsao (1954), so that the test may be applied to truncated samples. That is, perhaps only the X's and Y's less than $X^{(r)}$ are observed, as sometimes happens in life testing experiments. The Smirnov test may then be applied to the truncated samples, with the aid of tables derived recursively by Tsao (1954). The distribution functions of Tsao's statistics were derived analytically by Conover (1967a). Extensions of the Smirnov test to three or more samples are presented in the next section.

Just as the Smirnov test is a two-sample version of the Kolmogorov goodness of fit test, the next test is a two-sample version of the Cramér-von Mises goodness of fit test. This test is two-sided only, and involves slightly more calculations than the Smirnov test does.

The Cramér-von Mises Two-Sample Test

DATA. The data consist of two independent random samples, $X_1, \ldots, X_n$ and $Y_1, \ldots, Y_m$, with unknown distribution functions $F(x)$ and $G(x)$ respectively.

ASSUMPTIONS
1. The samples are random samples, independent of each other.
2. The measurement scale is at least ordinal.
3. The random variables are continuous. If the random variables are actually discrete the test is likely to be conservative.

HYPOTHESES

$$H_0 : F(x) = G(x) \text{ for all } x \text{ from } -\infty \text{ to } +\infty.$$

$$H_1 : F(x) \neq G(x) \text{ for at least one value of } x.$$

TEST STATISTIC. Let $S_1(x)$ and $S_2(x)$ be the empirical distribution functions of the two samples. The test statistic T_2 is defined as

$$(5) \qquad T_2 = \frac{mn}{(m+n)^2} \sum_{\substack{x=X_i \\ x=Y_j}} [S_1(x) - S_2(x)]^2$$

where the squared difference in the summation is computed at each X_i and at each Y_j. Perhaps it is clearer to write the test statistic as

$$(6) \qquad T_2 = \frac{mn}{(m+n)^2} \left\{ \sum_{i=1}^{n} [S_1(X_i) - S_2(X_i)]^2 + \sum_{j=1}^{m} [S_1(Y_j) - S_2(Y_j)]^2 \right\}$$

An equivalent form for T_2 may be obtained by letting $R(X^{(i)})$ and $R(Y^{(j)})$ be the ranks, in the combined ordered sample, of the ith smallest of the X's denoted by $X^{(i)}$, and the jth smallest of the Y's denoted by $Y^{(i)}$, respectively. Then we can write (6) as

$$(7) \qquad T_2 = \frac{mn}{(m+n)^2} \left\{ \sum_{i=1}^{n} \left[\frac{R(X^{(i)})}{m} - i \frac{n+m}{nm} \right]^2 + \sum_{j=1}^{m} \left[\frac{R(Y^{(i)})}{n} - j \frac{n+m}{nm} \right]^2 \right\}$$

If $n = m$, (7) reduces to

$$(8) \qquad T_2 = \frac{1}{4n^2} \left\{ \sum_{i=1}^{n} [R(X^{(i)}) - 2i]^2 + \sum_{j=1}^{m} [R(Y^{(j)}) - 2j]^2 \right\}$$

DECISION RULE. Reject H_0 at the approximate level α if T_2 exceeds the $1 - \alpha$ quantile $w_{1-\alpha}$ as given below. These quantiles are approximations based on the asymptotic distribution, valid for large m and n, but are considered fairly accurate even if the sample sizes are small (Burr, 1964).

$$w_{.10} = .046 \qquad w_{.50} = .119 \qquad w_{.90} = .347$$
$$w_{.20} = .062 \qquad w_{.60} = .147 \qquad w_{.95} = .461$$
$$w_{.30} = .079 \qquad w_{.70} = .184 \qquad w_{.99} = .743$$
$$w_{.40} = .097 \qquad w_{.80} = .241 \qquad w_{.999} = 1.168$$

These values were taken from Anderson and Darling (1952), and are the same quantiles that are used in the Cramér-von Mises goodness of fit test given in the previous section. Exact quantiles for $n + m \leq 17$ are given by Burr (1964).

Example 2. From the data of Example 1 the test statistic T_2 may be computed by first finding

$$\sum_{i=1}^{9} [S_1(X_i) - S_2(X_i)]^2 = .459$$

and

$$\sum_{j=1}^{15} [S_1(Y_j) - S_2(Y_j)]^2 = .657$$

Then from (6) we have

$$T_2 = \frac{mn}{(m+n)^2} \left\{ \sum_{i=1}^{n} [S_1(X_i) - S_2(X_i)]^2 + \sum_{j=1}^{m} [S_1(Y_j) - S_2(Y_j)]^2 \right\}$$
$$= \frac{(15)(9)}{(24)^2} (.459 + .657)$$
$$= .262$$

The null hypothesis of identical distribution functions is accepted at $\alpha = .05$, because $T = .262$ is less than $w_{.95} = .461$ as given above. The critical level is seen to be about $\hat{\alpha} = 0.18$. This is slightly smaller than the critical level for the Smirnov test on the same data.

Theory. The exact distribution of the Cramér-von Mises two sample test statistic may be obtained in the same way as with the Smirnov test statistic. The different ordered arrangements of the combined sample are equally likely under the null hypothesis, and the statistic T_2 may be computed from the ordered combined sample. Exact quantiles of T_2 were obtained by Anderson (1962) and Burr (1963 and 1964) for small samples, in essentially the manner just described, with some computational shortcuts.

The statistic T_2 was apparently introduced by Fisz (1960). He credits the statistic to Lehmann (1951) and the asymptotic distribution of the statistic to Rosenblatt (1952), but the statistic studied by Lehmann and

Rosenblatt is

$$(9) \quad T_3 = \frac{m}{2(m+n)} \sum_{i=1}^{n} \left[\frac{R(X^{(i)})}{m} - i \frac{m+n}{mn} \right]^2$$

$$+ \frac{n}{2(m+n)} \sum_{j=1}^{m} \left[\frac{R(Y^{(j)})}{n} - j \frac{m+n}{mn} \right]^2$$

which differs from Fisz's statistic T_2 unless m equals n. Fisz showed that T_2 has the same asymptotic distribution as T_3, which was shown by Rosenblatt to be the same as the asymptotic distribution of the Cramér-von Mises goodness of fit statistic defined in the previous section. Therefore the asymptotic distribution of T_2 was actually obtained by Anderson and Darling (1952) in their paper on the Cramér-von Mises goodness of fit statistic. That is why T_2 is called the Cramér-von Mises two-sample test statistic, even though neither Cramér nor von Mises is credited with its invention.

For a two-sample test designed for points on a circle see Stephens (1965b), Maag (1966), and Maag and Stephens (1968). A multivariate Smirnov test is described by Bickel (1969). Fine (1966) is concerned with the Cramér-von Mises statistic, while Csörgö (1965) and Percus and Percus (1970) work with variations of the Smirnov test. Recent papers on the asymptotic efficiency of the Smirnov test are by Capon (1965), Ramachandramurty (1966), Andel (1967), and Klotz (1967).

PROBLEMS

1. Test the null hypothesis $F(x) \leq G(x)$, where the observations from $F(x)$ are .6, .8, .8, 1.2 and 1.4, and the observations from $G(x)$ are 1.3, 1.3, 1.8, 2.4, and 2.9.

2. A random sample of five sixth grade boys in one section of town were given a literacy test with the following results; 82, 74, 87, 86, 75. A random sample of eight sixth grade boys from a different section of town were given the same literacy test with these scores resulting; 88, 77, 91, 88, 94, 93, 83, 94. Is there a difference in literacy, as measured by this test, in the two populations of sixth grade boys? (Use the Smirnov test).

3. Use the Cramér-von Mises test on the data in Problem 2, and compare results with the Smirnov test.

4. Find the .80, .90, and .95 quantiles of T_1 for $n = 3$ and $m = 2$ from the exact distribution obtained in the text. Compare these with the quantiles in Table 17 and explain any differences.

5. Find the exact distribution of T_1, T_1^+, T_1^-, and T_2 when $n = 3$ and $m = 3$.

6. Compare the exact .95 quantiles of T_1 with the approximation based on the asymptotic distribution, for $n = m = 30$ and $n = m = 10$.

6.3. TESTS FOR SEVERAL INDEPENDENT SAMPLES

The tests presented in this section are multisample analogues of the Smirnov test for two samples. These tests may be applied to several independent samples and, as such, they may be compared with the Kruskal-Wallis test and the Bell-Doksum test for several independent samples. The Kruskal-Wallis and Bell-Doksum tests are intended to be sensitive to differences in means or medians of the various populations, but may be insensitive to other differences, in particular differences in variances. In fact a wide disparity of variances may tend to hide the differences in means. These Smirnov-type tests are not as sensitive to differences only in means, but they are consistent against a wider variety of differences, and therefore they are often more powerful than the Kruskal-Wallis and Bell-Doksum tests if the differences in means are accompanied by differences in variances and other differences, as is often the case. A major drawback of these tests is that they may be applied only to samples of equal sizes, because the tables for the case of unequal sample sizes have not been developed.

The first test presented is a direct analogue of the two-sided Smirnov test. The test statistic and the method of obtaining its distribution may be used for any number of samples, and any sample sizes, whether equal or unequal. But the distribution of the test statistic has been obtained only for the case of three samples of equal size, and so from a practical standpoint the test is only a three-sample test at present. Two other tests will also be presented because the distribution of the test statistics in those tests has been obtained for up to ten samples. Their test statistics are also of the Smirnov type, but the tests are not consistent against all alternatives, as is this first test proposed by Birnbaum and Hall (1960).

The Birnbaum-Hall Test

DATA. The data consist of three independent random samples, each of size n. Denote the empirical distribution functions of the three samples by $S_1(x)$, $S_2(x)$, and $S_3(x)$, and denote their unknown distribution functions by $F_1(x)$, $F_2(x)$, and $F_3(x)$, respectively.

ASSUMPTIONS
1. The samples are random samples, mutually independent of each other.
2. The measurement scale is at least ordinal.
3. In order for the test to be exact the random variables need to be continuous. Otherwise the test is conservative.

HYPOTHESES

H_0: $F_1(x)$, $F_2(x)$, and $F_3(x)$ are identical with each other.

H_1: At least two of the distribution functions are different from each other.

TEST STATISTIC. Consider the largest vertical distance between $S_1(x)$ and $S_2(x)$, between $S_2(x)$ and $S_3(x)$, and between $S_1(x)$ and $S_3(x)$, without regard to sign. The Birnbaum-Hall test statistic T_1 equals the largest of these distances. Mathematically T_1 may be stated as

$$(1) \qquad T_1 = \sup_{x,i,j} \left| S_i(x) - S_j(x) \right|$$

which is read, "T_1 equals the supremum, over all x, and all i and j (from 1 to 3), of the absolute difference between the empirical distribution functions $S_i(x)$ and $S_j(x)$."

DECISION RULE. Reject H_0 at the level of significance α if T_1 exceeds the $1 - \alpha$ quantile $w_{1-\alpha}$ as given by Table 18. Otherwise accept the null hypothesis that the three distribution functions are identical.

Example 1. Twelve volunteers were assigned to each of three weight-reducing plans. The assignment of the volunteers to the plans was at random, and it was assumed that the thirty-six volunteers in all would resemble a random sample of people who might try a weight-reducing program. The null hypothesis is that there is no difference in the probability distributions of the amount of weight lost under the three programs, and the alternative is that there is a difference. The results are given below as the number of pounds lost by each person

Plan A		Plan B		Plan C	
2	17	17	5	29	5
12	4	15	6	3	25
5	25	3	19	25	32
4	6	19	4	28	24
26	21	5	9	11	36
8	6	14	7	7	20

The empirical distribution functions appear in Figure 1. The greatest vertical distance between any two empirical distribution functions is seen from Figure 1 to occur at $x = 19$, between $S_2(x)$ and $S_3(x)$, plans B and C. This distance is 8/12. The critical region of size $\alpha = .05$ corresponds to all values of T_1 greater than 7/12, the .95 quantile as obtained from Table 18 for $n = 12$. Therefore the null hypothesis is rejected, and we may conclude that the different weight-reducing plans do in fact differ with regard to the probability distribution of the number of pounds lost. The critical level is estimated from Table 18 as slightly less than .05. The exact critical level may be obtained from the more extensive tables furnished by Birnbaum and Hall (1960), and equals .022.

Theory. The exact distribution of the Birnbaum-Hall statistic is found in the same way that the exact distribution functions of the Smirnov

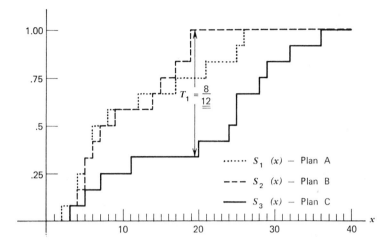

Figure 1. Graphs showing the three empirical distribution functions, and the Birnbaum-Hall test statistic T_1.

statistic and the Kruskal-Wallis statistic were found. That is, under the null hypothesis of identical distributions, each arrangement of the combined ordered sample is equally likely. The statistic T_1 is computed for each arrangement, and the distribution of T_1 is obtained. The procedure is simplified to a great extent by the difference equations used by Birnbaum and Hall (1960). The asymptotic quantiles are based on an unproved conjecture by the author, which states that the large sample distribution of T_1 is likely to be very similar to the large sample distribution of the one-sided statistic T_2 for seven samples, to be introduced next, because of similarities in the structure of their critical regions.

The following extension of the one-sided Smirnov statistic to the multi-sample case is appropriate if the alternative hypothesis of interest is one that not only says that differences exist, but states in which direction differences will exist, if indeed there are differences. For example, if the populations are identical except for one variable, such as dosage level on populations of plants or animals, or age levels in populations of humans, then the experimenter can often state that if differences exist among populations, those differences will be exhibited as a tendency for observations to be larger in older populations, or smaller as the dosage level increases, and so on. These are the one-sided alternatives that make this one-sided extension of the Smirnov test appropriate.

The One-Sided k-Sample Smirnov Test

DATA. The data consist of k random samples of equal size n. Let their respective empirical distribution functions be denoted by $S_1(x)$, $S_2(x)$, ..., $S_k(x)$, and let their respective, unknown, distribution functions be denoted by $F_1(x)$, $F_2(x)$, ..., $F_k(x)$.

ASSUMPTIONS
1. The samples are random samples, mutually independent of each other.
2. The measurement scale is at least ordinal.
3. In order for the test to be exact the random variables need to be continuous. Otherwise the test is likely to be conservative.

HYPOTHESES

$$H_0: F_1(x) \leq F_2(x) \leq \ldots \leq F_k(x) \text{ for all } x.$$

$$H_1: F_i(x) > F_j(x) \text{ for some } i < j, \text{ and for some } x.$$

These hypotheses are used when the alternative hypothesis is that the ith sample tends to have smaller values than the jth sample, for some i less than j. The null hypothesis is usually interpreted as, "All samples come from identical populations," because this one-sided test is usually appropriate if, for some physical reasons, differences between populations will only occur in the direction indicated by H_1.

TEST STATISTIC. The test statistic T_2 is defined as the largest vertical distance achieved by $S_i(x)$ above $S_{i+1}(x)$, where the adjacent samples are being compared as i ranges from 1 to $k - 1$. This may be stated mathematically as

$$(2) \qquad T_2 = \sup_{x, i < k} [S_i(x) - S_{i+1}(x)]$$

which is read, "T_2 equals the supremum, over all x, and over all i less than k (the number of samples) of the difference $S_i(x)$ minus $S_{i+1}(x)$." To evaluate T_2, one computes the usual one-sided Smirnov statistic, defined in the previous section, first for samples 1 and 2, then for samples 2 and 3, and so on to samples $k - 1$ and k, and lets T_2 equal the largest of these.

DECISION RULE. Reject H_0 at the level α if T_2 exceeds the $1 - \alpha$ quantile as given in Table 19. Table 19 is entered with k, the number of samples, and n, the size of each of the k samples. The entry in the column under $p = 1 - \alpha$ is divided by n to give the $1 - \alpha$ quantile. The approximation based on the asymptotic distribution, given at the bottom of each column, requires division by $\sqrt{n}$ as indicated.

Example 2.   As the human eye ages it loses its ability to focus on objects close to the eye. This is a well-recognized characteristic of people over 40 years old. In order to see if people in the 15 to 30 year old range also exhibit this loss of ability to focus on nearby objects as they get older, eight people were selected from each of four age groups; about 15 years old, about 20, about 25, and about 30 years old. It was assumed that these people would behave as a random sample from their age group populations would, with regard to the characteristic being measured. Each person held a printed paper in front of their right eye, with the left eye covered. The paper was moved closer to the eye until the person declared that the print began to look fuzzy. The closest distance at which the print was still sharp was measured once for each person.

The null hypothesis was that the distance measured was identically distributed for all populations. The alternative hypothesis was that the older groups tended to furnish greater distances measured. The samples were numbered from 1 to 4 in order of age group.

$$H_0 : F_1(x) = F_2(x) = F_3(x) = F_4(x) \text{ for all } x.$$

$$H_1 : F_i(x) > F_j(x) \text{ for some } x \text{ and some } i < j.$$

We are assuming that ability to focus on close objects does not improve with age, and therefore we are able to state the null hypothesis in the above slightly simpler form.

The distances, measured in inches, are given below. The samples are ordered within themselves for convenience.

1. 15 years old		2. 20 years old		3. 25 years old		4. 30 years old	
4.6	6.3	4.7	6.4	5.6	6.8	6.0	8.6
4.9	6.8	5.0	6.6	5.9	7.4	6.8	8.9
5.0	7.4	5.1	7.1	6.6	8.3	8.1	9.8
5.7	7.9	5.8	8.3	6.7	9.6	8.4	11.5

The largest differences between $S_i(x)$ above $S_{i+1}(x)$ occur at the jump points of $S_i(x)$, which are the values in the ith random sample. Therefore the differences $S_i(x) - S_{i+1}(x)$ need to be computed only at the $n = 8$ numbers in the ith sample.

$S_1(x) - S_2(x)$	$S_2(x) - S_3(x)$	$S_3(x) - S_4(x)$
$\frac{1}{8} - 0 = \frac{1}{8}$	$\frac{1}{8} - 0 = \frac{1}{8}$	$\frac{1}{8} - 0 = \frac{1}{8}$
$\frac{2}{8} - \frac{1}{8} = \frac{1}{8}$	$\frac{2}{8} - 0 = \frac{2}{8}$	$\frac{2}{8} - 0 = \frac{2}{8}$
$\frac{3}{8} - \frac{2}{8} = \frac{1}{8}$	$\frac{3}{8} - 0 = \frac{3}{8}$	$\frac{3}{8} - \frac{1}{8} = \frac{2}{8}$
$\frac{4}{8} - \frac{3}{8} = \frac{1}{8}$	$\frac{4}{8} - \frac{1}{8} = \frac{3}{8}$	$\frac{4}{8} - \frac{1}{8} = \frac{3}{8}$
$\frac{5}{8} - \frac{4}{8} = \frac{1}{8}$	$\frac{5}{8} - \frac{2}{8} = \frac{3}{8}$	$\frac{5}{8} - \frac{2}{8} = \frac{3}{8}$
$\frac{6}{8} - \frac{6}{8} = 0$	$\frac{6}{8} - \frac{3}{8} = \frac{3}{8}$	$\frac{6}{8} - \frac{2}{8} = \frac{4}{8}$
$\frac{7}{8} - \frac{7}{8} = 0$	$\frac{7}{8} - \frac{5}{8} = \frac{2}{8}$	$\frac{7}{8} - \frac{3}{8} = \frac{4}{8}$
$\frac{8}{8} - \frac{7}{8} = \frac{1}{8}$	$\frac{8}{8} - \frac{7}{8} = \frac{1}{8}$	$\frac{8}{8} - \frac{6}{8} = \frac{2}{8}$
$\sup_x [S_1(x) - S_2(x)]$ $= \frac{1}{8}$	$\sup_x [S_2(x) - S_3(x)]$ $= \frac{3}{8}$	$\sup_x [S_3(x) - S_4(x)]$ $= \frac{4}{8}$

The test statistic T_2 equals 4/8, the largest of the largest differences given at the bottom of the columns above. The critical region corresponds to values of T_2 greater than $w_{.95}$, at $\alpha = .05$, where $w_{.95}$ is given by Table 19 as

$$w_{.95} = \frac{5}{n} = \frac{5}{8}$$

for $k = 4$ samples and $n = 8$ observations per sample. Because T_2 does not exceed 5/8, the null hypothesis is accepted. The critical level is seen from Table 19 to be somewhat greater than .10. An examination of the data reveals a slight tendency of the observations to increase in the direction predicted by the alternative hypothesis, but the difference, if it is real, is too slight to be detected with such a small sample size.

Theory. As with the Birnbaum-Hall test, the exact distribution of the one-sided k-sample Smirnov test statistic may be obtained by considering each ordered arrangement of the combined k samples to be equally likely, evaluating the test statistic for each arrangement, and then tabulating the distribution function of the test statistic. However, this tedious task is not necessary, because the distribution function of T_2 was derived by Conover (1967b) as a mathematical function of k and n. The asymptotic distribution was also derived, and these distribution functions were used to obtain Table 19. The derivations of these distribution functions are beyond the mathematical level of this book, and are therefore omitted.

Another k-sample Smirnov test was suggested by Conover (1965). This differs from the previous test in that it is a two-sided test. It differs from the Birnbaum-Hall test in that it may be applied to more than three samples because of the more extensive tables available. This is a two-sided test for use with k independent samples, necessarily of the same size, where the alternative of interest indicates that some populations may tend to yield larger values than other populations. It is particularly suited to biological and agricultural type data, or other types of data that are bounded below by some number such as zero but not bounded above by any number, where larger means in a population are usually accompanied by other differences, such as larger variances.

The Two-Sided k-Sample Smirnov Test

DATA. The data consist of k random samples of equal size n. Denote the respective unknown distribution functions by $F_1(x), F_2(x), \ldots, F_k(x)$.

ASSUMPTIONS
1. The samples are random samples, mutually independent of each other.
2. The measurement scale is at least ordinal.
3. In order for the test to be exact the random variables need to be continuous. Otherwise the test is likely to be conservative.

HYPOTHESES

$$H_0: F_1(x) = F_2(x) = \ldots = F_k(x) \text{ for all } x.$$

(The population distribution functions are identical)

$$H_1: F_i(x) \neq F_j(x) \text{ for some } i, j \text{ and } x.$$

(The population distribution functions are not identical.)

TEST STATISTIC. The test statistic is evaluated by comparing the "largest" sample with the "smallest" sample. That is, find the largest observation in each sample and denote these by $Z_1, Z_2, \ldots, Z_k$. The sample with the *largest* Z_i (i.e., the largest observation of all) is called the "sample of rank k," or the largest sample, and its empirical distribution function is denoted by $S^{(k)}(x)$. The sample with the *smallest* Z_i is called the "sample of rank 1," or the smallest sample, and its empirical distribution function is denoted by $S^{(1)}(x)$. The test statistic T_3 is defined as the greatest vertical distance attained by $S^{(1)}(x)$ above $S^{(k)}(x)$. Mathematically this may be stated as

(3) $$T_3 = \sup_x \, [S^{(1)}(x) - S^{(k)}(x)]$$

DECISION RULE. Reject H_0 at the level α if T_3 exceeds the $1 - \alpha$ quantile as given by Table 20. To obtain the $1 - \alpha$ quantile $w_{1-\alpha}$ from Table 20, enter the row corresponding to the correct sample size n, and the column corresponding to $p = 1 - \alpha$. Then choose the table entry listed with the correct number of samples k. This table entry is then divided by n to obtain the desired quantile. The approximate quantiles for the asymptotic distribution require only division by $\sqrt{n}$ as indicated, and do not depend on k.

Example 3. For this example we shall use the same data as given in Example 1, where twelve volunteers were assigned to each of three weight-reducing plans. The null hypothesis is that there is no difference among plans, and the alternative is that some difference exists. The data were as follows.

Plan A		Plan B		Plan C	
2	17	17	5	29	5
12	4	15	6	3	25
5	25	3	19	25	32
4	6	19	4	28	24
26	21	5	9	11	36
8	6	14	7	7	20

The largest observation for each sample is underlined.

$$Z_1 = 26$$
$$Z_2 = 19$$
$$Z_3 = 36$$

The largest Z_i is 36, so Plan C is called the sample of rank 3. The smallest Z_i is 19, so Plan B is called the sample of rank 1. The test statistic is computed from those two samples. The difference $S^{(1)}(x) - S^{(k)}(x)$ needs to be computed only at the numbers listed in the sample of rank 1, because these are where the jumps of $S^{(1)}(x)$ occur, and therefore these are where $S^{(1)}(x)$ will achieve its largest value above $S^{(k)}(x)$. The samples are ordered below for convenience.

Plan B	Plan C	$S^{(1)}(x) - S^{(k)}(x)$	Plan B	Plan C	$S^{(1)}(x) - S^{(k)}(x)$
3	3	$\frac{1}{12} - \frac{1}{12} = 0$	9	25	
4	5	$\frac{2}{12} - \frac{1}{12} = \frac{1}{12}$	14	25	$\frac{7}{12} - \frac{3}{12} = \frac{4}{12}$
5	7		15	28	$\frac{8}{12} - \frac{4}{12} = \frac{4}{12}$
5	11	$\frac{4}{12} - \frac{2}{12} = \frac{2}{12}$	17	29	$\frac{9}{12} - \frac{4}{12} = \frac{5}{12}$
6	20	$\frac{5}{12} - \frac{2}{12} = \frac{3}{12}$	19	32	$\frac{10}{12} - \frac{4}{12} = \frac{6}{12}$
7	24	$\frac{6}{12} - \frac{3}{12} = \frac{3}{12}$	19	36	$\frac{12}{12} - \frac{4}{12} = \frac{8}{12}$

The test statistic T_3, defined by (3) as

$$T_3 = \sup_x [S^{(1)}(x) - S^{(k)}(x)]$$

equals 8/12 for these data. The critical region of size $\alpha = .05$ corresponds to values of T_3 greater than $w_{.95}$, which is given by Table 20 for $n = 12$ and $k = 3$ as

$$w_{.95} = \frac{6}{n} = \frac{6}{12}$$

Because T_3 exceeds 6/12, the null hypothesis is rejected. In fact the null hypothesis may be rejected at a level of significance as small as .01, so the critical level $\hat{\alpha}$ is about .01.

Theory.   As with all the tests of this section the exact distribution of the test statistic may be obtained by considering each ordered arrangement of the combined sample as equally likely, when the null hypothesis is true However, this tedious procedure is not necessary because the distribution function of T_3 is given as a mathematical function of n and k by Conover (1965). This simplifies tabulation, so that tables of quantiles may be easily obtained on a computer. The asymptotic distribution of T_3 is the same as that of the Smirnov one-sided test statistic for two samples, defined in the previous section and also defined as a special case of the previous test in this section.

The tests of this section were restricted to samples of equal size, and the Birnbaum-Hall test was further restricted to three samples. Actually any of these tests could be applied to any number of samples, and the samples could be of differing sizes, if tables of the distributions of the test statistics were available. Theoretically these tables are possible, and may be obtained by the enumeration method described following each example. From a practical standpoint, however, this enumeration method of considering all ordered arrangements of the combined sample is too exhaustive even for computers. At least that has been the feeling so far, except for the case of three samples of equal sizes. Only when the mathematical formulas for the distribution functions are known have the distributions been obtained for more than three samples, and those formulas are known only when the sample sizes are equal.

PROBLEMS

1. Do the following data indicate any difference due to gender in the lengths of Latin words? The observations represent the number of letters in Latin words selected at random from among those of the three genders, masculine, feminine, and neuter.

Masculine		Feminine		Neuter	
5	7	4	6	7	8
7	5	8	3	10	7
6	9	5	6	7	12

   Use both tests of this section which are appropriate, and compare results.

2. In order to see whether a longer time lapse between the last day of class and the time of the final exam tends to improve student performance on the final exam, a class of 48 students was divided at random into 4 groups of 12 students each. Group 1 took the final exam two days after the last class period. Group 2 took their exam four days after the last class period. Group 3 was given six days and Group 4 eight days. All groups were given comparable exams under otherwise comparable conditions. The final exam scores are as follows.

Group 1			Group 2			Group 3			Group 4		
48	71	80	42	70	77	38	73	83	49	77	84
61	74	82	48	71	81	58	74	87	58	79	93
67	75	87	62	73	89	70	75	90	73	80	94
68	79	89	67	75	92	71	79	94	74	84	97

   Does the increased time lapse tend to improve test performance?

3. The amount of iron present in the livers of white rats is measured after the animals had been fed one of five diets for a prescribed length of time. There were ten animals randomly assigned to each of the five diets.

Diet A	Diet B	Diet C	Diet D	Diet E
2.23	5.59	4.50	1.35	1.40
1.14	0.96	3.92	1.06	1.51
2.63	6.96	10.33	0.74	2.49
1.00	1.23	8.23	0.96	1.74
1.35	1.61	2.07	1.16	1.59
2.01	2.94	4.90	2.08	1.36
1.64	1.96	6.84	0.69	3.00
1.13	3.68	6.42	0.68	4.81
1.01	1.54	3.72	0.84	5.21
1.70	2.59	6.00	1.34	5.12

Do the different diets appear to affect the amount of iron present in the livers?

CHAPTER 7

Some Miscellaneous Tests

7.1. SOME QUICK TESTS

In this section we shall present some statistical procedures that were designed for simplicity and ease of application. The first is a location test for two independent samples, called Tukey's quick test. The second test is the Olmstead-Tukey test of association, for detecting whether or not two random variables are correlated.

Tukey's quick test was introduced by Tukey (1959) in response to a need for a test that would be much easier to use than the existing tests, in order to test the null hypothesis that two random variables were identically distributed, against the alternative that the two means were not equal to each other. This test probably has approximately the same power as the median test, but is likely to be less powerful than the Mann-Whitney test in most situations. However, Tukey argues for *practical power* as a basis for comparing tests, where practical power is defined as the product of the mathematical power by the probability that the procedure will be used. This simple test will presumably be used more often than the more complicated Mann-Whitney test. To emphasize the utility of his test, Tukey gave the following brief "complete description" of the test.

"Given two groups of measurements, taken under conditions (treatments, etc.) *A* and *B*, we feel the more confident of our identification of the direction of difference the less the two groups overlap one another. *If one group contains the highest value and the other the lowest value*, then we may choose (i) to count the number of values in the one group exceeding all numbers in the other, (ii) to count the number of values in the other group falling below all those in the one, and (iii) to sum these two counts (we *require* that neither count be zero). If the two groups are roughly the same size, then the critical values of the total count are, *roughly*, 7, 10, and 13, i.e. 7 for a two sided 5% level, 10 for a two sided 1% level, and 13 for a two sided 0.1% level." (*Technometrics*, Vol. 1, No. 1, p. 32.)

Of these three sentences in the above paragraph, the first sentence gives the intuitive justification of the test, the second sentence describes how to compute the test statistic, and the third sentence gives the critical values. Although we are not pretending to improve upon the description of the test, we shall reformulate the test in our usual format, for the sake of uniformity.

Tukey's Quick Test

DATA. The data consist of two independent random samples of sizes n and m; $X_1, X_2, \ldots, X_n$ and $Y_1, Y_2, \ldots, Y_m$.

ASSUMPTIONS
1. The two samples are random samples.
2. The two samples are mutually independent of each other.
3. The scale of measurement is at least ordinal.
4. Although some provision is made for ties, the test is based on the assumption that the random variables are continuous.
5. Either the two populations have identical distribution functions, or one population tends to yield larger observations than the other.

HYPOTHESES
A. (one-sided)

$$H_0 : E(X) \leq E(Y)$$
$$H_1 : E(X) > E(Y)$$

B. (one-sided)

$$H_0 : E(X) \geq E(Y)$$
$$H_1 : E(X) < E(Y)$$

C. (two-sided)

$$H_0 : E(X) = E(Y)$$
$$H_1 : E(X) \neq E(Y)$$

TEST STATISTIC. The test statistic is defined differently for the three sets of hypotheses which may be tested.

A. (one-sided) If the largest observation from both samples happens to be an X, and the smallest is a Y, then the test statistic is the sum of two counts; T_1 equals the number of X's larger than the largest Y, plus the number of Y's smaller than the smallest X. In all other cases T_1 equals zero.

B. (one-sided) If the smallest observation from both samples happens to be an X, and the largest is a Y, then the test statistic is the sum of two counts; T_2 equals the number of X's smaller than the smallest Y, plus the number of Y's larger than the largest X. In all other cases T_2 equals zero.

C. (two-sided) The two-sided test statistic T_3 equals the larger of T_1 and T_2. That is, if the largest of all of the observations is from one sample while the smallest observation is from the other sample, then T_3 equals the sum of two counts; T_3 equals the number of observations from the one sample larger than the largest observation from the other sample, plus the number of observations from the other sample which are smaller than the smallest observation from the one sample. If both the largest observation and the smallest observation happen to come from the same sample, then T_3 equals zero.

Ties may cause some confusion in the calculation of the test statistic, so the following procedures are recommended. If observations from *both* samples qualify as the largest observation, because they are tied with each other and no other observations are larger than they are, then define the test statistic to be zero. The same rule may be stated concerning ties among the smallest observations. The only other ties that may cause difficulty occur when the largest or smallest values from one sample are tied with "interior" values from the other sample. Then count those tied observations as 1/2 each when counting to compute the test statistic, as recommended by the author of the test.

DECISION RULE. As a quick approximate procedure, reject H_0 at the levels .05, .01, or .001 for the two-sided test, or .025, .005, or .0005 for the one-sided test, if the test statistic is greater than or equal to 7, 10, or 13 respectively. Otherwise accept H_0.

The values 7, 10, and 13 are strictly valid for fairly small sample sizes approximately equal to each other. For more precise values see Table 21 for samples which differ in size by 20 or less; that is, $|n - m| \le 20$. For other sample sizes the approximation at the bottom of Table 21 may be used.

Example 1.   In an eighth grade class of ten girls and eight boys, the students were timed as they ran over a measured distance to test the null hypothesis

H_0: The average speed of eighth grade boys equals the average speed of eighth grade girls

against the alternative

H_1: Their average speeds are not equal

It is assumed that the boys and girls in the class being studied resemble a random sample of the intended population, at least with respect to running speed.

The speed, recorded as time in seconds required to run the measured distance, is given for each student.

X_i (girls):	14.7	15.3	16.1	14.9	15.1	14.8
$n = 10$		16.7	17.3*	14.6*	15.0	
Y_j (boys):	13.9	14.6	14.2	15.0*	14.3	
$m = 8$		13.8*	14.7	14.4		

The highest and lowest values in each sample are noted with an asterisk. The largest value is an X and the smallest value is a Y. Therefore we find

$5\frac{1}{2}X$'s above 15.0 (15.3, 16.1, 15.1, 16.7, 17.3, and half of 15.0)

$5\frac{1}{2}Y$'s below 14.6 (13.9, 14.2, 14.3, 13.8, 14.4, and half of 14.6)

for a total count of 11. The two-sided test statistic T_3 is thus

$$T_3 = 11$$

At $\alpha = .05$, H_0 is rejected using either the approximate critical value of 7, or the exact critical value of 7 from Table 22 for $n = 10$ and $m = 8$. From Table 22 the critical level $\hat{\alpha}$ is seen to be less than .01.

Tukey's quick test may be used as the basis for a confidence interval for the difference between two means, just as the Mann-Whitney test was used as a basis for finding a confidence interval as was described in Section 5.4. The basic philosophy is the same as before; the confidence interval with coefficient $1 - \alpha$ consists of those values of c which would result in acceptance of

$$H_0: E(X) - E(Y) = c$$

when using the two-sided Tukey quick test at the level α. Actually we described Tukey's test only for the case where c equals zero, but it may be used to test the above null hypothesis merely by adding the constant c to all of the Y_j's before applying Tukey's quick test as we described it. Now we shall describe two convenient methods of deciding which values of c may be added to the Y_j's and still result in acceptance of H_0. Those values of c will form the confidence interval.

Confidence Interval for the Difference Between Two Means

DATA. The data consist of two random samples $X_1, \ldots, X_n$ and $Y_1, \ldots, Y_m$.

ASSUMPTIONS
1. The two samples are random samples, mutually independent of each other.
2. The random variables are continuous.

3. The difference between X and Y is solely one of location. That is, if X is not distributed the same as Y, then $X - c$ is distributed the same as Y, for some unknown constant c.

METHOD A (graphic). The following procedure is credited to Sandelius (1968).
1. The two samples, X and Y, say, are plotted on separate slips of paper with the same scale on both.
2. The slips are placed with the scales horizontal and with equal scale values opposite each other.
3. The Y slip is moved along the X slip until the left-most value is a Y and the right-most value is an X, and is then adjusted back and forth until T_3, the count of nonoverlapping values, is 7 or as little as possible above 7 (for a 95% confidence interval). The distance the Y slip was moved relative to the X slip is measured and is called L. This distance is negative if the Y slip was moved to the left relative to the X slip, and positive otherwise.
4. Now the Y slip is moved to the right of the position in step 3, until the left-most value is an X and the right-most value is a Y. The slips are adjusted until T_3 is 7 or as little as possible above 7. The distance the Y slip was moved, from its original position in step 2, is called U, and is negative if the Y slip is to the left of its original position relative to X, and positive otherwise.
5. The 95% confidence interval for $E(X) - E(Y)$ is between L and U, inclusive. More precisely,

(1) $$P(L \leq E(X) - E(Y) \leq U) \geq 95\%$$

Note that for a 99% or a 99.9% confidence interval the numbers 10 or 13 should be used instead of 7. More precise values may be obtained from Table 21.

Example 2. For the data in Example 1, we shall find a 95% confidence interval for the difference between the two population means. First the two samples are plotted on separate slips of paper, and then the slips are placed opposite each other as shown below. In the third step the Y slip is moved .3 units to the right, so that $3\frac{1}{2} Y$'s are to the left of the smallest X, and $3\frac{1}{2} X$'s are to the right of the largest Y, giving $T_3 = 7$. Thus L equals $+.3$. In the fourth step the Y slip is moved further to the right, a total of 2.3 units from its original position, until the largest Y value barely exceeds the largest X value. At this point T_3 goes from zero to nine in value. U equals $+2.3$.
The 95% confidence interval on $E(X) - E(Y)$ is from .3 to 2.3.

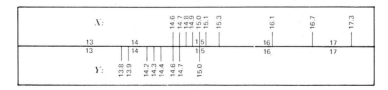

Step 2

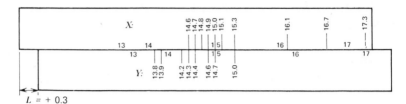

Step 3

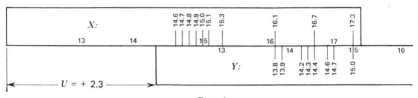

Step 4

METHOD B (algebraic) The algebraic method of finding a confidence interval is given by Tukey (1959) as follows.

1. Using 7 as a 5% critical value, subtract the largest Y from the 6 ($= 7 - 1$) largest X's and call these the highest $X - Y$ differences.

2. Subtract the 6 smallest Y values from the smallest X value to obtain the 6 "lowest $X - Y$ differences."

3. Of the 12 differences obtained in steps 1 and 2, let L equal the 7th largest (or 6th smallest) difference.

4. To obtain U, subtract the 6 largest Y values from the largest X value to obtain the "highest $Y - X$ differences," and subtract the smallest Y value from the 6 smallest X values to get the "lowest $Y - X$ differences."

5. Of the 12 differences found in step 4, let U equal the 7th smallest difference.

6. The interval from L to U, inclusive, is the 95% confidence interval for $E(X) - E(Y)$.

The 99% (or 99.9%) confidence interval is found by using 10 (or 13) instead of 7, working with the 9 (or 12) differences at each end, and letting L and U equal the 10th (or 13th) difference from the top or bottom respectively in the combined 18 (or 24) differences. More precise critical values may be obtained from Table 21.

Example 3.   In continuation of the two previous examples, we have the six largest X's and the largest Y yielding the highest $X - Y$ differences.

$\quad\quad$ $6 = 7 - 1$ largest X's $= 17.3, 16.7, 16.1, 15.3, 15.1, 15.0$

$\quad\quad$ The largest $Y = 15.0$.

$\quad\quad$ The highest $X - Y$ differences $= 2.3, 1.7, 1.1, .3, .1, .0$

Similarly the lowest $X - Y$ differences are found.

$\quad\quad$ The smallest $X = 14.6$

$\quad\quad$ $6 = 7 - 1$ smallest Y's $= 13.8, 13.9, 14.2, 14.3, 14.4, 14.6$

$\quad\quad$ The lowest $X - Y$ differences $= .8, .7, .4, .3, .2, .0$

$\quad\quad$ The 7 greatest differences $= 2.3, 1.7, 1.1, .8, .7, .4, \underline{.3}$

$\quad\quad$ $L = .3 =$ the 7th greatest $X - Y$ difference

To find U the procedure is slightly different.

The largest $X = 17.3$

$6 = 7 - 1$ largest Y's $= 15.0, 14.7, 14.6, 14.4, 14.3, 14.2$

The highest $Y - X$ differences $= 2.3, 2.6, 2.7, 2.9, 3.0, 3.1$

$6 = 7 - 1$ smallest X's $= 14.6, 14.7, 14.8, 14.9, 15.0, 15.1$

The smallest $Y = 13.8$

The lowest $Y - X$ differences $= .8, .9, 1.0, 1.1, 1.2, 1.3$

The 7 least differences $= .8, .9, 1.0, 1.1, 1.2, 1.3, \underline{2.3}$

$U = 2.3 =$ the 7th smallest $Y - X$ difference

The 95 % confidence interval for $E(X) - E(Y)$ is from .3 to 2.3, as before in Example 2.

Theory. The exact distribution of the test statistic may be obtained in the usual manner, by considering all ordered arrangements of the combined sample to be equally likely when the null hypothesis is true. The test statistic is computed for each ordered arrangement, and the various values of the test statistic are accumulated to give the distribution function. This process may be greatly simplified by some simple combinatorial shortcuts described by Tukey (1959).

The asymptotic approximation

$$(2) \qquad P(T_3 \geq h) \cong 2 \, \frac{nm}{m^2 - n^2} \left[\left(\frac{m}{n + m} \right)^h - \left(\frac{n}{n + m} \right)^h \right]$$

which is algebraically equivalent to the equation given at the bottom of Table 21, may be justified as follows. If the sample sizes of the X's and Y's are n and m, respectively, the probability of any particular value (in the ordered combined sample) being an X is $n/(n + m)$, and the probability of it being a Y is $m/(n + m)$. If the sample sizes are quite large, then one chosen value being an X or a Y will be nearly independent of any other value being an X or a Y. Then we have, approximately,

$$(3) \qquad P(\text{exactly } k \text{ highest are } X\text{'s}) = \left(\frac{n}{n + m} \right)^k \left(\frac{m}{n + m} \right)$$

since each of the k highest must be an X and the next highest value must be a Y. Similarly

$$(4) \qquad P(\text{at least } j \text{ lowest are } Y\text{'s}) = \left(\frac{m}{n + m} \right)^j$$

because each of the j lowest must be Y's and the others may be anything.

How can a total end count of h or more be attained? First, we shall consider only those cases where the highest value is an X. All of the situations fall into one of the following mutually exclusive categories:

(exactly $1X$ high) and (at least $h-1\,Y$'s low)
(exactly $2X$'s high) and (at least $h-2\,Y$'s low)

$\cdots$

(exactly $h-2X$'s high) and (at least $2\,Y$'s low)
(exactly $h-1X$'s high) and (at least $1\,Y$ low)
(*at least* hX's high) and (at least $1\,Y$ low)

Because those events are mutually exclusive, the probability of $T_3 \geq h$, with X high, is the sum of the probabilities of the individual events. Those probabilities are listed below.

(5) $P(T_3 \geq h,$ and X high$)$

$$\cong \left(\frac{n}{n+m}\right)^1 \left(\frac{m}{n+m}\right) \cdot \left(\frac{m}{n+m}\right)^{h-1}$$

$$+ \left(\frac{n}{n+m}\right)^2 \left(\frac{m}{n+m}\right) \cdot \left(\frac{m}{n+m}\right)^{h-2} + \cdots$$

$$+ \left(\frac{n}{n+m}\right)^{h-2} \left(\frac{m}{n+m}\right) \cdot \left(\frac{m}{n+m}\right)^2$$

$$+ \left(\frac{n}{n+m}\right)^{h-1} \left(\frac{m}{n+m}\right) \cdot \left(\frac{m}{n+m}\right)^1 + \left(\frac{n}{n+m}\right)^h \cdot \left(\frac{m}{n+m}\right)$$

$$\cong \left(\frac{1}{n+m}\right)^{h+1} (n^1m^h + n^2m^{h-1} + \cdots + n^{h-2}m^3 + n^{h-1}m^2 + n^hm^1)$$

This expression may be simplified by multiplying each term within the second set of parentheses by $(m-n)/(m-n)$, as follows.

(6) $P(T_3 \geq h,$ and X high$)$

$$\cong \left(\frac{1}{n+m}\right)^{h+1} \left[\frac{n^1m^h(m-n)}{m-n} + \frac{n^2m^{h-1}(m-n)}{m-n} + \cdots\right.$$

$$\left. + \frac{n^{h-2}m^3(m-n)}{m-n} + \frac{n^{h-1}m^2(m-n)}{m-n} + \frac{n^hm^1(m-n)}{m-n}\right]$$

$$\cong \left(\frac{1}{n+m}\right)^{h+1} \left(\frac{1}{m-n}\right)(n^1m^{h+1} - n^2m^h + n^2m^h - n^3m^{h-1} + \cdots$$

$$+ n^{h-2}m^4 - n^{h-1}m^3 + n^{h-1}m^3 - n^hm^2 + n^hm^2 - n^{h+1}m^1)$$

Most of the terms within the parentheses cancel each other, leaving

$$(7) \quad P(T_3 \geq h, \text{ and } X \text{ high}) \simeq \left(\frac{1}{n+m}\right)^{h+1}\left(\frac{1}{m-n}\right)(n^1 m^{h+1} - n^{h+1} m^1)$$

$$\simeq \frac{nm}{(n+m)(m-n)}\left[\left(\frac{m}{n+m}\right)^h - \left(\frac{n}{n+m}\right)^h\right]$$

The case where the highest value is a Y results in the same probability as given in (7), so the total probability of $T_3 \geq h$ is twice the probability given in (7).

$$(8) \quad P(T_3 \geq h) \simeq P(T_3 \geq h, \text{ and } X \text{ high}) + P(T_3 \geq h, \text{ and } Y \text{ high})$$

$$\simeq 2\frac{nm}{m^2 - n^2}\left[\left(\frac{m}{n+m}\right)^h - \left(\frac{n}{n+m}\right)^h\right]$$

and the derivation is complete. Incidentally, the probability given in (7) is actually the probability of $T_1 \geq h$, because T_1 is equal to T_3 if X is high, and equals zero if Y is high.

Tukey's test qualifies as a quick test because of the ease in which the test statistic may be computed. An important feature of this test is the lack of necessity to rank all of the observations from smallest to largest. Only relatively few observations actually need to be ranked. This is especially welcome when the sample sizes are quite large, because ranking the entire sample can be difficult with large samples. It is often true that when the samples are large the ease of calculation is more important than even the power of a test. That is, with very large sample sizes almost any test will detect the differences if they exist, so the test selected may as well be an easy one to apply. Some modifications of Tukey's quick test may be found in Rosenbaum (1965) and Neave (1966). See also the paper by Neave and Granger (1968).

The next test is another quick test, only this test is designed to detect correlation, if it exists, between two random variables, X and Y, say, where the observations occur in pairs (X_i, Y_i). This test was developed by Olmstead and Tukey (1947) who called it both a "corner test of association" and a "quadrant sum test." This test places heavy reliance on the very large and very small observations to detect the presence of correlation. And in practice these are often the most sensitive indicators of a dependency between two variables. So even though the Olmstead-Tukey test is a quick test to apply, it is often more powerful than other, more conventional, tests such as the Kendall test for correlation or the Spearman test for correlation given in Section 5.5, particularly when the correlation shows up most strongly in the very large and very small observations.

The Olmstead-Tukey Test of Association

DATA. The data consist of a bivariate random sample $(X_1, Y_1), (X_2, Y_2), \ldots,$ (X_n, Y_n) of size n.

ASSUMPTIONS

1. The bivariate sample is a random sample; that is, the (X_i, Y_i) all have the same bivariate distribution and are mutually independent. This does not imply that X_i has the same distribution as Y_i or is independent of Y_i.
2. The measurement scale is at least ordinal.
3. Although procedures for handling ties are given, the test assumes that the random variables are continuous.

HYPOTHESES

$$H_0: X_i \text{ and } Y_i \text{ are independent}$$

$$H_1: X_i \text{ and } Y_i \text{ are correlated}$$

TEST STATISTIC. First make a scatter diagram of the points; that is, plot each of the n data points (X_i, Y_i) as a point on the (x, y) coordinate axes by letting $x = X_i$ and $y = Y_i$ for each point. Draw a horizontal line through the median $Y_{\text{med.}}$ of the Y values, $y = Y_{\text{med.}}$, and draw a vertical line $x = X_{\text{med.}}$ through the median of the X values, as shown below. The upper right and lower left quadrants thus formed are called the $+$ quadrants, and the other two quadrants are the $-$ quadrants.

Beginning at the right-hand side of the scatter diagram, count (in order of abscissae) the observations until forced to cross the horizontal median. Write down the number of observations counted before this crossing, with a $+$ sign if they lay in the $+$ quadrant and a $-$ sign if they lay in the $-$ quadrant. In the following diagram the only dot counted is the solid dot on the right of the diagram, and this count is recorded as $+1$. Repeat this process moving up from below ($+6$ in the following diagram), to the right from the left ($+6$, counting ties as zero each), and downward from above ($+3$). The quadrant sum equals the sum of these four counts, signs attached, which is $+16$ in the following diagram. The Olmstead-Tukey test statistic T_4 equals the absolute value of the quadrant sum;

(9) $T_4 = |\text{quadrant sum}|$

As indicated above, if the extreme point on one side of the median line is tied with some points on the other side of the median line, none of the tied points contribute to the count because only points reached *before* crossing the line are counted. We also recommend, for simplicity, that points lying exactly on one of the median lines be completely ignored in the count, and that the count proceed as if those points were not there.

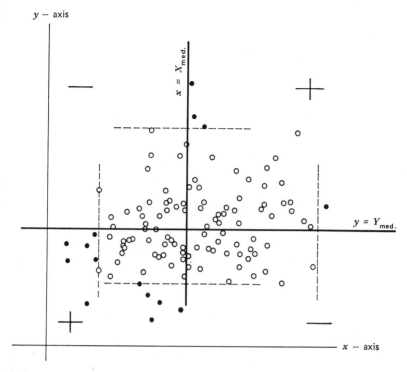

DECISION RULE. Reject H_0 at the level of significance α if T_4 *exceeds* the $1 - \alpha$ quantile, $x_{1-\alpha}$, as given below (from Olmstead and Tukey, 1947).

Sample Size	$x_{.80}$	$x_{.90}$	$x_{.95}$	$x_{.99}$	$x_{.999}$
$n = 6$	6	10	—	—	—
$n = 8$	6	8	10	—	—
$n = 10$	6	8	10	14	—
$n = 14$	6	9	10	14	21
$n = \infty$ (use for $n > 14$)	6	8	10	13	18

Example 4. We shall use the same data we used in Section 5.5 to illustrate usage of the rank correlation coefficients. The first-born twin and the second-born twin, in twelve sets of twins, were each given a measure of aggressiveness as follows

Twin set, $i =$	1	2	3	4	5	6	7	8	9	10	11	12
First-born $X_i =$	86	71	77	68	91	72	77	91	70	71	88	87
Second-born $Y_i =$	88	77	76	64	96	72	65	90	65	80	81	72

The hypotheses are:

H_0: Measures of aggressiveness of two twins are independent of each other.

H_1: The measures of aggressiveness of two twins are correlated.

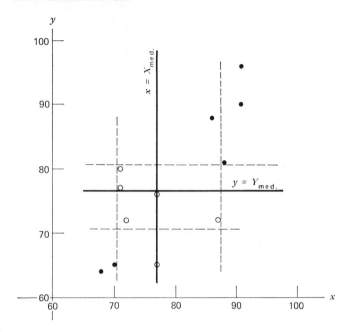

First a scatter diagram of the data is made. The lines through the median of the Y values ($Y_{med.} = 76.5$) and through the median of the X values ($X_{med.} = 77$) are drawn. Counting from the right, there are 3 points in the $+$ quadrant before the first point in the $-$ quadrant is encountered. Counting upwards from below, there are 2 points in the $+$ quadrant, ignoring the point on the median line, before the lowest point on the right side is met. The point on the right side happens to be tied with a point on the left side, so the latter point is not counted. Similar counts give $+2$ coming in from the left, and $+4$ counting down from the top. The algebraic sum of these counts is the quadrant sum:

$$\text{quadrant sum} = +3 + 2 + 2 + 4 = 11$$

The absolute value of the quadrant sum gives the test statistic:

$$T_4 = |\text{quadrant sum}| = 11$$

At $\alpha = .05$ we reject H_0 if T_4 exceeds the .95 quantile, given above as $x_{.95} = 10$. Because T_4 exceeds 10, the null hypothesis is rejected at $\alpha = .05$, just as it was in Section 5.5. The critical level is slightly less than .05, as compared with critical levels of about .01 obtained using Spearman's ρ and Kendall's τ. The computations were much simpler here, although normally one would use a rank correlation coefficient to test data with so few sample points.

Theory. As with the rank correlation coefficients of Section 5.5, the distribution of T_4, under the null hypothesis of independence, may be found by considering each possible pairing of the ranks of the X's with the ranks of the Y's to be equally likely, because the Olmstead-Tukey test statistic T_4 actually depends only on the relative sizes, or ranks, of the observations. For example, suppose there are four pairs of observations ($n = 4$), and the observations are perfectly matched so that the largest X is paired with the largest Y, the second largest X with the second largest Y, and so on. Then the count, on the scatter diagram, from the right includes the two points in the upper right quadrant, before the third point terminates the count, and the result is $+2$. Coming in from each direction the count is $+2$. So the quadrant sum is $+8$ and T_4 is $+8$.

As an illustration of how the distribution of T_4 may be obtained, all of the possible pairings for $n = 4$ are listed below. For brevity the notation $(3, 1)$ represents the pairing of the X value having rank 3 among the X's, with the Y value having rank 1 among the Y's. The pairings are grouped according to the *quadrant sums* resulting from each pairing.

Quadrant Sum	*Pairings of Ranks*			
-8	1. $(1, 4)$	$(2, 3)$	$(3, 2)$	$(4, 1)$
	2. $(1, 4)$	$(2, 3)$	$(3, 1)$	$(4, 2)$
	3. $(1, 3)$	$(2, 4)$	$(3, 2)$	$(4, 1)$
	4. $(1, 3)$	$(2, 4)$	$(3, 1)$	$(4, 2)$
-4	1. $(1, 4)$	$(2, 2)$	$(3, 3)$	$(4, 1)$
-2	1. $(1, 4)$	$(2, 2)$	$(3, 1)$	$(4, 3)$
	2. $(1, 3)$	$(2, 2)$	$(3, 4)$	$(4, 1)$
	3. $(1, 4)$	$(2, 1)$	$(3, 3)$	$(4, 2)$
	4. $(1, 2)$	$(2, 4)$	$(3, 3)$	$(4, 1)$
0	1. $(1, 1)$	$(2, 4)$	$(3, 3)$	$(4, 2)$
	2. $(1, 2)$	$(2, 3)$	$(3, 4)$	$(4, 1)$
	3. $(1, 2)$	$(2, 4)$	$(3, 1)$	$(4, 3)$
	4. $(1, 3)$	$(2, 1)$	$(3, 4)$	$(4, 2)$
	5. $(1, 4)$	$(2, 1)$	$(3, 2)$	$(4, 3)$
	6. $(1, 3)$	$(2, 2)$	$(3, 1)$	$(4, 4)$
$+2$	1. $(1, 1)$	$(2, 3)$	$(3, 4)$	$(4, 2)$
	2. $(1, 1)$	$(2, 4)$	$(3, 2)$	$(4, 3)$
	3. $(1, 2)$	$(2, 3)$	$(3, 1)$	$(4, 4)$
	4. $(1, 3)$	$(2, 1)$	$(3, 2)$	$(4, 4)$
$+4$	1. $(1, 1)$	$(2, 3)$	$(3, 2)$	$(4, 4)$
$+8$	1. $(1, 1)$	$(2, 2)$	$(3, 3)$	$(4, 4)$
	2. $(1, 1)$	$(2, 2)$	$(3, 4)$	$(4, 3)$
	3. $(1, 2)$	$(2, 1)$	$(3, 3)$	$(4, 4)$
	4. $(1, 2)$	$(2, 1)$	$(3, 4)$	$(4, 3)$

Under the null hypothesis, X and Y are independent, and therefore each of the $4_a = 24$ pairings is equally likely. The Olmstead-Tukey statistic equals the absolute value of the quadrant sum, so the distribution of T_4 is given below.

Probability Function	*Distribution Function*
$P(T_4 = 0) = 6/24 = .2500$	$P(T_4 \leq 0) = \ \ .2500$
$P(T_4 = 2) = 8/24 = .3333$	$P(T_4 \leq 2) = \ \ .5833$
$P(T_4 = 4) = 2/24 = .0833$	$P(T_4 \leq 4) = \ \ .6667$
$P(T_4 = 8) = 8/24 = .3333$	$P(T_4 \leq 8) = 1.0000$

Usage of T_4 as a test statistic effectively prevents the Olmstead-Tukey test from being presented in the one-sided version. However, if the quadrant sum is used as a test statistic, then the one-sided alternative hypothesis of positive correlation may be accepted if the quadrant sum is greater than its $1 - \alpha$ quantile, which equals the $1 - 2\alpha$ quantile of T_4. Also the null hypothesis of independence may be rejected in favor of the alternative "negative correlation" if the quadrant sum is less than its α quantile, which equals the negative of the $1 - 2\alpha$ quantile of T_4.

The asymptotic distribution of T_4 is derived by Olmstead and Tukey (1947), but uses probability tools we have not introduced in this book, so we shall omit the derivation. Also presented in their paper is an extension of this test to the case of three or more random variables, to test for three-way (or more) association, and an extension to test for serial correlation (nonrandomness in a sequence of random variables). However, the distribution theory necessary for these extensions is actually not fully developed.

In this section we have presented quick tests for testing for equal means with two independent samples, and for testing the independence of paired random variables. The quick test for testing whether paired random variables have equal means would be the sign test of Section 3.4. The next section presents a quick test for testing whether several populations have the same mean, on the basis of independent random samples from those populations.

PROBLEMS

1. Do these two samples come from populations with the same mean?

X_i			Y_j		
1.06	2.20	1.46	3.00	3.00	1.74
1.24	1.76	.78	2.91	2.65	2.70
2.65	2.75	2.37	3.23	3.01	2.84
3.46	1.35	3.31	2.41	2.01	.99
2.80	2.60	2.03	1.94	2.20	3.17
2.54	1.91	2.87	4.79	1.26	3.88

Assume that the samples are random, and independent of each other.

2. Twenty-four volunteers in a psychology experiment were asked to adjust the amount of sand in a pail until they thought the total weight of sand and the pail equaled five pounds. They were blindfolded, and they gave verbal instructions to a lab assistant who did the actual adjusting of the sand. First the subject used his right hand to lift the pail, and when he was sure the weight was exactly five pounds, the experiment was repeated with the left hand. The actual weights at the end of each phase of the experiment are given below. Is there a positive correlation between the actual weights recorded for the left hand compared with the right hand?

Subject	Right	Left	Subject	Right	Left	Subject	Right	Left
1.	7.3	6.6	9.	3.7	4.0	17.	5.4	5.6
2.	6.2	5.9	10.	6.5	5.5	18.	6.3	6.3
3.	5.2	5.4	11.	6.3	4.6	19.	3.9	4.4
4.	5.4	4.6	12.	4.5	5.3	20.	3.6	4.1
5.	5.3	5.8	13.	5.3	5.3	21.	4.8	5.1
6.	5.0	5.4	14.	5.7	6.1	22.	4.9	4.4
7.	6.4	5.9	15.	6.7	6.3	23.	3.5	3.8
8.	7.3	5.4	16.	4.6	4.8	24.	5.3	4.7

3. Fifteen apple trees were sprayed with a fungicide at the beginning of the summer, and twenty trees were not sprayed. Do the sprayed trees tend to yield more apples (measured in bushels) than the trees which were not sprayed?

Sprayed Trees			Unsprayed Trees			
2.6	.7	2.0	1.8	1.8	2.9	1.5
4.7	2.5	2.0	.8	1.8	2.9	1.2
1.5	1.6	2.9	2.1	1.0	1.7	2.1
2.5	2.0	4.5	1.1	2.2	1.9	1.6
3.7	1.5	3.2	0.2	2.9	2.0	1.6

4. In a simulation experiment in statistics, five samples of twenty random numbers per sample were generated on a computer, and both the F-statistic (for a parametric test) and the Kruskal-Wallis statistic (for a nonparametric test) were computed. This procedure was repeated for twenty sets of data. Does there appear to be a correlation between the two statistics?

Data Set	F	$K - W$	Data Set	F	$K - W$	Data Set	F	$K - W$
1.	.80	1.12	8.	5.47	16.24	15.	2.39	8.89
2.	1.75	5.92	9.	2.46	9.18	16.	1.63	5.26
3.	1.34	6.57	10.	.82	4.02	17.	3.71	14.33
4.	1.36	6.51	11.	1.55	7.41	18.	1.65	6.37
5.	.12	.46	12.	1.38	4.62	19.	.91	3.64
6.	2.27	9.22	13.	1.19	7.74	20.	1.70	5.67
7.	.79	3.28	14.	4.66	15.92			

5. Find a confidence interval for the increase in yield of the sprayed trees, from the data in Problem 3.

7.2. A SLIPPAGE TEST FOR SEVERAL INDEPENDENT SAMPLES

The name "slippage test" is given to any statistical procedure which tests the null hypothesis of equal means, where the data consist of several independent samples. Thus the Kruskal-Wallis test, discussed in Section 5.6, and the Bell-Doksum test for several independent samples, introduced in Section 5.9, are examples of slippage tests. They test whether one or more population distribution functions may have slipped to the right of the others, that is, whether they tend to yield larger numbers than the others.

This particular slippage test is distinguished from the others by its ease of application. The entire analysis, from raw data to the decision to reject or accept the null hypothesis of equal means, usually takes about two minutes, once the method of analysis is learned. Also this test places heavy reliance on the extreme (largest) values from each sample, which may be good or bad. If the populations having the larger means also possess larger variances, which is usually the situation with ratio type data, then the extreme values of a sample are sensitive indicators of any differences in means, and the test is quite powerful. However, if the variances of the populations remain equal even though some means may be larger than others, or if the variances tend to decrease as the means get larger then this test may be relatively insensitive to differences in means, and a rank test such as the Kruskal-Wallis test would probably be better to use.

This test may be used only when all samples have the same number of observations. However, one advantage of this test is that ties, caused by several observations appearing to be exactly equal to each other, are not likely to cause much difficulty when using this test.

A k-Sample Slippage Test

DATA. The data consist of k random samples, mutually independent of each other, with each sample containing the same number n of observations.

ASSUMPTIONS
1. Each sample is a random sample from some population.
2. The k samples are mutually independent of each other.
3. For the test to be exact the random variables are assumed to be continuous. Otherwise the test is conservative.
4. Either the k population distribution functions are identical, or else some of the populations tend to yield larger observations than other populations.
5. The data are at least ordinal in scale.

HYPOTHESES

H_0: All k population distribution functions are identical with each other.

H_1: Some populations tend to furnish larger observed values than other populations.

TEST STATISTIC. The test statistic is evaluated by comparing the "largest" sample with the "smallest" sample, as follows. In each of the k samples, find the largest of the n observations, and denote that extreme (largest) value by Z_i, $i = 1$, $\ldots, k$. Compare the Z_i's, the extremes, and find the largest Z_i, denoted by $Z^{(k)}$, and the smallest Z_i, denoted by $Z^{(1)}$. The sample containing $Z^{(k)}$, the largest of all the observed values, is called the sample of rank k, or the largest sample. The sample from which $Z^{(1)}$ came is called the sample of rank 1, or the smallest sample. The test statistic T equals the number of observations from the sample of rank k which is greater than $Z^{(1)}$, the largest observation in the sample of rank 1.

If several samples contain observations which are equal to the largest observation $Z^{(k)}$, thus causing difficulty in the selection of the one sample to be designated as the "largest" sample, it is recommended that the second largest observations be compared in those tied samples, and that the sample whose second largest observation is largest be selected as the sample of rank k. If these are also tied, the third largest observations may then be used. No other ties will cause any ambiguity in the computation of the test statistic.

DECISION RULE. Reject H_0 at the level of significance α if T exceeds the $1 - \alpha$ quantile $w_{1-\alpha}$, as given by Table 22. Quantiles are given for up to 10 samples, and for selected sample sizes ranging from $n = 3$ upwards to infinity. The test statistic is discrete, so the actual size of the critical region may be smaller than the apparent α used, but will never be larger than the apparent α.

Example 1. Ten students are selected at random from the senior classes of each of five high schools, and are given a standardized achievement examination. The null hypothesis to be tested is

H_0: The senior scores are identically distributed in the five different high schools against the alternative

H_1: Some of the schools tend to be associated with larger scores than at least one of the other schools

The hypothetical results are as follows.

High Schools

High School Student	A	B	C	D	E
1	143	210	159	113	98
2	152	183	69	128	106
3	216	258*	64	240*	66
4	236	204	85	134	177*
5	120	252	126	205	93
6	191	132	148	234	68
7	244*	171	98	188	81
8	104	231	90	149	143
9	208	226	202*	223	160
10	179	104	182	217	128

In this example k equals 5 and n equals 10. The following steps were followed in analyzing the data.

(1) An asterisk was placed beside the extreme (greatest) value in each sample.

(2) The smallest extreme was underlined once and the greatest extreme was underlined twice.

(3) The number of values in the sample with the greatest extreme, high school B, that exceeded the smallest extreme, 177, was determined to be 7. Thus $T = 7$.

(4) From Table 22 it is seen that the critical region of size $\alpha = .05$ (or smaller, perhaps) corresponds to values of T greater than 5. Hence the null hypothesis is rejected and it is concluded that at least one of the high schools tends to be associated with larger scores than at least one other school.

The critical level may be seen from Table 22 to be slightly less than .02. The entire analysis, as the reader may verify for himself, may be performed in less than two minutes.

It becomes apparent with this example that much emphasis is placed on the ability of the extremes from each sample to reflect the magnitude of the sample values. This slippage test should not be considered in cases where a shift in the population means does not affect the upper tail of the distribution in the same way. However, in many situations the upper tail of the distribution is shifted farther than the mean when the mean is shifted, and in these situations this slippage test is a good one to use.

Theory.   This slippage test is basically a rank test in the sense that the test statistic depends only on the order of the observations and not on their actual numerical value. Therefore the distribution function of the test statistic may be found by an enumeration of the various possible orderings which the observations might take on. For illustration, suppose k equals 3 and n equals 2. Then the two observations in any particular sample are just as likely to take the ranks 4 and 5, as they are to take the ranks 1 and 6, or any of the other ranks from 1 to 6 in the combined ordered sample, under the null hypothesis that all the random variables are independent and identically distributed. Each assignment of the ranks to the samples is just as likely, and has the same probability, as each other possible assignment. Therefore the distribution function of the test statistic may be found by listing the possible arrangements, and computing the test statistic for each arrangement. The possible arrangements of ranks when $k = 3$ and $n = 2$ are given below. For convenience the samples are arranged in columns, with the sample of rank 1 on the left and the sample of rank 3 on the right. Also the ranks within each column are ordered. In this way the number of possible arrangements is only 15, each with probability 1/15, while the total number of arrangements without this simplifying procedure is $6! = 720$, each with probability 1/720, and that would be too many arrangements to list here. The fifteen arrangements of ranks are as follows.

2 4 6	2 5 6	2 5 6	3 5 6	3 5 6
1 3 5	1 3 4	1 4 3	1 4 2	2 4 1
$T = 2$	$T = 2$	$T = 2$	$T = 1$	$T = 1$
3 4 6	3 5 6	4 5 6	4 5 6	4 5 6
1 2 5	1 2 4	1 2 3	1 3 2	2 3 1
$T = 2$	$T = 2$	$T = 1$	$T = 1$	$T = 1$
3 4 6	3 5 6	4 5 6	4 5 6	4 5 6
2 1 5	2 1 4	2 1 3	3 1 2	3 2 1
$T = 2$	$T = 2$	$T = 1$	$T = 1$	$T = 1$

There are 8 arrangements that give $T = 1$, and 7 that give $T = 2$. Therefore the distribution of T is given as follows.

Probability Function	*Distribution Function*
$P(T = 1) = 8/15$	$P(T \leq 1) = 8/15$
$P(T = 2) = 7/15$	$P(T \leq 2) = 1$

Although it is possible to find the quantiles of T in the manner just described, for all values of n and k, it is not necessary to use such a time consuming procedure. The distribution function of T, as a mathematical function of n and k, and the asymptotic distribution function appropriate for large values of n are much easier to use and are given by Conover (1968). Additional quantiles, not given in Table 22, may be found there also.

This slippage test is a modification of one presented by Mosteller (1948). The method of ranking the samples is similar to a method described by Bofinger (1965). Bofinger (1965) also presents a test, similar in some ways to the above test, but designed to be sensitive against the alternative that p populations have "slipped" to the right of the other $k - p$ populations, where p is a prespecified constant. Investigation of the power of the slippage test was undertaken by Cash (1967), and indicated that the slippage test was more powerful than the parametric F test in situations where the populations followed the normal distribution and the populations having larger means also had larger variances than the other populations. Other power comparisons were made by Lohrding (1969a, 1969b).

An interesting sidelight of this slippage test is the unusual way in which multiple comparisons may be made following the test. If the slippage test results in rejection of the null hypothesis, then the "largest" sample is discarded, and the population from which that sample was drawn is considered to have "slipped to the right." Then the analysis is repeated on the remaining $k - 1$ samples, using the portion of Table 22 appropriate for the now smaller number of samples. The same α level is used now as before. If the null hypothesis is again rejected, the second population is also considered to have slipped to the right, and the procedure is repeated on the remaining $k - 2$ samples. This process continues until H_0 is accepted.

Usually the experimenter is not only interested in knowing whether differences exist among the populations, but also where some of the differences might be. If the experimenter wishes to detect the populations which may be considered to be better than at least one of the other populations, the method of multiple comparisons just described is appropriate and simple. The following example illustrates the procedure.

Example 2. The same data given in Example 1, where ten students' scores were examined in each of five high schools, will be further examined to illustrate the multiple comparison technique associated with the slippage test.

The scores are repeated below, with an asterisk placed beside each extreme value, and with the smallest extreme and the largest extreme underlined once and twice respectively. In addition, ranks are assigned to all of the samples according to how their extremes rank with each other.

High Schools

High School Student	A	B	C	D	E
1	143	210	159	113	98
2	152	183	69	128	106
3	216	258*	64	240*	66
4	236	204	85	134	177*
5	120	252	126	205	93
6	191	132	148	234	68
7	244*	171	98	188	81
8	104	231	90	149	143
9	208	226	202*	223	160
10	179	104	182	217	128
The Sample Rank:	4	5	2	3	1

In Example 1 the high school of rank $k = 5$, high school B, was compared with the high school of rank 1, high school E, and the null hypothesis of identical populations was subsequently rejected at $\alpha = .05$. For purposes of making multiple comparisons, high school B is considered to yield higher scores than at least one other high school and is removed from further analysis.

The same slippage test is applied to the samples of ranks 1 through 4, from high schools, A, C, D, and E. There are 6 scores from the sample of rank 4, high school A, that exceed the highest score from the sample of rank 1, 177. The .95 quantile is found from Table 22, for $k = 4$ and $n = 10$ under $p = .95$, to be 5. Since the value $T = 6$ exceeds 5, high school A is also considered to have a tendency to yield larger scores than at least one other high school, and is removed from further analysis.

In a comparison of the samples of ranks 1, 2, and 3, the sample of rank 3 is seen to contain 6 values larger than $Z^{(1)} = 177$, and this number 6 exceeds the .95 quantile 5, obtained from Table 22 for $k = 3$ and $n = 10$. Therefore high school D is also considered to have slipped to the right relative to at least one other school.

A comparison of the samples of ranks 1 and 2 reveals only 2 observations in the sample of rank 2 which exceed $Z^{(1)} = 177$. The .95 quantile for $k = 2$, $n = 10$ is 4, so high school C is not considered better than the other one and the multiple comparisons procedure terminates. The result is that high schools B, A and D are selected as schools which tend to furnish higher senior achievement scores than other high schools studied.

Although we do not prove it here, it is shown in Conover (1968) that the probability of m or more values, in the sample of rank i, exceeding $Z^{(1)}$, is the same as if the samples of ranks $(i + 1)$, $(i + 2)$, . . . , k were not there at all. Therefore the same tables may be used to find the desired quantiles, by consulting the tables as if there were only i samples instead of k samples. This statement is true only for the *unconditional* probabilities. In the multiple comparisons procedure just described, the sample of rank i is not compared with the sample of rank 1 unless there is first a significant number of observations exceeding $Z^{(1)}$ in the samples of ranks k, $(k - 1)$, . . . , and $(i + 1)$. Then the probability of having a significant number of observations in the sample of rank i exceeding $Z^{(1)}$ is conditional, because we are given that the previous comparisons all were significant. This conditional probability of significance will in general be larger than the unconditional α indicated by the $1 - \alpha$ quantile used. Nevertheless, this multiple comparisons procedure, as is true with all multiple comparisons procedures, is still useful as an objective yardstick for determining which of the populations to separate from the others, after it has been determined by a statistical hypothesis test that some differences probably exist.

PROBLEMS

1. A certain large factory has many inspectors who sit at a moving belt and watch for defective items moving by. It is believed that the method of lighting has an appreciable effect on the percent efficiency of the inspectors. Therefore five different lighting schemes were tried by randomly assigning twelve inspectors

to each different lighting system and recording the percentage of the defective items which each inspector detected on the moving belt. Which lighting schemes, if any, could be considered to be better than others?

	Lighting Schemes				
	A	*B*	*C*	*D*	*E*
	46	50	69	48	44
	48	46	47	60	40
	32	50	46	54	59
Percentages of	42	48	65	47	44
defective items	39	37	49	50	55
detected by the	48	58	59	68	50
various	49	50	42	58	47
inspectors:	30	44	63	46	71
	48	40	47	46	43
	34	39	47	37	55
	38	45	53	40	59
	43	48	54	47	62

2. Six diets for chickens are tested over a fixed period of time to see if some diets might result in more eggs laid. The numbers of eggs laid by the different hens are given below.

Diet *A*	*B*	*C*	*D*	*E*	*F*
52	94	54	57	61	88
52	64	60	80	62	52
53	74	71	70	72	51
73	51	59	56	59	52
68	62	89	55	84	52
57	57	97	52	65	66
56	52	84	73	61	54
77	94	81	78	88	67
89	92	72	75	56	64
54	83	82	67	63	63
53	77	81	56	61	93
86	62	62	56	70	62
53	52	93	80	93	94
80	67	76	78	53	72

Are some diets more effective than others?

3. Find the distribution function of the slippage test statistic for two samples of three elements each, under the null hypothesis. Compare the quantiles given in Table 22 with the quantiles you obtained.

4. Compare the five populations from which the following random samples were

obtained to see which, if any, of the populations tend to yield larger observations than the others.

Sample 1		Sample 2		Sample 3		Sample 4		Sample 5	
.78	1.89	1.80	2.04	3.92	.93	.15	1.69	.64	4.15
2.67	1.82	1.75	1.86	1.78	2.38	3.85	2.15	2.57	2.62
3.25	3.90	2.25	1.47	3.56	5.20	3.03	2.42	1.50	1.28
.28	1.66	.91	.99	3.17	1.61	3.85	1.16	2.20	2.82
.84	2.81	1.20	1.65	4.83	1.90	.07	3.46	3.63	3.62
3.49	1.00	3.72	2.59	4.13	4.17	3.84	2.19	1.61	1.35
1.01	1.69	2.48	2.87	.63	2.18	2.41	1.82	.03	1.93
2.05	.34	2.04	.90	5.67	2.35	4.81	3.21	.30	4.03
2.86	.56	2.69	1.29	7.06	5.04	2.03	1.77	3.89	1.86
2.98	3.34	1.20	2.08	4.22	3.18	4.75	2.40	3.80	4.59

7.3. TESTS BASED ON RUNS

In the same way that a gambler speaks of a "run of bad luck," we may speak of runs of one type or another. In statistics, any sequence of like observations, bounded by observations of a different type, is called a *run*. The number of observations in the run is called the *length* of the run. Suppose a coin is tossed twenty times and the results H (heads) or T (tails) are recorded in the order in which they occur, as follows.

$$T\ HHHHHH\ T\ H\ T\ H\ TT\ HHH\ T\ H\ T\ H$$

The series begins with a run of tails of length 1, followed by a run of heads of length 6, followed by another run of tails of length 1, and so on. In all, there are six runs of tails and six runs of heads. In fact, with only two kinds of observations as we have with H and T, the number of runs of the one kind will always be within one run of the number of runs of the other kind, because each run of the one kind is preceded and followed by a run of the other kind, except at the beginning or end of the sequence.

In a sequence of two kinds of observations as we have above, the total number of runs may be used as a measure of the randomness of the sequence; too many runs may indicate that each observation tends to follow, and be followed by, an observation of the other type, while too few runs might indicate a tendency for like observations to follow like observations. In either case the sequence would indicate that the process generating the sequence (the method of obtaining "heads" or "tails" for the above sequence) was not random. By a random process, or a random sequence, we mean a sequence of independent and identically distributed random variables; that is, a sequence of heads and tails is random if the probability of any particular toss resulting in a head is the same as for any other particular toss in the sequence, independently of one another.

The same statistic, the total number of runs, may be used in a different situation, to test the null hypothesis that two independent random samples came from populations with identical distribution functions. The two samples are simply combined and ordered from smallest to largest. A run consists of a sequence of consecutive X values (from the one population) or Y values (from the other population) in the combined ordered sample. Too few runs indicates that there is a difference between the two population distribution functions. When the total number of runs is used in this fashion it is called the Wald-Wolfowitz runs test, and may be used as an alternative to other two sample tests such as the Mann-Whitney test, or the Smirnov test. The runs test is not very powerful in the two-sample problem, but it is quick and easy to apply. The runs test is presented below both as a test of randomness and as a two-sample test for differences of any type between two populations.

The Wald-Wolfowitz Runs Test

DATA

A. (test for randomness) The data consist of a sequence of observations, taken in order of occurrence. The observations are of two types, or can be reduced to data of two types, denoted by a and b in this presentation. Let n denote the number of a's and m the number of b's in the observed sequence.

B. (two-sample test) The data consist of two random samples, mutually independent of each other. One sample $X_1, \ldots, X_n$ is of size n and the other $Y_1, \ldots, Y_m$ is of size m. Arrange the two samples into one combined ordered sample, as follows:

$$X < X < Y < X < \ldots < Y < X$$

It is necessary to retain the identity of the observations as X's or Y's as above, although the actual numerical value is unimportant after the ordering. If several observations are equal to each other, causing an ambiguity in the total number of runs, it is recommended that two combined ordered arrangements be made. Let the first arrangement be the one that results in the fewest runs, let the second arrangement be the one that gives the most runs, and let the test statistic equal the average of those two numbers of runs.

ASSUMPTIONS. For the test of randomness the only assumption is that the observations be recordable as either one type (a) or the other (b). For the two-sample test we have the following assumptions.

1. The two samples are random samples, mutually independent of each other.
2. The random variables are continuous.

HYPOTHESES

A. (test for randomness)

H_0: The process which generates the sequence is a random process.

H_1: The random variables in the sequence are either dependent on other random variables in the sequence, or are distributed differently from one another.

B. (two-sample test)

H_0: The X's and the Y's have identical distribution functions.

H_1: The distribution function of the X's is different than the distribution function of the Y's.

TEST STATISTIC
A. (test for randomness) The test statistic T equals the total number of runs of like elements in the sequence of observations.
B. (two-sample test) The test statistic T equals the total number of runs of observations from the same population, both populations being considered, in the ordered combined sample.

DECISION RULE. Use Table 23 to obtain the quantiles w_p of T, under the assumption that H_0 is true. In the test for randomness use a two-tailed critical region, and reject H_0 at the level α if $T > w_{1-\alpha/2}$ or if $T < w_{\alpha/2}$. In the two-sample test only the lower tail is used, so reject H_0 at the level α if T is less than w_α. Table 23 furnishes some quantiles w_p for selected values of p (.005, .01, .025, .05, .10, .90, .95, .975, .99, .995) and for selected sample sizes m and n (2, 5, 8, 11, 14, 17, 20). For other values of m and n less than 20, use the nearest sample size given, and the quantiles will be exact in most cases. If m or n exceeds 20 use the large sample approximation given at the end of the table.

Example 1. (test for randomness) A meteorologist studies a 100 day record of precipitation, where each day is classified as W (wet, .01 inch of rain or more) or D (dry, less than .01 inch of rain). The null hypothesis is randomness; that is, the probability of a day being W is the same for all days, independently of whether the other days are W or D. The alternative hypothesis is non-randomness of one type or another. The record is as follows, with the runs bracketed for easy counting.

Day Number, and Precipitation Result

1-D⌉	11-D	21-D	31-D	41-D⌉	51-D	61-D	71-D⌉	81-D⌉	91-D
2-D	12-D⌋	22-D⌋	32-D	42-D⌋	52-D	62-D	72-D⌋	82-D⌋	92-D
3-D	13-W⌉	23-W⌉	33-D⌉	43-W⌉	53-D	63-D	73-W⌉	83-W⌉	93-D
4-D⌋	14-W	24-D⌉	34-D⌋	44-W	54-D	64-D	74-W⌋	84-W	94-D
5-W⌉	15-W	25-D	35-W⌉	45-W⌋	55-D⌋	65-D	75-D⌉	85-W⌋	95-D
6-W⌋	16-W⌋	26-D⌋	36-W⌋	46-D⌉	56-W⌉	66-D⌋	76-D	86-D⌉	96-D
7-D⌉	17-D⌉	27-W⌉	37-D⌉	47-D	57-W	67-W⌉	77-D	87-D	97-D⌋
8-D	18-D	28-W	38-D	48-D⌋	58-W	68-W⌋	78-D⌋	88-D	98-W⌉
9-D	19-D	29-W⌋	39-D	49-W⌉	59-W⌋	69-D⌉	79-W⌉	89-D	99-W
10-D	20-D	30-D⌉	40-D	50-D⌉	60-D⌉	70-D	80-D⌉	90-D	100-W⌋

There are 69 dry days and 31 wet days in the sequence, so $n = 69$ and $m = 31$. The total number of runs is 26;

$$T = 26$$

The approximate .025 and .975 quantiles are computed, to result in $\alpha = .05$. The approximation at the end of Table 23 gives

$$w_{.025} \cong \frac{2mn}{m + n} + 1 + x_{.025}\sqrt{\frac{2mn(2mn - m - n)}{(m + n)^2(m + n - 1)}}$$

$$\cong 43.78 + (-1.96)(4.249)$$

$$\cong 35.5$$

In a similar manner we find

$$w_{.975} = 52.1$$

Because $T = 26$ is less than $w_{.025} = 35.5$, the null hypothesis is rejected at $\alpha = .05$, and it is concluded that the sequence of wet and dry days is not random. The reason for the rejection of H_0 appears to be the tendency for wet days to follow wet days, as is the case when weather is caused primarily from frontal systems. Further examination shows the critical level $\hat{\alpha}$ to be somewhat smaller than .0002, the smallest $\hat{\alpha}$ which may be determined from Table 1 in a two tailed test.

Example 2.   (two-sample test) The same data are used here that were used in Example 5.3.2 to illustrate the Mann-Whitney test. Nine pieces of flint were collected, four from area A and five from area B. It is assumed that these pieces of flint resemble random samples from those areas, at least with respect to hardness. The pieces of flint were ranked in order of hardness by scratching each against the other, to test

H_0: The flints from areas A and B are of equal hardness against the alternative

H_1: The flints are not of equal hardness.

In this case we are actually testing whether the characteristic of hardness has the same probability distribution in the two areas, while the Mann-Whitney test is designed more for testing whether the average hardness is the same in both areas.

The combined ordered sample is ranked as follows, and the runs are underlined.

Origin of piece	A	A	A	B	A	B	B	B	B
Rank	1	2	3	4	5	6	7	8	9

The total number of runs is 4. From Table 23 we see that the .05 quantile, under $N_1 = 5$ and $N_2 = 5$ (which is as close to $n = 4$, $m = 5$ as we can get) for $\alpha = .05$ is given by $w_{.05} = 4$. Because T equals 4, the null hypothesis is accepted. The critical level is seen from Table 23 to be somewhat larger than .10, which says that 4 runs is not a very unlikely occurrence with sample sizes of 4 and 5. In the Mann-Whitney test illustration, the critical level was about .05.

Theory.   The exact distribution of the number of runs may be found in the same way that we have been finding distributions for statistics based on ranks. That is, under the null hypothesis that the two samples (in the two-sample test) came from identical populations, each ordered arrangement is equally likely, and has probability $1/\binom{m+n}{n}$, because there are $\binom{m+n}{n}$ ordered arrangements of the combined sample. The probability distribution of the test statistic is found by counting the number of arrangements that result in a specified number r, say, of runs, and dividing that number by $\binom{m+n}{n}$.

To facilitate the counting of arrangements, the key fact to consider is that each arrangement of X's and Y's in the combined ordered sample may be described completely merely by stating which X's begin runs of X's and which Y's begin runs of Y's, and whether the sequence begins with an X or a Y if r is an even number. For instance, if there are 4 X's and 5 Y's, and the runs of X's start with $X^{(1)}$, $X^{(3)}$, and $X^{(4)}$, and the runs of Y's start with $Y^{(1)}$, $Y^{(2)}$, and $Y^{(3)}$, then the only possible ordered arrangements are

$$\underline{X^{(1)} < X^{(2)}} < \underline{Y^{(1)}} < \underline{X^{(3)}} < \underline{Y^{(2)}} < \underline{X^{(4)}} < \underline{Y^{(3)} < Y^{(4)} < Y^{(5)}}$$

and

$$\underline{Y^{(1)}} < \underline{X^{(1)} < X^{(2)}} < \underline{Y^{(2)}} < \underline{X^{(3)}} < \underline{Y^{(3)} < Y^{(4)} < Y^{(5)}} < \underline{X^{(4)}}$$

each containing six runs. This key fact enables us to count easily the number of arrangements producing exactly r runs, in the following manner.

If r is an even number, then there are $r/2$ runs of X's and $r/2$ runs of Y's. The first run of X's begins with $X^{(1)}$, the smallest X, while the remaining $r/2 - 1$ runs of X's may begin with any $r/2 - 1$ of the remaining $n - 1$ X's. There are $\binom{n-1}{r/2-1}$ ways of selecting these $r/2 - 1$ X's from the $n - 1$ X's available. The same statement may be made of the Y's; that is, the first Y run begins with $Y^{(1)}$, but the remaining $r/2 - 1$ Y runs may begin with any $r/2 - 1$ of the remaining $m - 1$ Y's. These Y's may be selected in $\binom{m - 1}{r/2 - 1}$ ways. So, when r is even, we have

(1) $\binom{n - 1}{\frac{r}{2} - 1}$ = number of ways of specifying which X order statistics begin runs of X's

(2) $\binom{m - 1}{\frac{r}{2} - 1}$ = number of ways of specifying which Y order statistics begin runs of Y's

(3) $\binom{n - 1}{\frac{r}{2} - 1}\binom{m - 1}{\frac{r}{2} - 1}$ = number of ways of specifying which X's and which Y's begin runs

Because each arrangement may begin with an X or a Y when r is even, the above number of ways of specifying which X's and which Y's begin runs needs to be doubled to get the number of arrangements of the ordered combined sample which result in r runs.

(4) $2\binom{n - 1}{\frac{r}{2} - 1}\binom{m - 1}{\frac{r}{2} - 1}$ = number of arrangements of X's and Y's which yield r runs, r even.

Division by the total number of arrangements gives the desired probability.

(5) $P(T = r \mid H_0 \text{ is true}) = \dfrac{2\binom{n - 1}{\frac{r}{2} - 1}\binom{m - 1}{\frac{r}{2} - 1}}{\binom{n + m}{n}}$,   when r is even

This gives the probability of T equaling r when r is even.

If r is odd, either there is one more run of X's, $(r + 1)/2$ runs, than there are runs of Y's, $(r - 1)/2$, in which case the sequence begins and ends with an X, or else there is one more run of Y's than X's and the sequence begins and ends with a Y. The same reasoning used above when r is even may be used when r is odd to get the following.

(6) $\binom{n - 1}{\frac{(r - 1)}{2}}\binom{m - 1}{\frac{(r - 3)}{2}}$ = number of arrangements that give $(r + 1)/2$ X runs and $(r - 1)/2$ Y runs

(7) $\binom{n - 1}{\frac{(r - 3)}{2}}\binom{m - 1}{\frac{(r - 1)}{2}}$ = number of arrangements that give $(r - 1)/2$ runs of X's and $(r + 1)/2$ runs of Y's

(8) $\dbinom{n-1}{\frac{(r-1)}{2}}\dbinom{m-1}{\frac{(r-3)}{2}} + \dbinom{n-1}{\frac{(r-3)}{2}}\dbinom{m-1}{\frac{(r-1)}{2}}$ = the total number of arrangements that yield r runs, when r is odd

(9) $P(T = r \mid H_0 \text{ is true}) = \dfrac{\dbinom{n-1}{\frac{(r-1)}{2}}\dbinom{m-1}{\frac{(r-3)}{2}} + \dbinom{n-1}{\frac{(r-3)}{2}}\dbinom{m-1}{\frac{(r-1)}{2}}}{\dbinom{n+m}{2}}$

where r is odd

The last equation gives the probability of T equaling r runs when r is odd. Thus the probability function of T is specified by (5) when r is even and (9) when r is odd.

There is a slight difference in theory between the two-sample test and the test for randomness. In the two-sample test the sample sizes m and n are known in advance, and the probabilities are found by considering each ordered arrangement to be equally likely for a given m and n. In the test for randomness the number of a's and b's are not known until the entire sequence of observations is known, and m and n are part of the uncertainty of the data. It is still true that each arrangement of a's and b's is equally likely when m and n are given, so the probability function derived above still holds true for the test for randomness, although it is actually a conditional probability function, conditional on knowing the values for m and n. The test as described is legitimate also, because the critical region depends on m and n. That is, after the sequence is observed, m and n are noted, and then Table 23 is entered with those values of m and n to determine the critical region. As long as a critical region of size α is used for each actual m and n obtained, the overall level of significance for all possible m and n is also equal to the same α. Therefore the Wald-Wolfowitz runs test may be used as a test for randomness in a sequence. The unconditional probability distribution of T, given only $m + n$, is the subject of a recent paper by Dunn (1969).

The probability function for the total number of runs was given by Ising (1925). Wald and Wolfowitz (1940) applied this distribution to the two-sample problem and thus invented the two-sample test described above. They also showed the asymptotic normality of T, which is used to get the approximate quantiles at the end of Table 23. A basic paper by Mood (1940) gave the history of the theory of runs of various types up to that time, and developed the theory further, along other lines. Extensive tables were given by Swed and Eisenhart (1943).

The above test for randomness may be used as a test for trend by calling the observations a's if they fall above some point, such as the median, or b's if they fall below or on the point. The null hypothesis of randomness may be rejected in favor of the alternative, trend, if there are too few runs.

Mosteller (1941) suggested using the length of the longest run above (or below) the median as a test for trend. Wallis and Moore (1941) recommended using a goodness of fit test on the number of runs of length 1, 2, or more than 2 as a test for randomness. Later they (Moore and Wallis, 1943) introduced the theory of runs up and down, which is covered well by Bradley (1968), Chapter 12. Other types of runs tests have been introduced by Kruskal (1952), Mood (1954), Dixon (1945), Ferguson and Kraft (1955), Goodman (1957), Weiss (1960), and others.

Some recent papers present a runs test for circular distributions (Asano, 1965), for time series (Sen, 1965), for acceptance sampling (Prairie, Zimmer, and Brookhouse, 1962), for sequences of random digits (Bofinger and Bofinger, 1961), and for comovements between time series (Goodman and Grunfeld, 1961). See Barton (1967) for a recent discussion of runs above and below the median.

PROBLEMS

1. A professional baseball team had the following sequence of wins and losses for the month of July: $W L W W L W L W L L L W L L W W L W W L W L W L W L L$ $W L W L$. Does their win-loss record appear to be random?

2. Test the following data for randomness by dichotomizing the data at some convenient point.

61,	21,	89,	31,	81,	65,	67,	34,	64,	45,	97,	97,	29,	20,	98,
20,	22,	3,	61,	42,	92,	98,	14,	59,	69,	19,	36,	83,	71,	80,
17,	72,	28,	23,	91,	15,	14,	42,	49,	75,	58,	73,	57,	20,	77,
09,	24,	73,	67,	48,	32,	68,	65,	18,	12,	14,	65,	80,	23,	80,
52,	28,	91,	84,	26,	30,	66,	42,	43,	28,	23,	6,	96,	5,	94,
27,	83,	70,	18,	16,	10,	84,	94,	79,	50,	23,	9,	34,	22,	66,
81,	78,	76,	60,	14,	95,	25,	47,	47,	36.					

3. Use the Wald-Wolfowitz runs test to test for differences between the distribution functions of the two groups sampled in Problem 5.3.6.

4. Use the Wald-Wolfowitz runs test to test for differences between the comfortable temperatures for men and for women, using the data in Problem 5.3.2.

7.4. FISHER'S METHOD OF RANDOMIZATION

This final section of the book introduces the simple concept of randomization of the data, credited to Fisher (1935). This method of randomization is flexible enough to be used for many different types of hypothesis tests. It may be used to make inferences concerning the median in the one sample problem, or the median difference in the matched pairs problem, in much the same way that the Wilcoxon signed ranks test was used in Section 5.1. Fisher's method of randomization also furnishes a test for two independent samples, as did the Mann-Whitney test of Section 5.3. It may be used to test the hypothesis of independence in the matched pairs case, or equal means in the one way layout or the randomized complete block design. It is flexible enough to test almost any hypothesis in almost any situation for which another test exists.

The big drawback to Fisher's method of randomization is that the testing procedure is usually very long and tedious. This is because it is not possible to make tables of the critical regions, or the quantiles of the test statistics, and so each time a randomization test is applied the critical region must be determined on the basis of the data which were collected. The best way to describe the degree of difficulty involved is to imagine that instead of a randomization test we are going to use a rank test, such as the Mann-Whitney test for two independent samples, but no tables are available, not even a large sample approximation. Imagine that two samples are drawn, one of size 15 and the other of size 20. In order to find a critical region of size $\alpha = .05$, say, the $\binom{35}{15}$ arrangements in the combined ordered sample are assumed to be equally likely and the ones resulting in the largest (or smallest) values of the test statistic are counted. The job is difficult enough by hand when the sample sizes are 2 and 3, or with a computer for sample sizes of 10 and 15. For samples of size 15 and 20 the task is indeed monumental. The task of finding the critical region in a randomization test is accomplished in much the same way as in a rank test, but all of the convenient shortcuts available in the rank tests are not available in the randomization tests, so the task is much more difficult.

This all adds up to saying that tests based on Fisher's method of randomization are almost impossible to apply unless the sample sizes are very small. One reason we are presenting some of these tests is to point out that the rank tests of Chapter 5 are merely simplified versions of these older randomization tests. A second reason we feel these tests are important in any study of nonparametric statistics is that they may be applied equally well with discrete or continuous data, while the rank tests are designed for continuous data and are weakened somewhat by the presence of ties. A third reason for their importance is that their asymptotic relative efficiency (A.R.E.) is 1.0 when compared to the most powerful parametric tests in some specific situations described by Lehmann and Stein (1949) and Hoeffding (1952).

We shall now describe a test using matched pairs data, based on Fisher's method of randomization. According to Kempthorne and Doerfler (1969) the randomization test for matched pairs is always to be preferred over the Wilcoxon test or the sign test.

The Randomization Test for Matched Pairs

DATA. The data consist of observations on n' bivariate random variables $(X_1, Y_1), (X_2, Y_2), \ldots, (X_{n'}, Y_{n'})$. Omit from further consideration all pairs (X_i, Y_i) whose difference $Y_i - X_i$ is zero, and let the remaining number of pairs be denoted by n. Denote the nonzero differences $Y_i - X_i$ by $D_1, D_2, \ldots, D_n$.

ASSUMPTIONS

1. The distribution of each D_i is symmetric about zero.
2. The D_i's are mutually independent.
3. The D_i's all have the same median.
4. The measurement scale of the D_i's is at least interval.

HYPOTHESES. Only the two-tailed version of this test is presented, although the one-tailed versions may be obtained by comparison with the Wilcoxon signed ranks test of Section 5.1. Let the common median of the D_i's be denoted by $d_{.50}$.

$$H_0 : d_{.50} = 0$$
$$H_1 : d_{.50} \neq 0$$

TEST STATISTIC. The test statistic T_1 equals the sum of the positive differences.

(1) $$T_1 = \sum D_i \qquad \text{only for those } D_i > 0$$

DECISION RULE. Reject H_0 at the level α if $T_1 > w_{1-\alpha/2}$ or if $T_1 < w_{\alpha/2}$, where the quantiles $w_{1-\alpha/2}$ and $w_{\alpha/2}$ are found as follows.

Consider only the absolute values of the D_i's, $|D_i|$, without regard for whether they were originally positive or negative. There are 2^n ways of assigning $+$ or $-$ signs to the set of absolute differences obtained; that is, we might assign $+$ signs to all n of the $|D_i|$, or we might assign a $+$ to $|D_1|$ but $-$ signs to $|D_2|$ through $|D_n|$, and so on. To find the p quantile w_p, $0 \leq p \leq 1$, first find the $(2^n)(p)$ assignments of signs that give the smallest values for T_1, the sum of the "positive" absolute differences. [If $(2^n)(p)$ is not an integer, use the next larger integer.] The largest value of T_1 thus obtained is the p quantile w_p of T_1 under the null hypothesis. [If $(2^n)(p)$ is an integer, use the average of the largest value of T_1 thus obtained, and the largest value of T_1 possible if $(2^n)(p) + 1$ arrangements had been considered instead, according to our usual convention.]

The above method of finding w_p works for all values of p from 0 to 1, but in practice it should be used only for small values of p, such as $p = \alpha/2$. For large values of p, such as $p = 1 - \alpha/2$ the relationship

(2)
$$w_{1-\alpha/2} = \sum_{i=1}^{n} |D_i| - w_{\alpha/2}$$

should be used. The relationship in (2) is apparent if one considers that for every assignment of signs that results in a small value of T_1, a complete reversal of signs (pluses replaced by minuses, and vice versa) results in a large value of T_1. The latter value of T_1, the sum of the "positive" $|D_i|$'s, plus the former value of T_1, the sum of the now "negative" $|D_i|$'s, add up to the sum of all of the $|D_i|$'s, as indicated by (2).

The critical value $\hat{\alpha}$ is obtained by counting the number of assignments of signs which result in a smaller $\left(\text{or larger, if the observed } T_1 > 1/2 \sum_{i=1}^{n} |D_i|\right)$ value of T_1, or the same value for T_1, as the one obtained from the data. This number is doubled and divided by 2^n to get $\hat{\alpha}$.

Example 1. Suppose that eight matched pairs resulted in the following differences: $-16, -4, -7, -3, 0, +5, +1, -10$. The zero is discarded, and we have

$$D_1 = -16, D_2 = -4, D_3 = -7, D_4 = -3, D_5 = +5, D_6 = +1, D_7 = -10$$

and $n = 7$. The null hypothesis
$$H_0 : d_{.50} = 0$$
is tested against the alternative
$$H_1 : d_{.50} \neq 0$$

using the randomization test at the level $\alpha = .05$.

The quantile $w_{.025}$ is found by considering the 4 [because $(2^7)(.025)$ equals 3.2] ways of assigning signs which result in the lowest sum of the "positive" absolute differences. These are given as follows.

Assignment of Signs	$\sum$ "positive" $|D_i|$
$-16, -4, -7, -3, -5, -1, -10$	$T_1 = 0$
$-16, -4, -7, -3, -5, +1, -10$	$T_1 = 1$
$-16, -4, -7, +3, -5, -1, -10$	$T_1 = 3$
$-16, -4, -7, +3, -5, +1, -10$	$T_1 = 4$
(also $-16, +4, -7, -3, -5, -1, -10$ gives	$T_1 = 4$)

The largest of these T_1 values is 4, so
$$w_{.025} = 4$$

From Equation (2) we have
$$w_{.975} = \sum_{i=1}^{7} |D_i| - w_{.025}$$
$$= 46 - 4 = 42$$

The value of the test statistic obtained from the data is

$$T_1 = \sum \text{positive } D_i$$
$$= 5 + 1 = 6$$

which is neither less than 4 nor greater than 42, so H_0 is accepted.

The critical level $\hat{\alpha}$ is found by listing the assignments of signs that result in $T_1 \leq 6$, in addition to the five listed above.

Assignment of Signs	$\sum$ "positive" $|D_i|$
−16, +4, −7, −3, −5, +1, −10	$T = 5$
−16, −4, −7, −3, +5, −1, −10	$T = 5$
−16, −4, −7, −3, +5, +1, −10	$T = 6$

Thus there are eight arrangements that give values of T_1 less than or equal to the observed value of 6. Because this is a two-tailed test this number is doubled, and $\hat{\alpha}$ is given by

$$\hat{\alpha} = \frac{2(8)}{2^7} = \frac{16}{128} = .125$$

The previous randomization test is not as typical of randomization tests in general as is the next test. In the previous test the critical region was dependent upon the actual values observed, so that a different set of observations would generally result in a different critical region, and this is a characteristic of all tests using Fisher's method of randomization. However, the previous test also relied upon the assumption of symmetry in the distribution of the D_i's. In general the tests based on Fisher's method of randomization do not rely on assumptions which are often difficult to verify, such as the assumption of symmetry. For this reason the following test is more typical of the randomization tests.

The Randomization Test for Two Independent Samples

DATA. The data consists of two random samples $X_1, X_2, \ldots, X_n$ and $Y_1, Y_2, \ldots, Y_m$ of sizes n and m respectively.

ASSUMPTIONS

1. Both samples are random samples from their respective populations.

2. In addition to independence within each sample there is mutual independence between the two samples.

3. The measurement scale is at least interval.

4. Either the two population distribution functions are identical, or else one population has a larger mean than the other. (Without this assumption the test is still valid, but might lack consistency.)

HYPOTHESES. Only the two-tailed test is presented; the one-tailed tests may be surmised by direct analogy with the Mann-Whitney test in Section 5.3.

$$H_0 : E(X) = E(Y)$$

$$H_1 : E(X) \neq E(Y)$$

TEST STATISTIC. The test statistic T_2 is the sum of the X observations:

(3)
$$T_2 = \sum_{i=1}^{n} X_i$$

DECISION RULE. Reject H_0 at the level α if either $T_2 > w_{1-\alpha/2}$ or $T_2 < w_{\alpha/2}$, where the quantiles w_p are found as follows.

Consider the observed values of X_i and Y_j as merely a group of $n + m$ numbers, and consider the ways in which n of those numbers may be selected. There are $\binom{n + m}{n}$ such ways. To find the p quantile w_p consider the $\binom{n + m}{n}(p)$ selections which yield the smallest sums, which sum we shall call T_2. The largest T_2 thus obtained is w_p.

As before, if $\binom{n + m}{n}(p)$ is not an integer, round upwards to the next higher integer. If $\binom{n + m}{n}(p)$ is integer-valued, then w_p is the average of the largest T_2 thus obtained and the T_2 which would result from considering $\binom{n + m}{n}(p) + 1$ selections.

The critical value $\hat{\alpha}$ is obtained by counting the number of ways n of the $n + m$ observations may be selected so that their sum is smaller (or larger if the observed T_2 is in the upper tail) than, or equal to, the observed T_2 from the data. This number is doubled, because the test is two tailed, and divided by $\binom{n + m}{n}$ to get $\hat{\alpha}$.

Example 2. Suppose that a random sample yielded X_i's of 0, 1, 1, 0, -2, and an independent random sample of Y_j's gave 6, 7, 7, 4, -3, 9, 14. The null hypothesis

$$H_0 : E(X) = E(Y)$$

is tested against

$$H_1 : E(X) \neq E(Y)$$

at $\alpha = .05$, with the randomization test for two independent samples.

One sample is of size $n = 5$ and the other of size $m = 7$, so there are $\binom{12}{5} =$ 792 ways of forming a subset containing 5 of the 12 numbers. Not all of the subsets are distinguishable because there are two 0's, two 1's, and two 7's, so we shall distinguish between identical numbers with the aid of subscripts, such as $0_1, 0_2, 1_1$, etc. Because $(792)(.025) = 19.8$, we need to find the 20 groups which yield the lowest values of T_2, to obtain $w_{.025}$. These groups of numbers, and the corresponding values of T_2, are given as follows.

Observations	T_2	Observations	T_2
$-3, -2, 0_1, 0_2, 1_1$	-4	$-3, -2, 0_2, 1_2, 4$	0
$-3, -2, 0_1, 0_2, 1_2$	-4	$-3, -2, 1_1, 1_2, 4$	1
$-3, -2, 0_1, 1_1, 1_2$	-3	$-3, -2, 0_1, 0_2, 6$	1
$-3, -2, 0_2, 1_1, 1_2$	-3	$-3, -2, 0_1, 1_1, 6$	2
$-3, -2, 0_1, 0_2, 4$	-1	$-3, -2, 0_1, 1_2, 6$	2
$-3, 0_1, 0_2, 1_1, 1_2$	-1	$-3, -2, 0_2, 1_1, 6$	2
$-2, 0_1, 0_2, 1_1, 1_2$	0	$-3, -2, 0_2, 1_2, 6$	2
$-3, -2, 0_1, 1_1, 4$	0	$-3, 0_1, 0_2, 1_1, 4$	2
$-3, -2, 0_1, 1_2, 4$	0	$-3, 0_1, 0_2, 1_2, 4$	2
$-3, -2, 0_2, 1_1, 4$	0	$-3, -2, 0_1, 0_2, 7_1$	2

The largest T_2 thus obtained is

$$w_{.025} = 2$$

It is not necessary to find $w_{.975}$ even though this is a two-tailed test, because the observed T_2 is in the lower tail. The observed value of T_2 from the data is

$$T_2 = \sum_{i=1}^{5} X_i = 0 + 1 + 1 + 0 - 2 = 0$$

which is less than $w_{.025} = 2$, so H_0 is rejected. In fact H_0 could have been rejected at the level

$$\hat{\alpha} = \frac{2(11)}{792} = .028$$

because there are 11 possible arrangements of the numbers that yield values less than or equal to 0.

Theory.   The theory behind the randomization tests is partially explained by the method of finding the critical region. In the test for two independent samples, for instance, it is obvious that we are considering each selection of nX observations to be equally likely, from the $n + m$ observations available. It just remains to explain why we may consider the selections to be equally likely, and why we are working with the observations themselves as our "sample space," so to speak.

The selections may be considered to be equally likely because of the null hypothesis which states (along with the assumptions) that the X's and the Y's are all independent and identically distributed. Therefore the X's should have no more of a tendency to be low than the Y's have, or to be high, or to be in the middle. Given any group of $m + n$ numbers, whether they be observations or not, each subgroup of n of those numbers is just as likely to be the n values of X as any other subgroup of n of those numbers, because the numbers that are not X's have to be Y's and the overall probability attached to that group of numbers does not depend on which numbers are called X's and which numbers are called Y's. Now if the X's are distributed differently than the Y's, it will matter which numbers are called X's and which are called Y's, but for purposes of finding a critical region of size α we restrict our consideration to identically distributed random variables. So that is the intuitive argument for considering each selection of n observations as X's to be equally likely.

That also leads to the second question, "Why are we working with the observations themselves as our sample space?" We explained above that any set of $m + n$ numbers satisfies the "equally likely" criterion. But in a testing situation we need to identify the $m + n$ numbers used with the $m + n$ observations obtained. In a rank test the $m + n$ numbers used are the integers from 1 to $m + n$, and they are matched one for one with the observations by assigning ranks to the observations. In this case we are using the observations themselves as the numbers. This eliminates the problem of which numbers to assign to which observations, which occurs in the rank tests when ties confuse the ranking procedure. By using the observations themselves as the numbers, it is easy to identify one of the selections of n numbers as the one actually obtained in the data. Then, with the aid of the test statistic, all selections more extreme than the one obtained may also be identified, counted, and used to compute the critical level $\hat{\alpha}$.

The critical region is thus determined for individual subsets of the sample space, such as for the subset of all outcomes that have the same numerical values as the observed values in the data. These subsets are mutually exclusive, cover the entire sample space (given *any* set of observations we can find the critical region for that subset of the sample space), and each subset has a critical region of size α relative to the size of the entire subset. So the overall size of all the critical regions combined is also α, which shows that the test is indeed a valid one.

The principal difference between the test for two independent samples and the test for matched pairs, is that in the test for matched pairs the assumption of symmetry is used to justify the changing of algebraic signs without changing the probability. If a difference D_i can be a $+6$, say, then it can be a -6 with the same probability when its distribution is symmetric about zero. Again it does not matter which numbers are used. The Wilcoxon test used ranks. We use the observations themselves as numbers, so that we may easily identify one of the assignments of signs as corresponding to the one actually obtained.

The randomization test for matched pairs is discussed by Fisher (1935). The randomization test for two independent samples is presented by Pitman (1937/38) along with a randomization test for correlation and an analysis of variance test.

A randomization test for multivariate data is presented by Chung and Fraser (1958). Further discussions of randomization tests may be found in articles by Welch (1937), Scheffé (1943), Moses (1952), Smith (1953), and Kempthorne (1955), or in most of the books mentioned in the preface. A recent paper on multisample permutation tests is by Sen (1967b). Useful approximations to the distributions of the test statistics are discussed by Cleroux (1969).

PROBLEMS

1. Test whether the median difference $Y_i - X_i$ may be considered to be zero, where the observed differences are $+7$, $+3$, $+2$, $+8$, -1.

2. The paired observations on (X_i, Y_i) are $(17, 14)$, $(15, 14)$, $(12, 15)$, $(9, 7)$, $(17, 16)$, $(18, 18)$, $(14, 10)$. Is the median of $Y_i - X_i$ positive?

3. One random sample furnished the values 1, 4, 3, -2, 0, and the other gave 4, 6, 4. Can we conclude that the two population means are unequal?

4. Test $H_0 : E(X) \leq E(Y)$ against the one-sided alternative $H_1 : E(X) > E(Y)$, where the X values are 15, 17, 16, and the Y values are 12, 14, 15, 10, 12.

5. Suppose someone suggests subtracting a constant from all of the observations in the randomization test for two independent samples to make the calculations easier, such as subtracting 10 from each observation in Problem 4 before analyzing the data. Does this affect the results of the test? Explain. Would division of the observations by a constant affect the results?

6. Would the results of the randomization test for matched pairs be affected by subtracting of a constant from all of the observations, or by division of the data by a constant? Explain.

7. In the randomization test for correlation the critical region is determined, as in the rank correlation tests, by assuming that each pairing of X's with Y's is equally likely, where the data consist of a random bivariate sample (X_i, Y_i), $i = 1, \ldots, n$. Explain how to find the p quantile w_p of the test statistic $T_3 = \sum_{i=1}^{n} X_i Y_i$, under the null hypothesis of independence between the X's and the Y's.

Appendix

Table 1   Normal Distribution[a]

w_p	p	w_p	p	w_p	p
−3.7190	.0001	−.4677	.32	.5244	.70
−3.2905	.0005	−.4399	.33	.5534	.71
−3.0902	.001	−.4125	.34	.5828	.72
−2.5758	.005	−.3853	.35	.6128	.73
−2.3263	.01	−.3585	.36	.6433	.74
−2.1701	.015	−.3319	.37	.6745	.75
−2.0537	.02	−.3055	.38	.7063	.76
−1.9600	.025	−.2793	.39	.7388	.77
−1.8808	.03	−.2533	.40	.7722	.78
−1.7507	.04	−.2275	.41	.8064	.79
−1.6449	.05	−.2019	.42	.8416	.80
−1.5548	.06	−.1764	.43	.8779	.81
−1.4758	.07	−.1510	.44	.9154	.82
−1.4395	.075	−.1257	.45	.9542	.83
−1.4051	.08	−.1004	.46	.9945	.84
−1.3408	.09	−.0753	.47	1.0364	.85
−1.2816	.10	−.0502	.48	1.0803	.86
−1.2265	.11	−.0251	.49	1.1264	.87
−1.1750	.12	.0000	.50	1.1750	.88
−1.1264	.13	.0251	.51	1.2265	.89
−1.0803	.14	.0502	.52	1.2816	.90
−1.0364	.15	.0753	.53	1.3408	.91
−.9945	.16	.1004	.54	1.4051	.92
−.9542	.17	.1257	.55	1.4395	.925
−.9154	.18	.1510	.56	1.4758	.93
−.8779	.19	.1764	.57	1.5548	.94
−.8416	.20	.2019	.58	1.6449	.95
−.8064	.21	.2275	.59	1.7507	.96
−.7722	.22	.2533	.60	1.8808	.97
−.7388	.23	.2793	.61	1.9600	.975
−.7063	.24	.3055	.62	2.0537	.98
−.6745	.25	.3319	.63	2.1701	.985
−.6433	.26	.3585	.64	2.3263	.99
−.6128	.27	.3853	.65	2.5758	.995
−.5828	.28	.4125	.66	3.0902	.999
−.5534	.29	.4399	.67	3.2905	.9995
−.5244	.30	.4677	.68	3.7190	.9999
−.4959	.31	.4959	.69		

SOURCE.   Abridged from Tables 3 and 4, pp. 111–112, Pearson and Hartley (1962).

[a] The entries in this table are quantiles w_p of the standard normal random variable W, selected so $P(W \leq w_p) = p$ and $P(W > w_p) = 1 - p$.

Table 2   CHI-SQUARE DISTRIBUTION[a]

	$p = .750$	.900	.950	.975	.990	.995	.999
$k = 1$	1.323	2.706	3.841	5.024	6.635	7.879	10.83
2	2.773	4.605	5.991	7.378	9.210	10.60	13.82
3	4.108	6.251	7.815	9.348	11.34	12.84	16.27
4	5.385	7.779	9.488	11.14	13.28	14.86	18.47
5	6.626	9.236	11.07	12.83	15.09	16.75	20.51
6	7.841	10.64	12.59	14.45	16.81	18.55	22.46
7	9.037	12.02	14.07	16.01	18.48	20.28	24.32
8	10.22	13.36	15.51	17.53	20.09	21.96	26.13
9	11.39	14.68	16.92	19.02	21.67	23.59	27.88
10	12.55	15.99	18.31	20.48	23.21	25.19	29.59
11	13.70	17.28	19.68	21.92	24.73	26.76	31.26
12	14.85	18.55	21.03	23.34	26.22	28.30	32.91
13	15.98	19.81	22.36	24.74	27.69	29.82	34.53
14	17.12	21.06	23.68	26.12	29.14	31.32	36.12
15	18.25	22.31	25.00	27.49	30.58	32.80	37.70
16	19.37	23.54	26.30	28.85	32.00	34.27	39.25
17	20.49	24.77	27.59	30.19	33.41	35.72	40.79
18	21.60	25.99	28.87	31.53	34.81	37.16	42.31
19	22.72	27.20	30.14	32.85	36.19	38.58	43.82
20	23.83	28.41	31.41	34.17	37.57	40.00	45.32
21	24.93	29.62	32.67	35.48	38.93	41.40	46.80
22	26.04	30.81	33.92	36.78	40.29	42.80	48.27
23	27.14	32.01	35.17	38.08	41.64	44.18	49.73
24	28.24	33.20	36.42	39.37	42.98	45.56	51.18
25	29.34	34.38	37.65	40.65	44.31	46.93	52.62
26	30.43	35.56	38.89	41.92	45.64	48.29	54.05
27	31.53	36.74	40.11	43.19	46.96	49.64	55.48
28	32.62	37.92	41.34	44.46	48.28	50.99	56.89
29	33.71	39.09	42.56	45.72	49.59	52.34	58.30
30	34.80	40.26	43.77	46.98	50.89	53.67	59.70
40	45.62	51 81	55.76	59.34	63.69	66.77	73.40
50	56.33	63.17	67.50	71.42	76.15	79.49	86.66
60	66.98	74.40	79.08	83.30	88.38	91.95	99.61
70	77.58	85.53	90.53	95.02	100.4	104.2	112.3
80	88.13	96.58	101.9	106.6	112.3	116.3	124.8
90	98.65	107.6	113.1	118.1	124.1	128.3	137.2
100	109.1	118.5	124.3	129.6	135.8	140.2	149.4
x_p	.675	1.282	1.645	1.960	2.326	2.576	3.090

For $k > 100$ use the approximation $w_p = (1/2)x(_p + \sqrt{2k - 1})^2$, or the more accurate $w_p = k\left(1 - \dfrac{2}{9k} + x_p\sqrt{\dfrac{2}{9k}}\right)^3$, where x_p is the value from the standardized normal distribution shown in the bottom of the table.

SOURCE.   Abridged from Table 8, p. 131, Pearson and Hartley (1962).

[a] The entries in this table are quantiles w_p of a chi-square random variable W with k degrees of freedom, selected so $P(W \leq w_p) = p$ and $P(W > w_p) = 1 - p$.

Table 3 BINOMIAL DISTRIBUTION[a]

n	y	p = .05	.10	.15	.20	.25	.30	.35	.40	.45
1	0	.9500	.9000	.8500	.8000	.7500	.7000	.6500	.6000	.5500
	1	1.0000	1.0000	1.0000	1.0000	1.0000	1.0000	1.0000	1.0000	1.0000
2	0	.9025	.8100	.7225	.6400	.5625	.4900	.4225	.3600	.3025
	1	.9975	.9900	.9775	.9600	.9375	.9100	.8775	.8400	.7975
	2	1.0000	1.0000	1.0000	1.0000	1.0000	1.0000	1.0000	1.0000	1.0000
3	0	.8574	.7290	.6141	.5120	.4219	.3430	.2746	.2160	.1664
	1	.9928	.9720	.9392	.8960	.8438	.7840	.7182	.6480	.5748
	2	.9999	.9990	.9966	.9920	.9844	.9730	.9571	.9360	.9089
	3	1.0000	1.0000	1.0000	1.0000	1.0000	1.0000	1.0000	1.0000	1.0000
4	0	.8145	.6561	.5220	.4096	.3164	.2401	.1785	.1296	.0915
	1	.9860	.9477	.8905	.8192	.7383	.6517	.5630	.4752	.3910
	2	.9995	.9963	.9880	.9728	.9492	.9163	.8735	.8208	.7585
	3	1.0000	.9999	.9995	.9984	.9961	.9919	.9850	.9743	.9590
	4	1.0000	1.0000	1.0000	1.0000	1.0000	1.0000	1.0000	1.0000	1.0000
5	0	.7738	.5905	.4437	.3277	.2373	.1681	.1160	.0778	.0503
	1	.9774	.9185	.8352	.7373	.6328	.5282	.4284	.3370	.2562
	2	.9988	.9914	.9734	.9421	.8965	.8369	.7648	.6826	.5931
	3	1.0000	.9995	.9978	.9933	.9844	.9692	.9460	.9130	.8688
	4	1.0000	1.0000	.9999	.9997	.9990	.9976	.9947	.9898	.9815
	5	1.0000	1.0000	1.0000	1.0000	1.0000	1.0000	1.0000	1.0000	1.0000
6	0	.7351	.5314	.3771	.2621	.1780	.1176	.0754	.0467	.0277
	1	.9672	.8857	.7765	.6554	.5339	.4202	.3191	.2333	.1636
	2	.9978	.9842	.9527	.9011	.8306	.7443	.6471	.5443	.4415
	3	.9999	.9987	.9941	.9830	.9624	.9295	.8826	.9208	.7447
	4	1.0000	.9999	.9996	.9984	.9954	.9891	.9777	.9590	.9308
	5	1.0000	1.0000	1.0000	.9999	.9998	.9993	.9982	.9959	.9917
	6	1.0000	1.0000	1.0000	1.0000	1.0000	1.0000	1.0000	1.0000	1.0000
7	0	.6983	.4783	.3206	.2097	.1335	.0824	.0490	.0280	.0152
	1	.9556	.8503	.7166	.5767	.4449	.3294	.2338	.1586	.1024
	2	.9962	.9743	.9262	.8520	.7564	.6471	.5323	.4199	.3164
	3	.9998	.9973	.9879	.9667	.9294	.8740	.8002	.7102	.6083
	4	1.0000	.9998	.9988	.9953	.9871	.9812	.9444	.9037	.8471
	5	1.0000	1.0000	.9999	.9996	.9987	.9962	.9910	.9812	.9643
	6	1.0000	1.0000	1.0000	1.0000	.9999	.9998	.9994	.9984	.9963
	7	1.0000	1.0000	1.0000	1.0000	1.0000	1.0000	1.0000	1.0000	1.0000

[a] Y has the binomial distribution with parameters n and p. The entries are the values of $P(Y \leq y) = \sum_{i=0}^{y} \binom{n}{i} p^i (1 - p)^{n-i}$, for p ranging from .05 to .95.

Table 3 (CONTINUED)

n	y	p = .50	.55	.60	.65	.70	.75	.80	.85	.90	.95
1	0	.5000	.4500	.4000	.3500	.3000	.2500	.2000	.1500	.1000	.0500
	1	1.0000	1.0000	1.0000	1.0000	1.0000	1.0000	1.0000	1.0000	1.0000	1.0000
2	0	.2500	.2025	.1600	.1225	.0900	.0625	.0400	.0225	.0100	.0025
	1	.7500	.6975	.6400	.5775	.5100	.4375	.3600	.2775	.1900	.0975
	2	1.0000	1.0000	1.0000	1.0000	1.0000	1.0000	1.0000	1.0000	1.0000	1.0000
3	0	.1250	.0911	.0640	.0429	.0270	.0156	.0080	.0034	.0010	.0001
	1	.5000	.4252	.3520	.2818	.2160	.1562	.1040	.0608	.0280	.0072
	2	.8750	.8336	.7840	.7254	.6570	.5781	.4880	.3859	.2710	.1426
	3	1.0000	1.0000	1.0000	1.0000	1.0000	1.0000	1.0000	1.0000	1.0000	1.0000
4	0	.0625	.0410	.0256	.0150	.0081	.0039	.0016	.0005	.0001	.0000
	1	.3125	.2415	.1792	.1265	.0837	.0508	.0272	.0120	.0037	.0005
	2	.6875	.6090	.5248	.4370	.3483	.2617	.1808	.1095	.0523	.0140
	3	.9375	.9085	.8704	.8215	.7599	.6836	.5904	.4780	.3439	.1855
	4	1.0000	1.0000	1.0000	1.0000	1.0000	1.0000	1.0000	1.0000	1.0000	1.0000
5	0	.0312	.0185	.0102	.0053	.0024	.0010	.0003	.0001	.0000	.0000
	1	.1875	.1312	.0870	.0540	.0308	.0156	.0067	.0022	.0005	.0000
	2	.5000	.4069	.3174	.2352	.1631	.1035	.0579	.0266	.0086	.0012
	3	.8125	.7438	.6630	.5716	.4718	.3672	.2627	.1648	.0815	.0226
	4	.9688	.9497	.9222	.8840	.8319	.7627	.6723	.5563	.4095	.2262
	5	1.0000	1.0000	1.0000	1.0000	1.0000	1.0000	1.0000	1.0000	1.0000	1.0000
6	0	.0156	.0083	.0041	.0018	.0007	.0002	.0001	.0000	.0000	.0000
	1	.1094	.0692	.0410	.0223	.0109	.0046	.0016	.0004	.0001	.0000
	2	.3438	.2553	.1792	.1174	.0705	.0376	.0170	.0059	.0013	.0001
	3	.6562	.5585	.4557	.3529	.2557	.1694	.0989	.0473	.0158	.0022
	4	.8906	.8364	.7667	.6809	.5798	.4661	.3446	.2235	.1143	.0328
	5	.9844	.9723	.9533	.9246	.8824	.8220	.7379	.6229	.4686	.2649
	6	1.0000	1.0000	1.0000	1.0000	1.0000	1.0000	1.0000	1.0000	1.0000	1.0000
7	0	.0078	.0037	.0016	.0006	.0002	.0001	.0000	.0000	.0000	.0000
	1	.0625	.0357	.0188	.0090	.0038	.0013	.0004	.0001	.0000	.0000
	2	.2266	.1529	.0963	.0556	.0288	.0129	.0047	.0012	.0002	.0000
	3	.5000	.3917	.2898	.1998	.1260	.0706	.0333	.0121	.0027	.0002
	4	.7734	.6836	.5801	.4677	.3529	.2436	.1480	.0738	.0257	.0038
	5	.9375	.8976	.8414	.7662	.6706	.5551	.4233	.2834	.1497	.0444
	6	.9922	.9848	.9720	.9510	.9176	.8665	.7903	.6794	.5217	.3017
	7	1.0000	1.0000	1.0000	1.0000	1.0000	1.0000	1.0000	1.0000	1.0000	1.0000

Table 3 (CONTINUED)

n	y	p = .05	.10	.15	.20	.25	.30	.35	.40	.45
8	0	.6634	.4305	.2725	.1678	.1001	.0576	.0319	.0168	.0084
	1	.9428	.8131	.6572	.5033	.3671	.2553	.1691	.1064	.0632
	2	.9942	.9619	.8948	.7969	.6785	.5518	.4278	.3154	.2201
	3	.9996	.9950	.9786	.9437	.8862	.8059	.7064	.5941	.4770
	4	1.0000	.9996	.9971	.9896	.9727	.9420	.8939	.8263	.7396
	5	1.0000	1.0000	.9998	.9988	.9958	.9887	.9747	.9502	.9115
	6	1.0000	1.0000	1.0000	.9999	.9996	.9987	.9964	.9915	.9819
	7	1.0000	1.0000	1.0000	1.0000	1.0000	.9999	.9988	.9993	.9983
	8	1.0000	1.0000	1.0000	1.0000	1.0000	1.0000	1.0000	1.0000	1.0000
9	0	.6302	.3874	.2316	.1342	.0751	.0404	.0207	.0101	.0046
	1	.9288	.7748	.5995	.4362	.3003	.1960	.1211	.0705	.0385
	2	.9916	.9470	.8591	.7382	.6007	.4628	.3373	.2318	.1495
	3	.9994	.9917	.9661	.9144	.8343	.7297	.6089	.4826	.3614
	4	1.0000	.9991	.9944	.9804	.9511	.9012	.8283	.7334	.6214
	5	1.0000	.9999	.9994	.9969	.9900	.9747	.9464	.9006	.8342
	6	1.0000	1.0000	1.0000	.9997	.9987	.9957	.9888	.9750	.9502
	7	1.0000	1.0000	1.0000	1.0000	.9999	.9996	.9986	.9962	.9909
	8	1.0000	1.0000	1.0000	1.0000	1.0000	1.0000	.9999	.9997	.9992
	9	1.0000	1.0000	1.0000	1.0000	1.0000	1.0000	1.0000	1.0000	1.0000
10	0	.5987	.3487	.1969	.1074	.0563	.0282	.0135	.0060	.0025
	1	.9139	.7361	.5443	.3758	.2440	.1493	.0860	.0464	.0233
	2	.9885	.9298	.8202	.6778	.5256	.3828	.2616	.1673	.0996
	3	.9990	.9872	.9500	.8791	.7759	.6496	.5138	.3823	.2660
	4	.9999	.9984	.9901	.9672	.9219	.8497	.7515	.6331	.5044
	5	1.0000	.9999	.9986	.9936	.9803	.9527	.9051	.8338	.7384
	6	1.0000	1.0000	.9999	.9991	.9965	.9894	.9740	.9452	.8980
	7	1.0000	1.0000	1.0000	.9999	.9996	.9984	.9952	.9877	.9726
	8	1.0000	1.0000	1.0000	1.0000	1.0000	.9999	.9995	.9983	.9955
	9	1.0000	1.0000	1.0000	1.0000	1.0000	1.0000	1.0000	.9999	.9997
	10	1.0000	1.0000	1.0000	1.0000	1.0000	1.0000	1.0000	1.0000	1.0000
11	0	.5688	.3138	.1673	.0859	.0422	.0198	.0088	.0036	.0014
	1	.8981	.6974	.4922	.3221	.1971	.1130	.0606	.0302	.0139
	2	.9848	.9104	.7788	.6174	.4552	.3127	.2001	.1189	.0652
	3	.9984	.9815	.9306	.8389	.7133	.5696	.4256	.2963	.1911
	4	.9999	.9972	.9841	.9496	.8854	.7897	.6683	.5328	.3971
	5	1.0000	.9997	.9973	.9883	.9657	.9218	.8513	.7535	.6331
	6	1.0000	1.0000	.9997	.9980	.9924	.9784	.9499	.9006	.8262
	7	1.0000	1.0000	1.0000	.9998	.9988	.9957	.9878	.9707	.9390
	8	1.0000	1.0000	1.0000	1.0000	.9999	.9994	.9980	.9941	.9852
	9	1.0000	1.0000	1.0000	1.0000	1.0000	1.0000	.9998	.9993	.9978
	10	1.0000	1.0000	1.0000	1.0000	1.0000	1.0000	1.0000	1.0000	.9998
	11	1.0000	1.0000	1.0000	1.0000	1.0000	1.0000	1.0000	1.0000	1.0000

Table 3 (Continued)

n	y	p = .50	.55	.60	.65	.70	.75	.80	.85	.90	.95
8	0	.0039	.0017	.0007	.0002	.0001	.0000	.0000	.0000	.0000	.0000
	1	.0352	.0181	.0085	.0036	.0013	.0004	.0001	.0000	.0000	.0000
	2	.1445	.0885	.0498	.0253	.0113	.0042	.0012	.0002	.0000	.0000
	3	.3633	.2604	.1737	.1061	.0580	.0273	.0104	.0029	.0004	.0000
	4	.6367	.5230	.4059	.2936	.1941	.1138	.0563	.0214	.0050	.0004
	5	.8555	.7799	.6846	.5722	.4482	.3215	.2031	.1052	.0381	.0058
	6	.9648	.9368	.8936	.8309	.7447	.6329	.4967	.3428	.1869	.0572
	7	.9961	.9916	.9832	.9681	.9424	.8999	.8322	.7275	.5695	.3366
	8	1.0000	1.0000	1.0000	1.0000	1.0000	1.0000	1.0000	1.0000	1.0000	1.0000
9	0	.0020	.0008	.0003	.0001	.0000	.0000	.0000	.0000	.0000	.0000
	1	.0195	.0091	.0038	.0014	.0004	.0001	.0000	.0000	.0000	.0000
	2	.0898	.0498	.0250	.0112	.0043	.0013	.0003	.0000	.0000	.0000
	3	.2539	.1658	.0994	.0536	.0253	.0100	.0031	.0006	.0001	.0000
	4	.5000	.3786	.2666	.1717	.0988	.0489	.0196	.0056	.0009	.0000
	5	.7461	.6386	.5174	.3911	.2703	.1657	.0856	.0339	.0083	.0006
	6	.9102	.8505	.7682	.6627	.5372	.3993	.2618	.1409	.0530	.0084
	7	.9805	.9615	.9295	.8789	.8040	.6997	.5638	.4005	.2252	.0712
	8	.9980	.9954	.9899	.9793	.9596	.9249	.8658	.7684	.6126	.3698
	9	1.0000	1.0000	1.0000	1.0000	1.0000	1.0000	1.0000	1.0000	1.0000	1.0000
10	0	.0010	.0003	.0001	.0000	.0000	.0000	.0000	.0000	.0000	.0000
	1	.0107	.0045	.0017	.0005	.0001	.0000	.0000	.0000	.0000	.0000
	2	.0547	.0274	.0123	.0048	.0016	.0004	.0001	.0000	.0000	.0000
	3	.1719	.1020	.0548	.0260	.0106	.0035	.0009	.0001	.0000	.0000
	4	.3770	.2616	.1662	.0949	.0473	.0197	.0064	.0014	.0001	.0000
	5	.6230	.4956	.3669	.2485	.1503	.0781	.0328	.0099	.0016	.0001
	6	.8281	.7340	.6177	.4862	.3504	.2241	.1209	.0500	.0128	.0010
	7	.9453	.9004	.8327	.7384	.6172	.4744	.3222	.1798	.0702	.0115
	8	.9893	.9767	.9536	.9140	.8507	.7560	.6242	.4557	.2639	.0861
	9	.9990	.9975	.9940	.9865	.9718	.9437	.8926	.8031	.6513	.4013
	10	1.0000	1.0000	1.0000	1.0000	1.0000	1.0000	1.0000	1.0000	1.0000	1.0000
11	0	.0005	.0002	.0000	.0000	.0000	.0000	.0000	.0000	.0000	.0000
	1	.0059	.0022	.0007	.0002	.0000	.0000	.0000	.0000	.0000	.0000
	2	.0327	.0148	.0059	.0020	.0006	.0001	.0000	.0000	.0000	.0000
	3	.1133	.0610	.0293	.0122	.0043	.0012	.0002	.0000	.0000	.0000
	4	.2744	.1738	.0994	.0501	.0216	.0076	.0020	.0003	.0000	.0000
	5	.5000	.3669	.2465	.1487	.0782	.0343	.0117	.0027	.0003	.0000
	6	.7256	.6029	.4672	.3317	.2103	.1146	.0504	.0159	.0028	.0001
	7	.8867	.8089	.7037	.5744	.4304	.2867	.1611	.0694	.0185	.0016
	8	.9673	.9348	.8811	.7999	.6873	.5448	.3826	.2212	.0896	.0152
	9	.9941	.9861	.9698	.9394	.8870	.8029	.6779	.5078	.3026	.1019
	10	.9995	.9986	.9964	.9912	.9802	.9578	.9141	.8327	.6862	.4312
	11	1.0000	1.0000	1.0000	1.0000	1.0000	1.0000	1.0000	1.0000	1.0000	1.0000

Table 3 (CONTINUED)

n	y	p = .05	.10	.15	.20	.25	.30	.35	.40	.45
12	0	.5404	.2824	.1422	.0687	.0317	.0138	.0057	.0022	.0008
	1	.8816	.6590	.4435	.2749	.1584	.0850	.0424	.0424	.0083
	2	.9804	.8891	.7358	.5583	.3907	.2528	.1513	.0834	.0421
	3	.9978	.9744	.9078	.7946	.6488	.4925	.3467	.2253	.1345
	4	.9998	.9957	.9761	.9274	.8424	.7237	.5833	.4382	.3044
	5	1.0000	.9995	.9954	.9806	.9456	.8822	.7873	.6652	.5269
	6	1.0000	.9999	.9993	.9961	.9857	.9614	.9154	.8418	.7393
	7	1.0000	1.0000	.9999	.9994	.9972	.9905	.9745	.9427	.8883
	8	1.0000	1.0000	1.0000	.9999	.9996	.9983	.9944	.9847	.9644
	9	1.0000	1.0000	1.0000	1.0000	1.0000	.9998	.9992	.9972	.9921
	10	1.0000	1.0000	1.0000	1.0000	1.0000	1.0000	.9999	.9997	.9989
	11	1.0000	1.0000	1.0000	1.0000	1.0000	1.0000	1.0000	1.0000	.9999
	12	1.0000	1.0000	1.0000	1.0000	1.0000	1.0000	1.0000	1.0000	1.0000
13	0	.5133	.2542	.1209	.0550	.0238	.0097	.0037	.0013	.0004
	1	.8646	.6213	.3983	.2336	.1267	.0637	.0296	.0126	.0049
	2	.9755	.8661	.7296	.5017	.3326	.2025	.1132	.0579	.0269
	3	.9969	.9658	.9033	.7473	.5843	.4206	.2783	.1686	.0929
	4	.9997	.9935	.9740	.9009	.7940	.6543	.5005	.3530	.2279
	5	1.0000	.9991	.9947	.9700	.9198	.8346	.7159	.5744	.4268
	6	1.0000	.9999	.9987	.9930	.9757	.9376	.8705	.7712	.6437
	7	1.0000	1.0000	.9998	.9988	.9944	.9818	.9538	.9023	.8212
	8	1.0000	1.0000	1.0000	.9998	.9990	.9960	.9874	.9679	.9302
	9	1.0000	1.0000	1.0000	1.0000	.9999	.9993	.9975	.9922	.9797
	10	1.0000	1.0000	1.0000	1.0000	1.0000	.9999	.9997	.9987	.9959
	11	1.0000	1.0000	1.0000	1.0000	1.0000	1.0000	1.0000	.9999	.9995
	12	1.0000	1.0000	1.0000	1.0000	1.0000	1.0000	1.0000	1.0000	1.0000
	13	1.0000	1.0000	1.0000	1.0000	1.0000	1.0000	1.0000	1.0000	1.0000
14	0	.4877	.2288	.1028	.0440	.0178	.0068	.0024	.0008	.0002
	1	.8470	.5846	.3567	.1979	.1010	.0475	.0205	.0081	.0029
	2	.9699	.8416	.6479	.4481	.2811	.1608	.0839	.0398	.0170
	3	.9958	.9559	.8535	.6982	.5213	.3552	.2205	.1243	.0632
	4	.9996	.9908	.9533	.8702	.7415	.5842	.4227	.2793	.1672
	5	1.0000	.9985	.9885	.9561	.8883	.7805	.6405	.4859	.3373
	6	1.0000	.9998	.9978	.9884	.9617	.9067	.8164	.6925	.5461
	7	1.0000	1.0000	.9997	.9976	.9897	.9685	.9247	.8499	.7414
	8	1.0000	1.0000	1.0000	.9996	.9978	.9917	.9757	.9417	.8811
	9	1.0000	1.0000	1.0000	1.0000	.9997	.9983	.9940	.9825	.9574
	10	1.0000	1.0000	1.0000	1.0000	1.0000	.9998	.9989	.9961	.9886
	11	1.0000	1.0000	1.0000	1.0000	1.0000	1.0000	.9999	.9994	.9978
	12	1.0000	1.0000	1.0000	1.0000	1.0000	1.0000	1.0000	.9999	.9997
	13	1.0000	1.0000	1.0000	1.0000	1.0000	1.0000	1.0000	1.0000	1.0000
	14	1.0000	1.0000	1.0000	1.0000	1.0000	1.0000	1.0000	1.0000	1.0000

Table 3 (CONTINUED)

n	y	$p = .50$	.55	.60	.65	.70	.75	.80	.85	.90	.95
12	0	.0002	.0001	.0000	.0000	.0000	.0000	.0000	.0000	.0000	.0000
	1	.0032	.0011	.0003	.0001	.0000	.0000	.0000	.0000	.0000	.0000
	2	.0193	.0079	.0028	.0008	.0002	.0000	.0000	.0000	.0000	.0000
	3	.0730	.0356	.0153	.0056	.0017	.0004	.0001	.0000	.0000	.0000
	4	.1938	.1117	.0573	.0255	.0095	.0028	.0006	.0001	.0000	.0000
	5	.3872	.2607	.1582	.0846	.0386	.0143	.0039	.0007	.0001	.0000
	6	.6128	.4731	.3348	.2127	.1178	.0544	.0194	.0046	.0005	.0000
	7	.8062	.6956	.5618	.4167	.2763	.1576	.0726	.0239	.0043	.0002
	8	.9270	.8655	.7747	.6533	.5075	.3512	.2054	.0922	.0256	.0022
	9	.9807	.9579	.9166	.8487	.7472	.6093	.4417	.2642	.1109	.0196
	10	.9968	.9917	.9804	.9576	.9150	.8416	.7251	.5565	.3410	.1184
	11	.9998	.9992	.9978	.9943	.9862	.9683	.9313	.8578	.7176	.4596
	12	1.0000	1.0000	1.0000	1.0000	1.0000	1.0000	1.0000	1.0000	1.0000	1.0000
13	0	.0001	.0000	.0000	.0000	.0000	.0000	.0000	.0000	.0000	.0000
	1	.0017	.0005	.0001	.0000	.0000	.0000	.0000	.0000	.0000	.0000
	2	.0112	.0041	.0013	.0003	.0001	.0000	.0000	.0000	.0000	.0000
	3	.0461	.0203	.0078	.0025	.0007	.0001	.0000	.0000	.0000	.0000
	4	.1334	.0698	.0321	.0126	.0040	.0010	.0002	.0000	.0000	.0000
	5	.2905	.1788	.0977	.0462	.0182	.0056	.0012	.0002	.0000	.0000
	6	.5000	.3563	.2288	.1295	.0624	.0243	.0070	.0013	.0001	.0000
	7	.7095	.5732	.4256	.2841	.1654	.0802	.0300	.0053	.0009	.0000
	8	.8666	.7721	.6470	.4995	.3457	.2060	.0991	.0260	.0065	.0003
	9	.9539	.9071	.8314	.7217	.5794	.4157	.2527	.0967	.0342	.0031
	10	.9888	.9731	.9421	.8868	.7975	.6674	.4983	.2704	.1339	.0245
	11	.9983	.9951	.9874	.9704	.9363	.8733	.7664	.6017	.3787	.1354
	12	.9999	.9996	.9987	.9963	.9903	.9762	.9450	.8791	.7458	.4867
	13	1.0000	1.0000	1.0000	1.0000	1.0000	1.0000	1.0000	1.0000	1.0000	1.0000
14	0	.0000	.0000	.0000	.0000	.0000	.0000	.0000	.0000	.0000	.0000
	1	.0009	.0003	.0001	.0000	.0000	.0000	.0000	.0000	.0000	.0000
	2	.0065	.0022	.0006	.0001	.0000	.0000	.0000	.0000	.0000	.0000
	3	.0287	.0114	.0039	.0011	.0002	.0000	.0000	.0000	.0000	.0000
	4	.0898	.0462	.0175	.0060	.0017	.0003	.0000	.0000	.0000	.0000
	5	.2120	.1189	.0583	.0243	.0083	.0022	.0004	.0000	.0000	.0000
	6	.3953	.2586	.1501	.0753	.0315	.0103	.0024	.0003	.0000	.0000
	7	.6047	.4539	.3075	.1836	.0933	.0383	.0116	.0022	.0002	.0000
	8	.7880	.6627	.5141	.3595	.2195	.1117	.0439	.0115	.0015	.0000
	9	.9102	.8328	.7207	.5773	.4158	.2585	.1298	.0467	.0092	.0004
	10	.9713	.9368	.8757	.7795	.6448	.4787	.3018	.1465	.0441	.0042
	11	.9935	.9830	.9602	.9161	.8392	.7189	.5519	.3521	.1584	.0301
	12	.9991	.9971	.9919	.9795	.9525	.8990	.8021	.6433	.4154	.1530
	13	.9999	.9998	.9992	.9976	.9932	.9822	.9560	.8972	.7712	.5123
	14	1.0000	1.0000	1.0000	1.0000	1.0000	1.0000	1.0000	1.0000	1.0000	1.0000

Table 3   (CONTINUED)

n	y	p = .05	.10	.15	.20	.25	.30	.35	.40	.45
15	0	.4633	.2059	.0874	.0352	.0134	.0047	.0016	.0005	.0001
	1	.8290	.5490	.3186	.1671	.0802	.0353	.0142	.0052	.0017
	2	.9638	.8159	.6042	.3980	.2361	.1268	.0617	.0271	.0107
	3	.9945	.9444	.8227	.6482	.4613	.2969	.1727	.0905	.0424
	4	.9994	.9873	.9383	.8358	.6865	.5155	.3519	.2173	.1204
	5	.9999	.9978	.9832	.9389	.8516	.7216	.5643	.4032	.2608
	6	1.0000	.9997	.9964	.9819	.9434	.8689	.7548	.6098	.4522
	7	1.0000	1.0000	.9994	.9958	.9827	.9500	.8868	.7869	.6535
	8	1.0000	1.0000	.9999	.9992	.9958	.9848	.9578	.9050	.8182
	9	1.0000	1.0000	1.0000	.9999	.9992	.9963	.9876	.9662	.9231
	10	1.0000	1.0000	1.0000	1.0000	.9999	.9993	.9972	.9907	.9745
	11	1.0000	1.0000	1.0000	1.0000	1.0000	.9999	.9995	.9981	.9937
	12	1.0000	1.0000	1.0000	1.0000	1.0000	1.0000	.9999	.9997	.9989
	13	1.0000	1.0000	1.0000	1.0000	1.0000	1.0000	1.0000	1.0000	.9999
	14	1.0000	1.0000	1.0000	1.0000	1.0000	1.0000	1.0000	1.0000	1.0000
	15	1.0000	1.0000	1.0000	1.0000	1.0000	1.0000	1.0000	1.0000	1.0000
16	0	.4401	.1853	.0743	.0281	.0100	.0033	.0010	.0003	.0001
	1	.8108	.5147	.2839	.1407	.0635	.0261	.0098	.0033	.0010
	2	.9571	.7892	.5614	.3518	.1971	.0994	.0451	.0183	.0066
	3	.9930	.9316	.7899	.5981	.4050	.2459	.1339	.0651	.0281
	4	.9991	.9830	.9209	.7982	.6302	.4499	.2892	.1666	.0853
	5	.9999	.9967	.9765	.9183	.8103	.6598	.4900	.3288	.1976
	6	1.0000	.9995	.9944	.9733	.9204	.8247	.6881	.5272	.3660
	7	1.0000	.9999	.9989	.9930	.9729	.9256	.8406	.7161	.5629
	8	1.0000	1.0000	.9998	.9985	.9925	.9743	.9329	.8577	.7441
	9	1.0000	1.0000	1.0000	.9998	.9984	.9929	.9771	.9417	.8759
	10	1.0000	1.0000	1.0000	1.0000	.9997	.9984	.9938	.9809	.9514
	11	1.0000	1.0000	1.0000	1.0000	1.0000	.9997	.9987	.9951	.9851
	12	1.0000	1.0000	1.0000	1.0000	1.0000	1.0000	.9998	.9991	.9965
	13	1.0000	1.0000	1.0000	1.0000	1.0000	1.0000	1.0000	.9999	.9994
	14	1.0000	1.0000	1.0000	1.0000	1.0000	1.0000	1.0000	1.0000	.9999
	15	1.0000	1.0000	1.0000	1.0000	1.0000	1.0000	1.0000	1.0000	1.0000
	16	1.0000	1.0000	1.0000	1.0000	1.0000	1.0000	1.0000	1.0000	1.0000

Table 3 (Continued)

n	y	p = .50	.55	.60	.65	.70	.75	.80	.85	.90	.95
15	0	.0000	.0000	.0000	.0000	.0000	.0000	.0000	.0000	.0000	.0000
	1	.0005	.0001	.0000	.0000	.0000	.0000	.0000	.0000	.0000	.0000
	2	.0037	.0011	.0003	.0001	.0000	.0000	.0000	.0000	.0000	.0000
	3	.0176	.0063	.0019	.0005	.0001	.0000	.0000	.0000	.0000	.0000
	4	.0592	.0255	.0093	.0028	.0007	.0001	.0000	.0000	.0000	.0000
	5	.1509	.0769	.0338	.0124	.0037	.0008	.0001	.0000	.0000	.0000
	6	.3036	.1818	.0950	.0422	.0152	.0042	.0008	.0001	.0000	.0000
	7	.5000	.3465	.2131	.1132	.0500	.0173	.0042	.0006	.0000	.0000
	8	.6964	.5478	.3902	.2452	.1311	.0566	.0181	.0036	.0003	.0000
	9	.8491	.7392	.5968	.4357	.2784	.1484	.0611	.0168	.0022	.0001
	10	.9408	.8796	.7827	.6481	.4845	.3135	.1642	.0617	.0127	.0006
	11	.9824	.9576	.9095	.8273	.7031	.5387	.3518	.1773	.0556	.0055
	12	.9963	.9893	.9729	.9383	.8732	.7639	.6020	.3958	.1841	.0362
	13	.9995	.9983	.9948	.9858	.9647	.9198	.8329	.6814	.4510	.1710
	14	1.0000	.9999	.9995	.9984	.9953	.9866	.9648	.9126	.7941	.5367
	15	1.0000	1.0000	1.0000	1.0000	1.0000	1.0000	1.0000	1.0000	1.0000	1.0000
16	0	.0000	.0000	.0000	.0000	.0000	.0000	.0000	.0000	.0000	.0000
	1	.0003	.0001	.0000	.0000	.0000	.0000	.0000	.0000	.0000	.0000
	2	.0021	.0006	.0001	.0000	.0000	.0000	.0000	.0000	.0000	.0000
	3	.0106	.0035	.0009	.0002	.0000	.0000	.0000	.0000	.0000	.0000
	4	.0384	.0149	.0049	.0013	.0003	.0000	.0000	.0000	.0000	.0000
	5	.1051	.0486	.0191	.0062	.0016	.0003	.0000	.0000	.0000	.0000
	6	.2272	.1241	.0583	.0229	.0071	.0016	.0002	.0000	.0000	.0000
	7	.4018	.2559	.1423	.0671	.0257	.0075	.0015	.0002	.0000	.0000
	8	.5982	.4371	.2839	.1594	.0744	.0271	.0070	.0011	.0001	.0000
	9	.7228	.6340	.4728	.3119	.1753	.0796	.0267	.0056	.0005	.0000
	10	.8949	.8024	.6712	.5100	.3402	.1897	.0817	.0235	.0033	.0001
	11	.9616	.9147	.8334	.7108	.5501	.3698	.2018	.0791	.0170	.0009
	12	.9894	.9719	.9349	.8661	.7541	.5950	.4019	.2101	.0684	.0070
	13	.9979	.9934	.9817	.9549	.9006	.8729	.6482	.4386	.2108	.0429
	14	.9997	.9990	.9967	.9902	.9739	.9365	.8593	.7161	.4853	.1892
	15	1.0000	.9999	.9997	.9990	.9967	.9900	.9719	.9257	.8147	.5599
	16	1.0000	1.0000	1.0000	1.0000	1.0000	1.0000	1.0000	1.0000	1.0000	1.0000

Table 3 (CONTINUED)

n	y	p = .05	.10	.15	.20	.25	.30	.35	.40	.45
17	0	.4181	.1668	.0631	.0225	.0075	.0023	.0007	.0002	.0000
	1	.7922	.4818	.2525	.1182	.0501	.0193	.0067	.0021	.0006
	2	.9497	.7618	.5198	.3096	.1637	.0774	.0327	.0123	.0041
	3	.9912	.9174	.7556	.5489	.3530	.2019	.1028	.0464	.0184
	4	.9988	.9779	.9013	.7582	.5739	.3887	.2348	.1260	.0596
	5	.9999	.9953	.9681	.8943	.7653	.5968	.4197	.2639	.1471
	6	1.0000	.9992	.9917	.9623	.8929	.7752	.6188	.4478	.2902
	7	1.0000	.9999	.9983	.9891	.9598	.8954	.7872	.6405	.4743
	8	1.0000	1.0000	.9997	.9974	.9876	.9597	.9006	.8011	.6626
	9	1.0000	1.0000	1.0000	.9995	.9969	.9873	.9617	.9081	.8166
	10	1.0000	1.0000	1.0000	.9999	.9994	.9968	.9880	.9652	.9174
	11	1.0000	1.0000	1.0000	1.0000	.9999	.9993	.9970	.9894	.9699
	12	1.0000	1.0000	1.0000	1.0000	1.0000	.9999	.9994	.9975	.9914
	13	1.0000	1.0000	1.0000	1.0000	1.0000	1.0000	.9999	.9995	.9981
	14	1.0000	1.0000	1.0000	1.0000	1.0000	1.0000	1.0000	.9999	.9997
	15	1.0000	1.0000	1.0000	1.0000	1.0000	1.0000	1.0000	1.0000	1.0000
	16	1.0000	1.0000	1.0000	1.0000	1.0000	1.0000	1.0000	1.0000	1.0000
	17	1.0000	1.0000	1.0000	1.0000	1.0000	1.0000	1.0000	1.0000	1.0000
18	0	.3972	.1501	.0536	.0180	.0056	.0016	.0004	.0001	.0000
	1	.7735	.4503	.2241	.0991	.0395	.0142	.0046	.0013	.0003
	2	.9419	.7338	.4797	.2713	.1353	.0600	.0236	.0082	.0025
	3	.9891	.9018	.7202	.5010	.3057	.1646	.0783	.0328	.0120
	4	.9985	.9718	.8794	.7164	.5187	.3327	.1886	.0942	.0411
	5	.9998	.9936	.9581	.8671	.7175	.5344	.3550	.2088	.1077
	6	1.0000	.9988	.9882	.9487	.8610	.7217	.5491	.3743	.2258
	7	1.0000	.9998	.9973	.9837	.9431	.8593	.7283	.5634	.3915
	8	1.0000	1.0000	.9995	.9957	.9807	.9404	.8609	.7368	.5778
	9	1.0000	1.0000	.9999	.9991	.9946	.9790	.9403	.8653	.7473
	10	1.0000	1.0000	1.0000	.9998	.9988	.9939	.9788	.9424	.8720
	11	1.0000	1.0000	1.0000	1.0000	.9998	.9986	.9938	.9797	.9463
	12	1.0000	1.0000	1.0000	1.0000	1.0000	.9997	.9986	.9942	.9817
	13	1.0000	1.0000	1.0000	1.0000	1.0000	1.0000	.9997	.9987	.9951
	14	1.0000	1.0000	1.0000	1.0000	1.0000	1.0000	1.0000	.9998	.9990
	15	1.0000	1.0000	1.0000	1.0000	1.0000	1.0000	1.0000	1.0000	.9999
	16	1.0000	1.0000	1.0000	1.0000	1.0000	1.0000	1.0000	1.0000	1.0000
	17	1.0000	1.0000	1.0000	1.0000	1.0000	1.0000	1.0000	1.0000	1.0000
	18	1.0000	1.0000	1.0000	1.0000	1.0000	1.0000	1.0000	1.0000	1.0000

Table 3 (Continued)

n	y	p = .50	.55	.60	.65	.70	.75	.80	.85	.90	.95
17	0	.0000	.0000	.0000	.0000	.0000	.0000	.0000	.0000	.0000	.0000
	1	.0001	.0000	.0000	.0000	.0000	.0000	.0000	.0000	.0000	.0000
	2	.0012	.0003	.0001	.0000	.0000	.0000	.0000	.0000	.0000	.0000
	3	.0064	.0019	.0005	.0001	.0000	.0000	.0000	.0000	.0000	.0000
	4	.0245	.0086	.0025	.0006	.0001	.0000	.0000	.0000	.0000	.0000
	5	.0717	.0301	.0106	.0030	.0007	.0001	.0000	.0000	.0000	.0000
	6	.1662	.0826	.0348	.0120	.0032	.0006	.0001	.0000	.0000	.0000
	7	.3145	.1834	.0919	.0383	.0127	.0031	.0005	.0000	.0000	.0000
	8	.5000	.3374	.1989	.0994	.0403	.0124	.0026	.0003	.0000	.0000
	9	.6855	.5257	.3595	.2128	.1046	.0402	.0109	.0017	.0001	.0000
	10	.8338	.7098	.5522	.3812	.2248	.1071	.0377	.0083	.0008	.0000
	11	.9283	.8529	.7361	.5803	.4032	.2347	.1057	.0319	.0047	.0001
	12	.9755	.9404	.8740	.7652	.6113	.4261	.2418	.0987	.0221	.0012
	13	.9936	.9816	.9536	.8972	.7981	.6470	.4511	.2444	.0826	.0088
	14	.9988	.9959	.9877	.9673	.9226	.8363	.6904	.4802	.2382	.0503
	15	.9999	.9994	.9979	.9933	.9807	.9499	.8818	.7475	.5182	.2078
	16	1.0000	1.0000	.9998	.9993	.9977	.9925	.9775	.9369	.8332	.5819
	17	1.0000	1.0000	1.0000	1.0000	1.0000	1.0000	1.0000	1.0000	1.0000	1.0000
18	0	.0000	.0000	.0000	.0000	.0000	.0000	.0000	.0000	.0000	.0000
	1	.0001	.0000	.0000	.0000	.0000	.0000	.0000	.0000	.0000	.0000
	2	.0007	.0001	.0000	.0000	.0000	.0000	.0000	.0000	.0000	.0000
	3	.0038	.0010	.0002	.0000	.0000	.0000	.0000	.0000	.0000	.0000
	4	.0154	.0049	.0013	.0003	.0000	.0000	.0000	.0000	.0000	.0000
	5	.0481	.0183	.0058	.0014	.0003	.0000	.0000	.0000	.0000	.0000
	6	.1189	.0537	.0203	.0062	.0014	.0002	.0000	.0000	.0000	.0000
	7	.2403	.1280	.0576	.0212	.0061	.0012	.0002	.0000	.0000	.0000
	8	.4073	.2527	.1347	.0597	.0210	.0054	.0009	.0001	.0000	.0000
	9	.5927	.4222	.2632	.1391	.0596	.0193	.0043	.0005	.0000	.0000
	10	.7597	.6085	.4366	.2717	.1407	.0569	.0163	.0027	.0002	.0000
	11	.8811	.7742	.6257	.4509	.2783	.1390	.0513	.0118	.0012	.0000
	12	.9519	.8923	.7912	.6450	.4656	.2825	.1329	.0419	.0064	.0002
	13	.9846	.9589	.9058	.8114	.6673	.4813	.2836	.1206	.0282	.0015
	14	.9962	.9880	.9672	.9217	.8354	.6943	.4990	.2798	.0982	.0109
	15	.9993	.9975	.9918	.9764	.9400	.8647	.7287	.5203	.2662	.0581
	16	.9999	.9997	.9987	.9954	.9858	.9605	.9009	.7759	.5497	.2265
	17	1.0000	1.0000	.9999	.9996	.9984	.9944	.9820	.9464	.8499	.6028
	18	1.0000	1.0000	1.0000	1.0000	1.0000	1.0000	1.0000	1.0000	1.0000	1.0000

Table 3 (CONTINUED)

n	y	p = .05	.10	.15	.20	.25	.30	.35	.40	.45
19	0	.3774	.1351	.0456	.0144	.0042	.0011	.0003	.0001	.0000
	1	.7547	.4203	.1985	.0829	.0310	.0104	.0031	.0008	.0002
	2	.9335	.7054	.4413	.2369	.1113	.0462	.0170	.0055	.0015
	3	.9869	.8850	.6841	.4551	.2631	.1332	.0591	.0230	.0077
	4	.9980	.9648	.8556	.6733	.4654	.2822	.1500	.0696	.0280
	5	.9998	.9914	.9463	.8369	.6678	.4739	.2968	.1629	.0777
	6	1.0000	.9983	.9837	.9324	.8251	.6655	.4812	.3081	.1727
	7	1.0000	.9997	.9959	.9767	.9225	.8180	.6656	.4878	.3169
	8	1.0000	1.0000	.9992	.9933	.9713	.9161	.8145	.6675	.4940
	9	1.0000	1.0000	.9999	.9984	.9911	.9674	.9125	.8139	.6710
	10	1.0000	1.0000	1.0000	.9997	.9977	.9895	.9653	.9115	.8159
	11	1.0000	1.0000	1.0000	1.0000	.9995	.9972	.9886	.9648	.9129
	12	1.0000	1.0000	1.0000	1.0000	.9999	.9994	.9969	.9884	.9658
	13	1.0000	1.0000	1.0000	1.0000	1.0000	.9999	.9993	.9969	.9891
	14	1.0000	1.0000	1.0000	1.0000	1.0000	1.0000	.9999	.9994	.9972
	15	1.0000	1.0000	1.0000	1.0000	1.0000	1.0000	1.0000	.9999	.9995
	16	1.0000	1.0000	1.0000	1.0000	1.0000	1.0000	1.0000	1.0000	.9999
	17	1.0000	1.0000	1.0000	1.0000	1.0000	1.0000	1.0000	1.0000	1.0000
	18	1.0000	1.0000	1.0000	1.0000	1.0000	1.0000	1.0000	1.0000	1.0000
	19	1.0000	1.0000	1.0000	1.0000	1.0000	1.0000	1.0000	1.0000	1.0000
20	0	.3585	.1216	.0388	.0115	.0032	.0008	.0002	.0000	.0000
	1	.7358	.3917	.1756	.0692	.0243	.0076	.0021	.0005	.0001
	2	.9245	.6769	.4049	.2061	.0913	.0355	.0121	.0036	.0009
	3	.9841	.8670	.6477	.4114	.2252	.1071	.0444	.0160	.0049
	4	.9974	.9568	.8298	.6296	.4148	.2375	.1182	.0510	.0189
	5	.9997	.9887	.9327	.8042	.6172	.4164	.2454	.1256	.0553
	6	1.0000	.9976	.9781	.9133	.7858	.6080	.4166	.2500	.1299
	7	1.0000	.9996	.9941	.9679	.8982	.7723	.6010	.4159	.2520
	8	1.0000	.9999	.9987	.9900	.9591	.8867	.7624	.5956	.4143
	9	1.0000	1.0000	.9998	.9974	.9861	.9520	.8782	.7553	.5914
	10	1.0000	1.0000	1.0000	.9994	.9961	.9829	.9468	.8725	.7507
	11	1.0000	1.0000	1.0000	.9999	.9991	.9949	.9804	.9435	.8692
	12	1.0000	1.0000	1.0000	1.0000	.9998	.9987	.9940	.9790	.9420
	13	1.0000	1.0000	1.0000	1.0000	1.0000	.9997	.9985	.9935	.9786
	14	1.0000	1.0000	1.0000	1.0000	1.0000	1.0000	.9997	.9984	.9936
	15	1.0000	1.0000	1.0000	1.0000	1.0000	1.0000	1.0000	.9997	.9985
	16	1.0000	1.0000	1.0000	1.0000	1.0000	1.0000	1.0000	1.0000	.9997
	17	1.0000	1.0000	1.0000	1.0000	1.0000	1.0000	1.0000	1.0000	1.0000
	18	1.0000	1.0000	1.0000	1.0000	1.0000	1.0000	1.0000	1.0000	1.0000
	19	1.0000	1.0000	1.0000	1.0000	1.0000	1.0000	1.0000	1.0000	1.0000
	20	1.0000	1.0000	1.0000	1.0000	1.0000	1.0000	1.0000	1.0000	1.0000

Table 3 (CONTINUED)

n	y	p = .50	.55	.60	.65	.70	.75	.80	.85	.90	.95
19	0	.0000	.0000	.0000	.0000	.0000	.0000	.0000	.0000	.0000	.0000
	1	.0000	.0000	.0000	.0000	.0000	.0000	.0000	.0000	.0000	.0000
	2	.0004	.0001	.0000	.0000	.0000	.0000	.0000	.0000	.0000	.0000
	3	.0022	.0005	.0001	.0000	.0000	.0000	.0000	.0000	.0000	.0000
	4	.0096	.0028	.0006	.0001	.0000	.0000	.0000	.0000	.0000	.0000
	5	.0318	.0109	.0031	.0007	.0001	.0000	.0000	.0000	.0000	.0000
	6	.0835	.0342	.0116	.0031	.0006	.0001	.0000	.0000	.0000	.0000
	7	.1796	.0871	.0352	.0114	.0028	.0005	.0000	.0000	.0000	.0000
	8	.3238	.1841	.0885	.0347	.0105	.0023	.0003	.0000	.0000	.0000
	9	.5000	.3290	.1861	.0875	.0326	.0089	.0016	.0001	.0000	.0000
	10	.6762	.5060	.3325	.1855	.0839	.0287	.0067	.0008	.0000	.0000
	11	.8204	.6831	.5122	.3344	.1820	.0775	.0233	.0041	.0003	.0000
	12	.9165	.8273	.6919	.5188	.3345	.1749	.0676	.0163	.0017	.0000
	13	.9682	.9223	.8371	.7032	.5261	.3322	.1631	.0537	.0086	.0002
	14	.9904	.9720	.9304	.8500	.7178	.5346	.3267	.1444	.0352	.0020
	15	.9978	.9923	.9770	.9409	.8668	.7369	.5449	.3159	.1150	.0132
	16	.9996	.9985	.9945	.9830	.9538	.8887	.7631	.5587	.2946	.0665
	17	1.0000	.9998	.9992	.9969	.9896	.9690	.9171	.8015	.5797	.2453
	18	1.0000	1.0000	.9999	.9997	.9989	.9958	.9856	.9544	.8649	.6226
	19	1.0000	1.0000	1.0000	1.0000	1.0000	1.0000	1.0000	1.0000	1.0000	1.0000
20	0	.0000	.0000	.0000	.0000	.0000	.0000	.0000	.0000	.0000	.0000
	1	.0000	.0000	.0000	.0000	.0000	.0000	.0000	.0000	.0000	.0000
	2	.0002	.0000	.0000	.0000	.0000	.0000	.0000	.0000	.0000	.0000
	3	.0013	.0003	.0000	.0000	.0000	.0000	.0000	.0000	.0000	.0000
	4	.0059	.0015	.0003	.0000	.0000	.0000	.0000	.0000	.0000	.0000
	5	.0207	.0064	.0016	.0003	.0000	.0000	.0000	.0000	.0000	.0000
	6	.0577	.0214	.0065	.0015	.0003	.0000	.0000	.0000	.0000	.0000
	7	.1316	.0580	.0210	.0060	.0013	.0002	.0000	.0000	.0000	.0000
	8	.2517	.1308	.0565	.0196	.0051	.0009	.0001	.0000	.0000	.0000
	9	.4119	.2493	.1275	.0532	.0171	.0039	.0006	.0000	.0000	.0000
	10	.5881	.4086	.2447	.1218	.0480	.0139	.0026	.0002	.0000	.0000
	11	.7483	.5857	.4044	.2376	.1133	.0409	.0100	.0013	.0001	.0000
	12	.8684	.7480	.5841	.3990	.2277	.1018	.0321	.0059	.0004	.0000
	13	.9423	.8701	.7500	.5834	.3920	.2142	.0867	.0219	.0024	.0000
	14	.9793	.9447	.8744	.7546	.5836	.3828	.1958	.0673	.0113	.0003
	15	.9941	.9811	.9490	.8818	.7625	.5852	.3704	.1702	.0432	.0026
	16	.9987	.9951	.9840	.9556	.8929	.7748	.5886	.3523	.1330	.0159
	17	.9998	.9991	.9964	.9879	.9645	.9087	.7939	.5951	.3231	.0755
	18	1.0000	.9999	.9995	.9979	.9924	.9757	.9308	.8244	.6083	.2642
	19	1.0000	1.0000	1.0000	.9998	.9992	.9968	.9885	.9612	.8784	.6415
	20	1.0000	1.0000	1.0000	1.0000	1.0000	1.0000	1.0000	1.0000	1.0000	1.0000

For n larger than 20, the rth quantile y_r of a binomial random variable may be approximated using $y_r = np + w_r \sqrt{np(1 - p)}$, where w_r is the rth quantile of a standard normal random variable, obtained from Table 1.

Table 4 Chart Providing Confidence Limits for p in Binomial Sampling, Given a Sample Fraction Y/n. Confidence Coefficient, $1 - 2\alpha = .95$

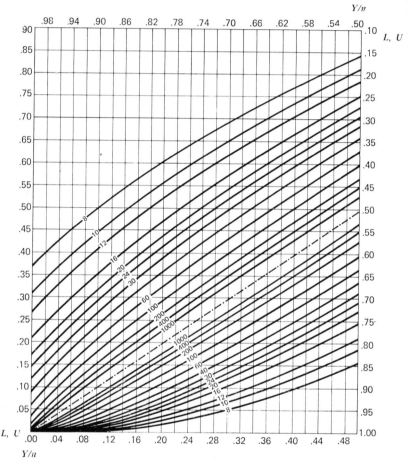

The numbers printed along the curves indicate the sample size n. If for a given value of the abscissa Y/n, L, and U are the ordinates read from (or interpolated between) the appropriate lower and upper curves, then

$$P(L \le p \le U) \ge 1 - 2\alpha$$

Table 4 (Continued) Confidence Coefficient, $1 - 2\alpha = .99$

The numbers printed along the curves indicate the sample size n.

SOURCE.   Adapted from Table 41, pp. 204–205, Pearson and Hartley (1962).

NOTE.   The process of reading from the curves can be simplified with the help of the right-angled corner of a loose sheet of paper or thin card, along the edges of which are marked off the scales shown in the top left-hand corner of each chart.

Table 5 Sample Sizes for One-Sided Nonparametric Tolerance Limits[a]

$1 - \alpha$	$q = .500$	.700	.750	.800	.850	.900	.950	.975	.980	.990
.500	1	2	3	4	5	7	14	28	35	69
.700	2	4	5	6	8	12	24	48	60	120
.750	2	4	5	7	9	14	28	55	69	138
.800	3	5	6	8	10	16	32	64	80	161
.850	3	6	7	9	12	19	37	75	94	189
.900	4	7	9	11	15	22	45	91	144	230
.950	5	9	11	14	19	29	59	119	149	299
.975	6	11	13	17	23	36	72	146	183	368
.980	6	11	14	18	25	38	77	155	194	390
.990	7	13	17	21	29	44	90	182	228	459
.995	8	15	19	24	33	51	104	210	263	528
.999	10	20	25	31	43	66	135	273	342	688

[a] The quantity tabled is the sample size n such that $q^n \leq \alpha$, for use in finding the tolerance limits

$$P(X^{(1)} \leq q \text{ of the population}) \geq 1 - \alpha$$

or

$$P(q \text{ of the population} \leq X^{(n)}) \geq 1 - \alpha$$

as described in Section 3.3.

Table 6 Sample Sizes for Nonparametric Tolerance Limits When $r + m = 2$[a]

$1 - \alpha$	$q = .500$	.700	.750	.800	.850	.900	.950	.975	.980	.990
.500	3	6	7	9	11	17	34	67	84	168
.700	5	8	10	12	16	24	49	97	122	244
.750	5	9	10	13	18	27	53	107	134	269
.800	5	9	11	14	19	29	59	119	149	299
.850	6	10	13	16	22	33	67	134	168	337
.900	7	12	15	18	25	38	77	155	194	388
.950	8	14	18	22	30	46	93	188	236	473
.975	9	17	20	26	35	54	110	221	277	555
.980	9	17	21	27	37	56	115	231	290	581
.990	11	20	24	31	42	64	130	263	330	662
.995	12	22	27	34	47	72	146	294	369	740
.999	14	27	33	42	58	89	181	366	458	920

[a] The quantity tabled is that sample size n such that $q^n + nq^{n-1}(1 - q) \leq \alpha$ for use in finding the tolerance limits

$$P(X^{(r)} \leq q \text{ of the population} \leq X^{(n+1-m)}) \geq 1 - \alpha$$

when $r + m$ equals 2.

Table 7   QUANTILES OF THE WILCOXON SIGNED RANKS TEST STATISTIC[a]

	$w_{.005}$	$w_{.01}$	$w_{.025}$	$w_{.05}$	$w_{.10}$	$w_{.20}$	$w_{.30}$	$w_{.40}$	$w_{.50}$	$\dfrac{n(n+1)}{2}$
$n = 4$	0	0	0	0	1	3	3	4	5	10
5	0	0	0	1	3	4	5	6	7.5	15
6	0	0	1	3	4	6	8	9	10.5	21
7	0	1	3	4	6	9	11	12	14	28
8	1	2	4	6	9	12	14	16	18	36
9	2	4	6	9	11	15	18	20	22.5	45
10	4	6	9	11	15	19	22	25	27.5	55
11	6	8	11	14	18	23	27	30	33	66
12	8	10	14	18	22	28	32	36	39	78
13	10	13	18	22	27	33	38	42	45.5	91
14	13	16	22	26	32	39	44	48	52.5	105
15	16	20	26	31	37	45	51	55	60	120
16	20	24	30	36	43	51	58	63	68	136
17	24	28	35	42	49	58	65	71	76.5	153
18	28	33	41	48	56	66	73	80	85.5	171
19	33	38	47	54	63	74	82	89	95	190
20	38	44	53	61	70	82	91	98	105	210

For n larger than 20, the pth quantile w_p of the Wilcoxon signed ranks test statistic may be approximated by $w_p = [n(n+1)/4] + x_p \sqrt{n(n+1)(2n+1)/24}$, where x_p is the pth quantile of a standard normal random variable, obtained from Table 1.

SOURCE.   Adapted from Table 1, McCornack (1965).

[a] The entries in this table are quantiles w_p of the Wilcoxon signed ranks test statistic T, given by Equation (5.1.4), for selected values of $p \leq .50$. Quantiles w_p for $p > .50$ may be computed from the equation

$$w_p = n(n+1)/2 - w_{1-p}$$

where $n(n+1)/2$ is given in the right hand column in the table. Note that $P(T < w_p) \leq p$ and $P(T > w_p) \leq 1 - p$ if H_0 is true. Critical regions correspond to values of T less than (or greater than) but not including the appropriate quantile.

Table 8 QUANTILES OF THE MANN-WHITNEY TEST STATISTIC[a]

n	p	m = 2	3	4	5	6	7	8	9	10	11	12	13	14	15	16	17	18	19	20
2	.001	0	0	0	0	0	0	0	0	0	0	0	0	0	0	0	0	0	0	0
	.005	0	0	0	0	0	0	0	0	0	0	0	0	0	0	0	0	0	1	1
	.01	0	0	0	0	0	0	0	0	0	0	0	1	1	1	1	1	1	2	2
	.025	0	0	0	0	0	0	1	1	1	1	2	2	2	2	2	3	3	3	3
	.05	0	0	0	1	1	1	2	2	2	2	3	3	4	4	4	4	5	5	5
	.10	0	1	1	2	2	2	3	3	4	4	5	5	5	6	6	7	7	8	8
3	.001	0	0	0	0	0	0	0	0	0	0	0	0	0	0	0	1	1	1	1
	.005	0	0	0	0	0	0	0	1	1	1	2	2	2	3	3	3	3	4	4
	.01	0	0	0	0	0	1	1	2	2	2	3	3	3	4	4	5	5	5	6
	.025	0	0	0	1	2	2	3	3	4	4	5	5	6	6	7	7	8	8	9
	.05	0	1	1	2	3	3	4	5	5	6	6	7	8	8	9	10	10	11	12
	.10	1	2	2	3	4	5	6	6	7	8	9	10	11	11	12	13	14	15	16
4	.001	0	0	0	0	0	0	0	0	1	1	1	2	2	2	3	3	4	4	4
	.005	0	0	0	0	1	1	2	2	3	3	4	4	5	6	6	7	7	8	9
	.01	0	0	0	1	2	2	3	4	4	5	6	6	7	9	8	9	10	10	11
	.025	0	0	1	2	3	4	5	5	6	7	8	9	10	11	12	12	13	14	15
	.05	0	1	2	3	4	5	6	7	8	9	10	11	12	13	15	16	17	18	19
	.10	1	2	4	5	6	7	8	10	11	12	13	14	16	17	18	19	21	22	23

SOURCE. Adapted from Table 1, Verdooren (1963).

[a] The entries in this table are the quantiles w_p of the Mann-Whitney test statistic T, given by Equation (5.3.2), for selected values of p. Note that $P(T < w_p) \le p$. Upper quantiles may be found from the equation

$$w_{1-p} = nm - w_p$$

Critical regions correspond to values less than (or greater than) but not including the appropriate quantile.

Table 8 (CONTINUED)

n	p	m = 2	3	4	5	6	7	8	9	10	11	12	13	14	15	16	17	18	19	20
5	.001	0	0	0	0	0	0	1	2	2	3	3	4	4	5	6	6	7	8	8
	.005	0	0	0	1	2	2	3	4	5	6	7	8	8	9	10	11	12	13	14
	.01	0	0	1	2	3	4	5	6	7	8	9	10	11	12	13	14	15	16	17
	.025	0	1	2	3	4	6	7	8	9	10	12	13	14	15	16	18	19	20	21
	.05	1	2	3	5	6	7	9	10	12	13	14	16	17	19	20	21	23	24	26
	.10	2	3	5	6	8	9	11	13	14	16	18	19	21	23	24	26	28	29	31
6	.001	0	0	0	0	0	0	2	3	4	5	5	6	7	8	9	10	11	12	13
	.005	0	0	1	2	3	4	5	6	7	8	10	11	12	13	14	16	17	18	19
	.01	0	0	2	3	4	5	7	8	9	10	12	13	14	16	17	19	20	21	23
	.025	0	2	3	4	6	7	9	11	12	14	15	17	18	20	22	23	25	26	28
	.05	1	3	4	6	8	9	11	13	15	17	18	20	22	24	26	27	29	31	33
	.10	2	4	6	8	10	12	14	16	18	20	22	24	26	28	30	32	35	37	39
7	.001	0	0	0	0	1	2	3	4	6	7	8	9	10	11	12	14	15	16	17
	.005	0	0	1	2	4	5	7	8	10	11	13	14	16	17	19	20	22	23	25
	.01	0	1	2	4	5	7	8	10	12	13	15	17	18	20	22	24	25	27	29
	.025	0	2	4	6	7	9	11	13	15	17	19	21	23	25	27	29	31	33	35
	.05	1	3	5	7	9	12	14	16	18	20	22	25	27	29	31	34	36	38	40
	.10	2	5	7	9	12	14	17	19	22	24	27	29	32	34	37	39	42	44	47
8	.001	0	0	0	1	2	3	5	6	7	9	10	12	13	15	16	18	19	21	22
	.005	0	0	2	3	5	7	8	10	12	14	16	18	19	21	23	25	27	29	31
	.01	0	1	3	5	7	8	10	12	14	16	18	21	23	25	27	29	31	33	35
	.025	1	3	5	7	9	11	14	16	18	20	23	25	27	30	32	35	37	39	42
	.05	2	4	6	9	11	14	16	19	21	24	27	29	32	34	37	40	42	45	48
	.10	3	6	8	11	14	17	20	23	25	28	31	34	37	40	43	46	49	52	55

Table 8 (CONTINUED)

n	p	m = 2	3	4	5	6	7	8	9	10	11	12	13	14	15	16	17	18	19	20
9	.001	0	0	0	2	3	4	6	8	9	11	13	15	16	18	20	22	24	26	27
	.005	0	1	2	4	6	8	10	12	14	17	19	21	23	25	28	30	32	34	37
	.01	0	2	4	6	8	10	12	15	17	19	22	24	27	29	32	34	37	39	41
	.025	1	3	5	8	11	13	16	18	21	24	27	29	32	35	38	40	43	46	49
	.05	2	5	7	10	13	16	19	22	25	28	31	34	37	40	43	46	49	52	55
	.10	3	6	10	13	16	19	23	26	29	32	36	39	42	46	49	53	56	59	63
10	.001	0	0	1	2	4	6	7	9	11	13	15	18	20	22	24	26	28	30	33
	.005	0	0	3	5	7	10	12	14	17	19	22	25	27	30	32	35	38	40	43
	.01	0	1	4	7	9	12	14	17	20	23	25	28	31	34	37	39	42	45	48
	.025	1	3	6	9	12	15	18	21	24	27	30	34	37	40	43	46	49	53	56
	.05	2	5	8	12	15	18	21	25	28	32	35	38	42	45	49	52	56	59	63
	.10	4	7	11	14	18	22	25	29	33	37	40	44	48	52	55	59	63	67	71
11	.001	0	0	1	3	5	7	9	11	13	16	18	21	23	25	28	30	33	35	38
	.005	0	0	3	6	8	11	14	17	19	22	25	28	31	34	37	40	43	46	49
	.01	0	1	5	8	10	13	16	19	23	26	29	32	35	38	42	45	48	51	54
	.025	1	4	7	10	14	17	20	24	27	31	34	38	41	45	48	52	56	59	63
	.05	2	5	9	13	17	20	24	28	32	35	39	43	47	51	55	58	62	66	70
	.10	4	8	12	16	20	24	28	32	37	41	45	49	53	58	62	66	70	74	79
12	.001	0	0	1	3	5	8	10	13	15	18	21	24	26	29	32	35	38	41	43
	.005	0	2	4	7	10	13	16	19	22	25	28	32	35	38	42	45	48	52	55
	.01	0	3	6	9	12	15	18	22	25	29	32	36	39	43	47	50	54	57	61
	.025	2	5	8	12	15	19	23	27	30	34	38	42	46	50	54	58	62	66	70
	.05	3	6	10	14	18	22	27	31	35	39	43	48	52	56	61	65	69	73	78
	.10	5	9	13	18	22	27	31	36	40	45	50	54	59	64	68	73	78	82	87

Table 8 (Continued)

n	P	m = 2	3	4	5	6	7	8	9	10	11	12	13	14	15	16	17	18	19	20
13	.001	0	0	2	4	6	9	12	15	18	21	24	27	30	33	36	39	43	46	49
	.005	0	2	4	8	11	14	18	21	25	28	32	35	39	43	46	50	54	58	61
	.01	1	3	6	10	13	17	21	24	28	32	36	40	44	48	52	56	60	64	68
	.025	2	5	9	13	17	21	25	29	34	38	42	46	51	55	60	64	68	73	77
	.05	3	7	11	16	20	25	29	34	38	43	48	52	57	62	66	71	76	81	85
	.10	5	10	14	19	24	29	34	39	44	49	54	59	64	69	75	80	85	90	95
14	.001	0	0	2	4	7	10	13	16	20	23	26	30	33	37	40	44	47	51	55
	.005	0	2	5	8	12	16	19	23	27	31	35	39	43	47	51	55	59	64	68
	.01	1	3	7	11	14	18	23	27	31	35	39	44	48	52	57	61	66	70	74
	.025	2	5	7	14	18	23	27	32	37	41	46	51	56	60	65	70	75	79	84
	.05	4	8	11	17	22	27	32	37	42	47	52	57	62	67	72	78	83	88	93
	.10	5	11	16	21	26	32	37	42	48	53	59	64	70	75	81	86	92	98	103
15	.001	0	0	2	5	8	11	15	18	22	25	29	33	37	41	44	48	52	56	60
	.005	0	3	6	9	13	17	21	25	30	34	38	43	47	52	56	61	65	70	74
	.01	1	4	8	12	16	20	25	29	34	38	43	48	52	57	62	67	71	76	81
	.025	2	6	11	15	20	25	30	35	40	45	50	55	60	65	71	76	81	86	91
	.05	4	9	14	19	24	29	34	40	45	51	56	62	67	73	78	84	89	95	101
	.10	6	11	17	23	28	34	40	46	52	58	64	69	75	81	87	93	99	105	111
16	.001	0	0	3	6	9	12	16	20	24	28	32	36	40	44	49	53	57	61	66
	.005	0	3	6	10	14	19	23	28	32	37	42	46	51	56	61	66	71	75	80
	.01	1	4	8	13	17	22	27	32	37	42	47	52	57	62	67	72	77	83	88
	.025	2	7	12	16	22	27	32	38	43	48	54	60	65	71	76	82	87	93	99
	.05	4	9	15	20	26	31	37	43	49	55	61	66	72	78	84	90	96	102	108
	.10	6	12	18	24	30	37	43	49	55	62	68	75	81	87	94	100	107	113	120

Table 8 (CONTINUED)

n	p	m = 2	3	4	5	6	7	8	9	10	11	12	13	14	15	16	17	18	19	20
17	.001	0	1	3	6	10	14	18	22	26	30	35	39	44	48	53	58	62	67	71
	.005	0	3	7	11	16	20	25	30	35	40	45	50	55	61	66	71	76	82	87
	.01	1	5	9	14	19	24	29	34	39	45	50	56	61	67	72	78	83	89	94
	.025	3	7	12	18	23	29	35	40	46	52	58	64	70	76	82	88	94	100	106
	.05	4	10	16	21	27	34	40	46	52	58	65	71	78	84	90	97	103	110	116
	.10	7	13	19	26	32	39	46	53	59	66	73	80	86	93	100	107	114	121	128
18	.001	0	1	4	7	11	15	19	24	28	33	38	43	47	52	57	62	67	72	77
	.005	0	3	7	12	17	22	27	32	38	43	48	54	59	65	71	76	82	88	93
	.01	1	5	10	15	20	25	31	37	42	48	54	60	66	71	77	83	89	95	101
	.025	3	8	13	19	25	31	37	43	49	56	62	68	75	81	87	94	100	107	113
	.05	5	10	17	23	29	36	42	49	56	62	69	76	83	89	96	103	110	117	124
	.10	7	14	21	28	35	42	49	56	63	70	78	85	92	99	107	114	121	129	136
19	.001	0	1	4	8	12	16	21	26	30	35	41	46	51	56	61	67	72	78	83
	.005	1	4	8	13	18	23	29	34	40	46	52	58	64	70	75	82	88	94	100
	.01	2	5	10	16	21	27	33	39	45	51	57	64	70	76	83	89	95	102	108
	.025	3	8	14	20	26	33	39	46	53	59	66	73	79	86	93	100	107	114	120
	.05	5	11	18	24	31	38	45	52	59	66	73	81	88	95	102	110	117	124	131
	.10	8	15	22	29	37	44	52	59	67	74	82	90	98	105	113	121	129	136	144
20	.001	0	1	4	8	13	17	22	27	33	38	43	49	55	60	66	71	77	83	89
	.005	1	4	9	14	19	25	31	37	43	49	55	61	68	74	80	87	93	100	106
	.01	2	6	11	17	23	29	35	41	48	54	61	68	74	81	88	94	101	108	115
	.025	3	9	15	21	28	35	42	48	56	63	70	77	84	91	99	106	113	120	128
	.05	5	12	19	26	33	40	48	56	63	70	78	85	93	101	108	116	124	131	139
	.10	8	16	23	31	39	47	55	63	71	79	87	95	103	111	120	128	136	144	152

For n or m greater than 20, the pth quantile w_p of the Mann-Whitney test statistic may be approximated by

$$w_p = \frac{nm}{2} + x_p \sqrt{\frac{nm(n+m+1)}{12}}$$

where x_p is the pth quantile of a standard normal random variable, obtained from Table 1.

Table 9   QUANTILES OF THE HOTELLING-PABST TEST STATISTIC[a]

n	$p = .001$	.005	.010	.025	.050	.100	$\frac{1}{3}n(n^2 - 1)$
4					2	2	20
5			2	2	4	6	40
6		2	4	6	8	14	70
7	2	6	8	14	18	26	112
8	6	12	16	24	32	42	168
9	12	22	28	38	50	64	240
10	22	36	44	60	74	92	330
11	36	56	66	86	104	128	440
12	52	78	94	120	144	172	572
13	76	110	130	162	190	226	728
14	106	148	172	212	246	290	910
15	142	194	224	270	312	364	1120
16	186	250	284	340	390	450	1360
17	238	314	356	420	480	550	1632
18	300	390	438	512	582	664	1938
19	372	476	532	618	696	790	2280
20	454	574	638	738	826	934	2660
21	546	686	758	870	972	1092	3080
22	652	810	892	1020	1134	1270	3542
23	772	950	1042	1184	1312	1464	4048
24	904	1104	1208	1366	1510	1678	4600
25	1050	1274	1390	1566	1726	1912	5200
26	1212	1462	1590	1786	1960	2168	5850
27	1390	1666	1808	2024	2216	2444	6552
28	1586	1890	2046	2284	2494	2744	7308
29	1800	2134	2306	2564	2796	3068	8120
30	2032	2398	2584	2868	3120	3416	8990

For n greater than 30, the quantiles of T may be approximated by

$$w_p \cong \tfrac{1}{6}n(n^2 - 1) + x_p \cdot \frac{1}{6} \frac{n(n^2 - 1)}{\sqrt{n - 1}}$$

where x_p is the pth quantile of a standard normal random variable given in Table 1.

SOURCE.   Adapted from Glasser and Winter (1961), with corrections.

[a] The entries in this table are the quantiles w_p of the Hotelling-Pabst test statistic T, defined by Equation (5.5.11), for selected values of p. Note that $P(T < w_p) \leq p$. Upper quantiles may be found from the equation

$$w_{1-p} = \tfrac{1}{3}n(n^2 - 1) - w_p$$

Critical regions correspond to values of T less than (or greater than) but not including the appropriate quantiles. Note that the median of T is given by

$$w_{.50} = \tfrac{1}{6}n(n^2 - 1)$$

Table 10 QUANTILES OF THE SPEARMAN TEST STATISTIC[a]

n	p = .900	.950	.975	.990	.995	.999
4	.8000	.8000				
5	.7000	.8000	.9000	.9000		
6	.6000	.7714	.8286	.8857	.9429	
7	.5357	.6786	.7450	.8571	.8929	.9643
8	.5000	.6190	.7143	.8095	.8571	.9286
9	.4667	.5833	.6833	.7667	.8167	.9000
10	.4424	.5515	.6364	.7333	.7818	.8667
11	.4182	.5273	.6091	.7000	.7455	.8364
12	.3986	.4965	.5804	.6713	.7273	.8182
13	.3791	.4780	.5549	.6429	.6978	.7912
14	.3626	.4593	.5341	.6220	.6747	.7670
15	.3500	.4429	.5179	.6000	.6536	.7464
16	.3382	.4265	.5000	.5824	.6324	.7265
17	.3260	.4118	.4853	.5637	.6152	.7083
18	.3148	.3994	.4716	.5480	.5975	.6904
19	.3070	.3895	.4579	.5333	.5825	.6737
20	.2977	.3789	.4451	.5203	.5684	.6586
21	.2909	.3688	.4351	.5078	.5545	.6455
22	.2829	.3597	.4241	.4963	.5426	.6318
23	.2767	.3518	.4150	.4852	.5306	.6186
24	.2704	.3435	.4061	.4748	.5200	.6070
25	.2646	.3362	.3977	.4654	.5100	.5962
26	.2588	.3299	.3894	.4564	.5002	.5856
27	.2540	.3236	.3822	.4481	.4915	.5757
28	.2490	.3175	.3749	.4401	.4828	.5660
29	.2443	.3113	.3685	.4320	.4744	.5567
30	.2400	.3059	.3620	.4251	.4665	.5479

For n greater than 30 the approximate quantiles of ρ may be obtained from

$$w_p \cong \frac{x_p}{\sqrt{n-1}}$$

where x_p is the p quantile of a standard normal random variable obtained from Table 1.

SOURCE.   Adapted from Glasser and Winter (1961), with corrections.

[a] The entries in this table are selected quantiles w_p of the Spearman rank correlation coefficient ρ when used as a test statistic. The lower quantiles may be obtained from the equation

$$w_p = -w_{1-p}$$

The critical region corresponds to values of ρ smaller than (or greater than) but not including the appropriate quantile. Note that the median of ρ is 0.

Table 11 QUANTILES OF THE KENDALL TEST STATISTIC[a]

n	$p = .900$	.950	.975	.990	.995
4	4	4	6	6	6
5	6	6	8	8	10
6	7	9	11	11	13
7	9	11	13	15	17
8	10	14	16	18	20
9	12	16	18	22	24
10	15	19	21	25	27
11	17	21	25	29	31
12	18	24	28	34	36
13	22	26	32	38	42
14	23	31	35	41	45
15	27	33	39	47	51
16	28	36	44	50	56
17	32	40	48	56	62
18	35	43	51	61	67
19	37	47	55	65	73
20	40	50	60	70	78
21	42	54	64	76	84
22	45	59	69	81	89
23	49	63	73	87	97
24	52	66	78	92	102
25	56	70	84	98	108
26	59	75	89	105	115
27	61	79	93	111	123
28	66	84	98	116	128
29	68	88	104	124	136
30	73	93	109	129	143

SOURCE.   Adapted from Table III, p. 53, of Kaarsemaker and van Wijngaarden (1953).

  [a] The entries in this table are selected quantiles w_p of the Kendall test statistic T, defined by Equation (5.5.13), for selected values of p. Only upper quantiles are given here, but lower quantiles may be obtained from the relationship

$$w_p = -w_{1-p}$$

Critical regions correspond to values of T greater than (or less than) but not including the appropriate quantile. Note that the median of T is 0.

Table 11   (CONTINUED)

n	$p = .900$	.950	.975	.990	.995
31	75	97	115	135	149
32	80	102	120	142	158
33	84	106	126	150	164
34	87	111	131	155	173
35	91	115	137	163	179
36	94	120	144	170	188
37	98	126	150	176	196
38	103	131	155	183	203
39	107	137	161	191	211
40	110	142	168	198	220

For n greater than 40, approximate quantiles of T may be obtained from

$$w_p \cong x_p \sqrt{\frac{n(n-1)(2n+5)}{18}}$$

where x_p is from the standard normal distribution given by Table 1.

Table 12   CRITICAL VALUES OF THE KRUSKAL-WALLIS TEST STATISTIC FOR THREE SAMPLES AND SMALL SAMPLE SIZES[a]

Sample sizes			Critical		Sample sizes			Critical	
n_1	n_2	n_3	Value	α	n_1	n_2	n_3	Value	α
2	1	1	2.7000	.500	4	2	2	6.0000	.014
2	2	1	3.6000	.200				5.3333	.033
2	2	2	4.5714	.067				5.1250	.052
			3.7143	.200				4.4583	.100
								4.1667	.105
3	1	1	3.2000	.300					
					4	3	1	5.8333	.021
3	2	1	4.2857	.100				5.2083	.050
			3.8571	.133				5.0000	.057
								4.0556	.093
3	3	2	5.3572	.029				3.8889	.129
			4.7143	.048					
			4.5000	.067	4	3	2	6.4444	.008
			4.4643	.105				6.3000	.011
3	3	1	5.1429	.043				5.4444	.046
			4.5714	.100				5.4000	.051
			4.0000	.129				4.5111	.098
3	3	2	6.2500	.011				4.4444	.102
			5.3611	.032					
			5.1389	.061	4	3	3	6.7455	.010
			4.5556	.100				6.7091	.013
			4.2500	.121				5.7909	.046
3	3	3	7.2000	.004				5.7273	.050
			6.4889	.001					
			5.6889	.029				4.7091	.092
			5.6000	.050				4.7000	.101
			5.0667	.086					
			4.6222	.100	4	4	1	6.6667	.010
4	1	1	3.5714	.200				6.1667	.022
								4.9667	.048
4	2	1	4.8214	.057				4.8667	.054
			4.5000	.076				4.1667	.082
			4.0179	.114				4.0667	.102

SOURCE.   Obtained from Kruskal and Wallis (1952).

[a] The null hypothesis may be rejected at the level α if the Kruskal-Wallis test statistic, given by Equation (5.6.3), is *equal to or greater than* the critical value given in the table. Use Table 2 if there are more than three samples or if the sample sizes exceed 5.

Table 12 (Continued)

| Sample sizes | | | Critical | | Sample sizes | | | Critical | |
n_1	n_2	n_3	Value	α	n_1	n_2	n_3	Value	α
4	4	2	7.0364	.006	5	3	2	6.9091	.009
			6.8727	.011				6.8281	.010
			5.4545	.046				5.2509	.049
			5.2364	.052				5.1055	.052
			4.5545	.098				4.6509	.091
			4.4455	.103				4.4121	.101
4	4	3	7.1439	.010	5	3	3	7.0788	.009
			7.1364	.011				6.9818	.011
			5.5985	.049				5.6485	.049
			5.5758	.051				5.5152	.051
			4.5455	.099				4.5333	.097
			4.4773	.102				4.4121	.109
4	4	4	7.6538	.008	5	4	1	6.9545	.008
			7.5385	.011				6.8400	.011
			5.6923	.049				4.9855	.044
			5.6538	.054				4.8600	.056
			4.6539	.097				3.9873	.098
			4.5001	.104				3.9600	.102
5	1	1	3.8571	.143	5	4	2	7.2045	.009
5	2	1	5.2500	.036				7.1182	.010
			5.0000	.048				5.2727	.049
			4.4500	.071				5.2682	.050
			4.2000	.095				4.5409	.098
			4.0500	.119				4.5182	.101
5	2	2	6.5333	.005	5	4	3	7.4449	.110
			6.1333	.013				7.3949	.011
			5.1600	.034				5.6564	.049
			5.0400	.056				5.6308	.050
			4.3733	.090				4.5487	.099
			4.2933	.112				4.5231	.103
5	3	1	6.4000	.012	5	4	4	7.7604	.009
			4.9600	.048				7.7440	.011
			4.8711	.052				5.6571	.049
			4.0178	.095				5.6176	.050
			3.8400	.123				4.6187	.100
								4.5527	.102

Table 12   (CONTINUED)

n_1	n_2	n_3	Critical Value	α	n_1	n_2	n_3	Critical Value	α
5	5	1	7.3091	.009				5.6264	.051
			6.8364	.011				4.5451	.100
			5.1273	.046				4.5363	.102
			4.9091	.053					
			4.1091	.086	5	5	4	7.8229	.010
			4.0364	.105				7.7914	.010
								5.6657	.049
5	5	2	7.3385	.010				5.6429	.050
			7.2692	.010				4.5229	.100
			5.3385	.047				4.5200	.101
			5.2462	.051					
			4.6231	.097	5	5	5	8.0000	.009
			4.5077	.100				7.9800	.010
								5.7800	.049
5	5	3	7.5780	.010				5.6600	.051
			7.5429	.010				4.5600	.100
			5.7055	.046				4.5000	.102

Table 13   Random Normal Deviates[a]

2.87	−.59	−.08	.74	.45	−.33	.70	−.76
−.51	1.13	−.71	1.14	−.84	−.31	−.72	−1.00
−.13	−.52	−.62	1.14	1.50	1.40	1.03	.57
−.35	−1.81	−.09	−.87	−.97	2.56	−1.18	1.09
−2.31	1.81	1.49	1.00	−1.00	.38	−.55	−1.65
−.18	.95	−1.10	.67	.76	.21	1.61	−1.02
−2.50	−.73	.52	−1.37	1.64	.53	−.60	−.22
−.32	.41	.52	1.08	−.59	−.53	−1.32	−.35
.45	−1.78	.77	−.37	−.94	.37	2.83	1.01
.82	−.16	−.05	−.18	−1.06	1.30	.84	.51
−.12	.05	−1.42	1.49	.79	1.05	−.11	−.25
1.11	−.82	−.67	1.16	.78	1.88	1.13	1.65
.94	−.15	−1.25	2.76	.05	−.75	−.68	.56
−1.09	−1.31	−.77	−1.31	.35	−.27	.88	1.88
.73	−.68	−.44	.65	−.38	−.68	.63	−.92
.09	.57	1.33	−.35	1.64	1.28	.42	−2.08
.64	−.65	.93	−.19	−1.25	.07	1.01	.63
−1.55	−1.40	.43	−.12	−.67	1.29	1.26	2.35
−.31	−.35	−.28	.20	−.83	.76	−.24	−2.01
−.30	−.55	1.61	−.24	1.37	.32	−.96	.63

For more random normal deviates, see the table of 100,000 random normal deviates furnished by the Rand Corporation (1955).

[a] The entries in this table are computer-generated numbers, which resemble observations on independent random variables having the standard normal distribution.

Table 14 Quantiles of the Kolmogorov Test Statistic[a]

One-Sided Test $p = .90$	.95	.975	.99	.995		$p = .90$	.95	.975	.99	.995
Two-Sided Test $p = .80$	.90	.95	.98	.99		$p = .80$	.90	.95	.98	.99
$n = 1$ .900	.950	.975	.990	.995	$n = 21$	.226	.259	.287	.321	.344
2 .684	.776	.842	.900	.929	22	.221	.253	.281	.314	.337
3 .565	.636	.708	.785	.829	23	.216	.247	.275	.307	.330
4 .493	.565	.624	.689	.734	24	.212	.242	.269	.301	.323
5 .447	.509	.563	.627	.669	25	.208	.238	.264	.295	.317
6 .410	.468	.519	.577	.617	26	.204	.233	.259	.290	.311
7 .381	.436	.483	.538	.576	27	.200	.229	.254	.284	.305
8 .358	.410	.454	.507	.542	28	.197	.225	.250	.279	.300
9 .339	.387	.430	.480	.513	29	.193	.221	.246	.275	.295
10 .323	.369	.409	.457	.489	30	.190	.218	.242	.270	.290
11 .308	.352	.391	.437	.468	31	.187	.214	.238	.266	.285
12 .296	.338	.375	.419	.449	32	.184	.211	.234	.262	.281
13 .285	.325	.361	.404	.432	33	.182	.208	.231	.258	.277
14 .275	.314	.349	.390	.418	34	.179	.205	.227	.254	.273
15 .266	.304	.338	.377	.404	35	.177	.202	.224	.251	.269
16 .258	.295	.327	.366	.392	36	.174	.199	.221	.247	.265
17 .250	.286	.318	.355	.381	37	.172	.196	.218	.244	.262
18 .244	.279	.309	.346	.371	38	.170	.194	.215	.241	.258
19 .237	.271	.301	.337	.361	39	.168	.191	.213	.238	.255
20 .232	.265	.294	.329	.352	40	.165	.189	.210	.235	.252
			Approximation for $n > 40$			$\dfrac{1.07}{\sqrt{n}}$	$\dfrac{1.22}{\sqrt{n}}$	$\dfrac{1.36}{\sqrt{n}}$	$\dfrac{1.52}{\sqrt{n}}$	$\dfrac{1.63}{\sqrt{n}}$

SOURCE. Adapted from Table 1 of Miller (1956).

[a] The entries in this table are selected quantiles w_p of the Kolmogorov test statistics T_1, T_1^+, and T_1^- as defined by (6.1.1) for two-sided tests and by (6.1.2) and (6.1.3) for one-sided tests. Reject H_0 at the level α if T exceeds the $1 - \alpha$ quantile given in this table. These quantiles are exact for $n \leq 20$ in the two-tailed test. The other quantiles are approximations which are equal to the exact quantiles in most cases.

Table 15 Quantiles of the Lilliefors Test Statistic[a]

	$p = .80$	.85	.90	.95	.99
Sample size $n = 4$	.300	.319	.352	.381	.417
5	.285	.299	.315	.337	.405
6	.265	.277	.294	.319	.364
7	.247	.258	.276	.300	.348
8	.233	.244	.261	.285	.331
9	.223	.233	.249	.271	.311
10	.215	.224	.239	.258	.294
11	.206	.217	.230	.249	.284
12	.199	.212	.223	.242	.275
13	.190	.202	.214	.234	.268
14	.183	.194	.207	.227	.261
15	.177	.187	.201	.220	.257
16	.173	.182	.195	.213	.250
17	.169	.177	.189	.206	.245
18	.166	.173	.184	.200	.239
19	.163	.169	.179	.195	.235
20	.160	.166	.174	.190	.231
25	.142	.147	.158	.173	.200
30	.131	.136	.144	.161	.187
Over 30	$\dfrac{.736}{\sqrt{n}}$	$\dfrac{.768}{\sqrt{n}}$	$\dfrac{.805}{\sqrt{n}}$	$\dfrac{.886}{\sqrt{n}}$	$\dfrac{1.031}{\sqrt{n}}$

SOURCE.   Adapted from Table 1 of Lilliefors (1967), with corrections.

[a] The entries in this table are the approximate quantiles w_p of the Lilliefors test statistic T_2 as defined by equation (6.1.11). Reject H_0 at the level α if T_2 exceeds $w_{1-\alpha}$ for the particular sample size n.

Table 16   QUANTILES OF THE SMIRNOV TEST STATISTIC FOR TWO SAMPLES OF EQUAL
SIZE n^a

	One-Sided Test: $p = .90$	.95	.975	.99	.995		One-Sided Test: $p = .90$	.95	.975	.99	.995
	Two-Sided Test: $p = .80$	.90	.95	.98	.99		Two-sided Test: $p = .80$	.90	.95	.98	.99
$n = 3$	2/3	2/3				$n = 20$	6/20	7/20	8/20	9/20	10/20
4	3/4	3/4	3/4			21	6/21	7/21	8/21	9/21	10/21
5	3/5	3/5	4/5	4/5	4/5	22	7/22	8/22	8/22	10/22	10/22
6	3/6	4/6	4/6	5/6	5/6	23	7/23	8/23	9/23	10/23	10/23
7	4/7	4/7	5/7	5/7	5/7	24	7/24	8/24	9/24	10/24	11/24
8	4/8	4/8	5/8	5/8	6/8	25	7/25	8/25	9/25	10/25	11/25
9	4/9	5/9	5/9	6/9	6/9	26	7/26	8/26	9/26	10/26	11/26
10	4/10	5/10	6/10	6/10	7/10	27	7/27	8/27	9/27	11/27	11/27
11	5/11	5/11	6/11	7/11	7/11	28	8/28	9/28	10/28	11/28	12/28
12	5/12	5/12	6/12	7/12	7/12	29	8/29	9/29	10/29	11/29	12/29
13	5/13	6/13	6/13	7/13	8/13	30	8/30	9/30	10/30	11/30	12/30
14	5/14	6/14	7/14	7/14	8/14	31	8/31	9/31	10/31	11/31	12/31
15	5/15	6/15	7/15	8/15	8/15	32	8/32	9/32	10/32	12/32	12/32
16	6/16	6/16	7/16	8/16	9/16	34	8/34	10/34	11/34	12/34	13/34
17	6/17	7/17	7/17	8/17	9/17	36	9/36	10/36	11/36	12/36	13/36
18	6/18	7/18	8/18	9/18	9/19	38	9/38	10/38	11/38	13/38	14/38
19	6/19	7/19	8/19	9/19	9/19	40	9/40	10/40	12/40	13/40	14/40
				Approximation for $n > 40$:			$\dfrac{1.52}{\sqrt{n}}$	$\dfrac{1.73}{\sqrt{n}}$	$\dfrac{1.92}{\sqrt{n}}$	$\dfrac{2.15}{\sqrt{n}}$	$\dfrac{2.30}{\sqrt{n}}$

SOURCE.   Adapted from Birnbaum and Hall (1960).

^{a} The entries in this tables are selected quantiles w_p of the Smirnov two-sample test statistic T defined by (6.2.2) and (6.2.3) for the one-tailed test and defined by (6.2.1) for the two-tailed test. Reject H_0 at the level α if T exceeds the $1 - \alpha$ quantile of T as given in this table. The test statistic is a discrete random variable, so the exact level of significance may be less than the apparent α used in this table.

Table 17   QUANTILES OF THE SMIRNOV TEST STATISTIC FOR TWO SAMPLES OF
DIFFERENT SIZE n AND m^a

One-Sided Test: Two-Sided Test:		$p = .90$ $p = .80$	.95 .90	.975 .95	.99 .98	.995 .99
$N_1 = 1$	$N_2 =$ 9	17/18				
	10	9/10				
$N_1 = 2$	$N_2 =$ 3	5/6				
	4	3/4				
	5	4/5	4/5			
	6	5/6	5/6			
	7	5/7	6/7			
	8	3/4	7/8	7/8		
	9	7/9	8/9	8/9		
	10	7/10	4/5	9/10		
$N_1 = 3$	$N_2 =$ 4	3/4	3/4			
	5	2/3	4/5	4/5		
	6	2/3	2/3	5/6		
	7	2/3	5/7	6/7	6/7	
	8	5/8	3/4	3/4	7/8	
	9	2/3	2/3	7/9	8/9	8/9
	10	3/5	7/10	4/5	9/10	9/10
	12	7/12	2/3	3/4	5/6	11/12
$N_1 = 4$	$N_2 =$ 5	3/5	3/4	4/5	4/5	
	6	7/12	2/3	3/4	5/6	5/6
	7	17/28	5/7	3/4	6/7	6/7
	8	5/8	5/8	3/4	7/8	7/8
	9	5/9	2/3	3/4	7/9	8/9
	10	11/20	13/20	7/10	4/5	4/5
	12	7/12	2/3	2/3	3/4	5/6
	16	9/16	5/8	11/16	3/4	13/16
$N_1 = 5$	$N_2 =$ 6	3/5	2/3	2/3	5/6	5/6
	7	4/7	23/35	5/7	29/35	6/7
	8	11/20	5/8	27/40	4/5	4/5
	9	5/9	3/5	31/45	7/9	4/5
	10	1/2	3/5	7/10	7/10	4/5
	15	8/15	3/5	2/3	11/15	11/15
	20	1/2	11/20	3/5	7/10	3/4
$N_1 = 6$	$N_2 =$ 7	23/42	4/7	29/42	5/7	5/6
	8	1/2	7/12	2/3	3/4	3/4
	9	1/2	5/9	2/3	13/18	7/9
	10	1/2	17/30	19/30	7/10	11/15
	12	1/2	7/12	7/12	2/3	3/4
	18	4/9	5/9	11/18	2/3	13/18
	24	11/24	1/2	7/12	5/8	2/3

Table 17   (CONTINUED)

One-Sided Test: Two-Sided Test:	$p = .90$ $p = .80$	$.95$ $.90$	$.975$ $.95$	$.99$ $.98$	$.995$ $.99$
$N_1 = 7$ $N_2 =$ 8	27/56	33/56	5/8	41/56	3/4
9	31/63	5/9	40/63	5/7	47/63
10	33/70	39/70	43/70	7/10	5/7
14	3/7	1/2	4/7	9/14	5/7
28	3/7	13/28	15/28	17/28	9/14
$N_1 = 8$ $N_2 =$ 9	4/9	13/24	5/8	2/3	3/4
10	19/40	21/40	23/40	27/40	7/10
12	11/24	1/2	7/12	5/8	2/3
16	7/16	1/2	9/16	5/8	5/8
32	13/32	7/16	1/2	9/16	19/32
$N_1 = 9$ $N_2 =$ 10	7/15	1/2	26/45	2/3	31/45
12	4/9	1/2	5/9	11/18	2/3
15	19/45	22/45	8/15	3/5	29/45
18	7/18	4/9	1/2	5/9	11/18
36	13/36	5/12	17/36	19/36	5/9
$N_1 = 10$ $N_2 =$ 15	2/5	7/15	1/2	17/30	19/30
20	2/5	9/20	1/2	11/20	3/5
40	7/20	2/5	9/20	1/2	
$N_1 = 12$ $N_2 =$ 15	23/60	9/20	1/2	11/20	7/12
16	3/8	7/16	23/48	13/24	7/12
18	13/36	5/12	17/36	19/36	5/9
20	11/30	5/12	7/15	31/60	17/30
$N_1 = 15$ $N_2 = 20$	7/20	2/5	13/30	29/60	31/60
$N_1 = 16$ $N_2 = 20$	27/80	31/80	17/40	19/40	41/80
Large-sample approximation	$1.07\sqrt{\dfrac{m+n}{mn}}$	$1.22\sqrt{\dfrac{m+n}{mn}}$	$1.36\sqrt{\dfrac{m+n}{mn}}$	$1.52\sqrt{\dfrac{m+n}{mn}}$	$1.63\sqrt{\dfrac{m+n}{mn}}$

SOURCE.   Adapted from Massey (1952).

[a] The entries in this table are selected quantities w_p of the Smirnov test statistic T for two samples, defined by Equations (6.2.1), (6.2.2), and (6.2.3). To enter the table let N_1 be the smaller sample size and let N_2 be the larger sample size. Reject H_0 at the level α if T exceeds $w_{1-\alpha}$ as given in the table. If n and m are not covered by this table, use the large sample approximation given at the end of the table.

Table 18 QUANTILES OF THE BIRNBAUM-HALL, OR
THREE-SAMPLE SMIRNOV, STATISTIC[a]

	$p = .80$	.90	.95	.98	.99
$n = 4$	3/4	3/4			
5	3/5	4/5	4/5		
6	4/6	4/6	5/6	5/6	5/6
7	4/7	5/7	5/7	6/7	6/7
8	5/8	5/8	5/8	6/8	6/8
9	5/9	5/9	6/9	6/9	7/9
10	5/10	6/10	6/10	7/10	7/10
11	5/11	6/11	7/11	7/11	8/11
12	6/12	6/12	7/12	8/12	8/12
13	6/13	7/13	7/13	8/13	8/13
14	6/14	7/14	8/14	8/14	9/14
15	6/15	7/15	8/15	9/15	9/15
16	7/16	7/16	8/16	9/16	9/16
17	7/17	8/17	8/17	9/17	10/17
18	7/18	8/18	9/18	9/18	10/18
19	7/19	8/19	9/19	10/19	10/19
20	7/20	8/20	9/20	10/20	11/20
22	8/22	9/22	10/22	11/22	11/22
24	8/24	9/24	10/24	11/24	12/24
26	9/26	10/26	10/26	11/26	12/26
28	9/28	10/28	11/28	12/28	13/28
30	9/30	10/30	11/30	12/30	13/30
32	10/32	11/32	12/32	13/32	14/32
34	10/34	11/34	12/34	13/34	14/34
36	10/36	11/36	12/36	14/36	14/36
38	10/38	12/38	13/38	14/38	15/38
40	11/40	12/40	13/40	14/40	15/40
Approximation for $n > 40$	$2.02/\sqrt{n}$	$2.18/\sqrt{n}$	$2.34/\sqrt{n}$	$2.53/\sqrt{n}$	$2.66/\sqrt{n}$

SOURCE. The entries for $n \leq 40$ were obtained from Table 1 of Birnbaum and Hall (1960). The large sample approximations result from a conjecture by the author.

[a] The entries in this table are selected quantiles w_p of the three-sample Smirnov statistic T_1 given by (6.3.1). All three samples are of size n. Reject H_0 at the level α if T_1 exceeds $w_{1-\alpha}$ as given in this table.

Table 19 QUANTILES OF THE ONE-SIDED k-SAMPLE SMIRNOV STATISTIC[a]

	$k = 2$					$k = 3$					$k = 4$				
$p =$	.90	.95	.975	.99	.995	.90	.95	.975	.99	.995	.90	.95	.975	.99	.995
$n = 2$															
3	2	2	3			2	3								
4	3	3	4	4	4	3	4	4	4						
5	3	3	4	5	5	3	4	5	5	5	3	4	4		5
6	3	4	5	5	5	4	5	5	6	6	4	4	5	5	6
7	3	4	5	5	6	4	5	5	6	7	4	5	5	6	6
8	4	4	5	6	6	4	5	6	7	7	4	5	6	6	7
9	4	5	6	6	7	5	6	6	7	8	5	6	6	7	7
10	4	5	6	7	7	5	6	7	7	8	5	6	6	7	8
12	4	5	6	7	8	5	6	7	8	9	5	6	7	8	9
14	5	6	7	8	9	6	7	7	8	9	6	7	8	8	9
16	5	6	7	8	9	6	7	8	9	10	6	7	8	9	9
18	5	7	8	9	10	7	8	8	9	10	7	8	8	9	10
20	6	7	8	9	11	7	8	9	10	10	7	8	9	10	11
25	6	7	9	10	12	8	9	10	11	12	8	9	10	11	12
30	6	8	10	11	13	9	10	11	12	13	9	10	11	12	13
35	7	9	11	12	14	10	11	12	13	14	10	11	12	14	14
40	8	9	11	13	15	10	12	13	14	15	10	12	13	14	15
45	8	10	12	14	16	11	12	14	15	16	11	13	14	15	16
50	10	12	13	15	16	12	13	14	16	17	12	14	15	16	17
Approximation for $n > 50$	$\dfrac{1.52}{\sqrt{n}}$	$\dfrac{1.73}{\sqrt{n}}$	$\dfrac{1.92}{\sqrt{n}}$	$\dfrac{2.15}{\sqrt{n}}$	$\dfrac{2.30}{\sqrt{n}}$	$\dfrac{1.73}{\sqrt{n}}$	$\dfrac{1.92}{\sqrt{n}}$	$\dfrac{2.09}{\sqrt{n}}$	$\dfrac{2.30}{\sqrt{n}}$	$\dfrac{2.45}{\sqrt{n}}$	$\dfrac{1.85}{\sqrt{n}}$	$\dfrac{2.02}{\sqrt{n}}$	$\dfrac{2.19}{\sqrt{n}}$	$\dfrac{2.39}{\sqrt{n}}$	$\dfrac{2.53}{\sqrt{n}}$

[a] The entries in this table, divided by n, are selected quantiles of the one-sided k-sample Smirnov statistic T_2 defined by (6.3.2). To obtain the p quantile, enter the table with the number of samples k and the sample size n. The value obtained from the table is then divided by n to become the p quantile. Reject H_0 if T_2 exceeds the q quantile. The approximate quantiles given at the bottom of the table require only division by $\sqrt{n}$ as indicated.

Table 19 (CONTINUED)

n	$k = 5$					$k = 6$					$k = 7$				
$p =$	.90	.95	.975	.99	.995	.90	.95	.975	.99	.995	.90	.95	.975	.99	.995
2															
3	3														
4	4	4	4			3					3				
5	4	5	5	5		4	4	5			4	4	4		
6	5	5	5	6	6	4	5	5	6	6	4	5	5		
7	5	5	6	6	6	5	5	5	6	7	5	5	5	5	6
8	5	6	6	7	7	5	5	6	7	7	5	6	6	6	6
9	5	6	6	7	7	5	6	6	7	8	5	6	6	7	7
10	6	6	7	8	8	6	6	7	8	8	6	7	7	7	8
12	6	7	8	8	9	6	7	7	9	9	6	7	7	8	8
14	7	7	8	9	10	7	7	8	9	10	7	8	8	9	9
16	7	8	8	10	10	7	8	9	10	11	8	8	9	9	10
18	8	8	9	10	11	8	9	9	10	11	8	9	9	10	11
20	8	9	9	11	12	8	9	10	11	12	9	9	10	11	11
25	9	10	11	12	13	9	10	11	12	13	10	10	11	12	13
30	10	11	12	13	14	10	11	12	13	14	11	11	12	13	14
35	11	12	13	14	15	11	12	13	14	15	11	12	13	14	15
40	12	13	14	15	16	12	13	14	15	16	12	13	14	15	16
45	12	13	15	16	17	13	14	15	16	17	13	14	15	16	17
50	13	14	15	17	18	13	15	16	17	18	14	15	16	17	18
Approximation for $n > 50$	$\dfrac{1.92}{\sqrt{n}}$	$\dfrac{2.09}{\sqrt{n}}$	$\dfrac{2.25}{\sqrt{n}}$	$\dfrac{2.45}{\sqrt{n}}$	$\dfrac{2.59}{\sqrt{n}}$	$\dfrac{1.97}{\sqrt{n}}$	$\dfrac{2.14}{\sqrt{n}}$	$\dfrac{2.30}{\sqrt{n}}$	$\dfrac{2.49}{\sqrt{n}}$	$\dfrac{2.63}{\sqrt{n}}$	$\dfrac{2.02}{\sqrt{n}}$	$\dfrac{2.18}{\sqrt{n}}$	$\dfrac{2.34}{\sqrt{n}}$	$\dfrac{2.53}{\sqrt{n}}$	$\dfrac{2.66}{\sqrt{n}}$

Table 19 (CONCLUDED)

n	k = 8					k = 9					k = 10				
p =	.90	.95	.975	.99	.995	.90	.95	.975	.99	.995	.90	.95	.975	.99	.995
2															
3	3														
4	4	4													
5	4	5	5			4	4					4			
6	5	5	6	5		5	5	5	5		4	5	5	5	
7	5	6	6	6	6	5	5	6	6	6	5	5	6	6	6
8	6	6	6	6	7	5	6	6	6	7	5	6	6	7	7
9	6	6	7	7	7	6	6	6	7	7	5	6	7	7	7
10	6	7	8	7	8	6	6	7	7	8	6	7	7	7	8
12	7	7	8	8	9	7	7	8	8	9	6	7	8	8	9
14	7	8	8	9	9	7	8	8	9	9	7	8	8	9	9
16	8	8	9	10	10	8	8	9	10	10	7	8	9	10	10
18	8	9	9	10	11	8	9	10	10	11	8	9	10	10	11
20	9	9	10	11	11	9	9	10	11	11	8	10	10	11	12
25	10	11	11	12	13	10	11	11	12	13	9	11	12	12	13
30	11	12	12	13	14	11	12	13	14	14	10	12	13	14	14
35	12	13	13	15	15	12	13	14	15	15	11	13	14	15	16
40	12	13	14	16	16	13	14	15	16	17	12	14	15	16	17
45	13	14	15	17	17	13	15	16	17	18	13	15	16	17	18
50	14	15	16	17	18	14	15	16	18	19	14	16	17	18	19
Approximation for n > 50	$\dfrac{2.05}{\sqrt{n}}$	$\dfrac{2.22}{\sqrt{n}}$	$\dfrac{2.37}{\sqrt{n}}$	$\dfrac{2.55}{\sqrt{n}}$	$\dfrac{2.69}{\sqrt{n}}$	$\dfrac{2.09}{\sqrt{n}}$	$\dfrac{2.25}{\sqrt{n}}$	$\dfrac{2.40}{\sqrt{n}}$	$\dfrac{2.58}{\sqrt{n}}$	$\dfrac{2.72}{\sqrt{n}}$	$\dfrac{2.11}{\sqrt{n}}$	$\dfrac{2.27}{\sqrt{n}}$	$\dfrac{2.42}{\sqrt{n}}$	$\dfrac{2.61}{\sqrt{n}}$	$\dfrac{2.74}{\sqrt{n}}$

Table 20 Quantiles of the Two-Sided k-Sample Smirnov Statistic[a]

	$p = .90$	$p = .95$	$p = .975$	$p = .99$	$p = .995$
$n = 3$	2 ($k = 2$)				
$n = 4$	3 ($2 \le k \le 6$)	3 ($k = 2$)			
$n = 5$	3 ($k = 2$) 4 ($3 \le k \le 10$)	4 ($2 \le k \le 10$)		4 ($k = 2$)	
$n = 6$	4 ($2 \le k \le 8$) 5 ($k = 9, 10$)	4 ($k = 2, 3$) 5 ($4 \le k \le 10$)	4 ($k = 2$) 5 ($3 \le k \le 10$)	5 ($2 \le k \le 6$)	5 ($k = 2, 3$)
$n = 7$	4 ($2 \le k \le 4$) 5 ($5 \le k \le 10$)	4 ($k = 2$) 5 ($3 \le k \le 10$)	5 ($2 \le k \le 5$) 6 ($6 \le k \le 10$)	5 ($k = 2$) 6 ($3 \le k \le 10$)	6 ($2 \le k \le 10$)
$n = 8$	4 ($k = 2$) 5 ($3 \le k \le 10$)	5 ($2 \le k \le 6$) 6 ($7 \le k \le 10$)	5 ($k = 2$) 6 ($3 \le k \le 10$)	6 ($2 \le k \le 7$) 7 ($8 \le k \le 10$)	6 ($k = 2, 3$) 7 ($4 \le k \le 10$)
$n = 9$	4 ($k = 2$) 5 ($3 \le k \le 10$)	5 ($k = 2, 3$) 6 ($4 \le k \le 10$)	6 ($2 \le k \le 9$) 7 ($k = 10$)	6 ($k = 2, 3$) 7 ($4 \le k \le 10$)	7 ($2 \le k \le 10$)
$n = 10$	5 ($2 \le k \le 6$) 6 ($7 \le k \le 10$)	5 ($k = 2$) 6 ($3 \le k \le 10$)	6 ($2 \le k \le 5$) 7 ($6 \le k \le 10$)	7 ($2 \le k \le 10$)	7 ($2 \le k \le 4$) 8 ($5 \le k \le 10$)
$n = 12$	5 ($k = 2, 3$) 6 ($4 \le k \le 10$)	6 ($2 \le k \le 4$) 7 ($5 \le k \le 10$)	6 ($k = 2$) 7 ($3 \le k \le 10$)	7 ($k = 2, 3$) 8 ($4 \le k \le 10$)	8 ($2 \le k \le 7$) 9 ($8 \le k \le 10$)
$n = 14$	6 ($2 \le k \le 7$) 7 ($8 \le k \le 10$)	6 ($k = 2$) 7 ($3 \le k \le 10$)	7 ($k = 2, 3$) 8 ($4 \le k \le 10$)	8 ($2 \le k \le 5$) 9 ($6 \le k \le 10$)	8 ($k = 2$) 9 ($3 \le k \le 10$)
$n = 16$	6 ($k = 2, 3$) 7 ($4 \le k \le 10$)	7 ($2 \le k \le 5$) 8 ($6 \le k \le 10$)	8 ($2 \le k \le 8$) 9 ($k = 9, 10$)	8 ($k = 2$) 9 ($3 \le k \le 10$)	9 ($2 \le k \le 4$) 10 ($5 \le k \le 10$)

[a] The entries in this table, divided by n, are selected quantiles of the two-sided k-sample Smirnov statistic T_3 defined by (6.3.3). To obtain the p quantile of T_3 enter the table with n, the number of observations in each of the k samples, and obtain the entry listed with the number of samples k and under the desired p. The value obtained from the table is then divided by n to become the p quantile. Reject H_0 if T_3 exceeds the p quantile. The approximate quantiles given at the end of the table require only division by $\sqrt{n}$ as indicated, and are valid for all values of k.

Table 20 (CONTINUED)

	$p = .90$	$p = .95$	$p = .975$	$p = .99$	$p = .995$
$n = 18$	6 ($k = 2$) 7 ($3 \leq k \leq 10$)	7 ($k = 2$) 8 ($3 \leq k \leq 10$)	8 ($2 \leq k \leq 4$) 9 ($5 \leq k \leq 10$)	9 ($2 \leq k \leq 4$) 10 ($5 \leq k \leq 10$)	10 ($2 \leq k \leq 9$) 11 ($k = 10$)
$n = 20$	7 ($2 \leq k \leq 6$) 8 ($7 \leq k \leq 10$)	8 ($2 \leq k \leq 7$) 9 ($8 \leq k \leq 10$)	8 ($k = 2$) 9 ($3 \leq k \leq 10$)	9 ($k = 2$) 10 ($3 \leq k \leq 10$)	10 ($k = 2, 3$) 11 ($4 \leq k \leq 10$)
$n = 25$	8 ($2 \leq k \leq 8$) 9 ($k = 9, 10$)	9 ($2 \leq k \leq 8$) 10 ($k = 9, 10$)	9 ($k = 2$) 10 ($3 \leq k \leq 9$) 11 ($k = 10$)	11 ($2 \leq k \leq 8$) 12 ($k = 9, 10$)	11 ($k = 2$) 12 ($3 \leq k \leq 10$)
$n = 30$	8 ($k = 2$) 9 ($3 \leq k \leq 10$)	9 ($k = 2$) 10 ($3 \leq k \leq 10$)	10 ($k = 2$) 11 ($3 \leq k \leq 10$)	12 ($2 \leq k \leq 8$) 13 ($k = 9, 10$)	12 ($k = 2$) 13 ($3 \leq k \leq 10$)
$n = 35$	9 ($2 \leq k \leq 4$) 10 ($5 \leq k \leq 10$)	10 ($k = 2, 3$) 11 ($4 \leq k \leq 10$)	11 ($k = 2$) 12 ($3 \leq k \leq 10$)	13 ($2 \leq k \leq 8$) 14 ($k = 9, 10$)	13 ($k = 2$) 14 ($3 \leq k \leq 10$)
$n = 40$	10 ($2 \leq k \leq 8$) 11 ($k = 9, 10$)	11 ($2 \leq k \leq 5$) 12 ($6 \leq k \leq 10$)	12 ($k = 2, 3$) 13 ($4 \leq k \leq 10$)	13 ($k = 2$) 14 ($3 \leq k \leq 10$)	14 ($k = 2$) 15 ($3 \leq k \leq 10$)
$n = 45$	10 ($k = 2, 3$) 11 ($4 \leq k \leq 10$)	12 ($2 \leq k \leq 8$) 13 ($k = 9, 10$)	13 ($2 \leq k \leq 5$) 14 ($6 \leq k \leq 10$)	14 ($k = 2$) 15 ($3 \leq k \leq 10$)	15 ($k = 2$) 16 ($3 \leq k \leq 10$)
$n = 50$	11 ($2 \leq k \leq 6$) 12 ($7 \leq k \leq 10$)	12 ($k = 2, 3$) 13 ($4 \leq k \leq 10$)	14 ($2 \leq k \leq 9$) 15 ($k = 10$)	15 ($k = 2, 3$) 16 ($4 \leq k \leq 10$)	16 ($k = 2, 3$) 17 ($4 \leq k \leq 10$)
Approximation for $n > 50$	$1.52/\sqrt{n}$	$1.73/\sqrt{n}$	$1.92/\sqrt{n}$	$2.15/\sqrt{n}$	$2.30/\sqrt{n}$

Table 21 CRITICAL VALUES OF THE TUKEY QUICK TEST STATISTIC[a]

$N_2 - N_1$	N_1	2.5%/.5%/.05% (one-sided) 5%/1%/.1% (two-sided)	$N_2 - N_1$	N_1	2.5%/.5%/.05% 5%/1%/.1%
0	4–8	7/9/13	6	3–4	9/11/–
	9–21	7/10/13		5–11	8/11/14
	22–24	7/10/14		12–	8/10/14
	25–	8/10/14			
			7	3–4	9/12/–
1	3–4	7/–/–		5	9/12/15
	5–6	7/9/–		6–8	8/11/15
	7	7/9/13		9–17	8/11/14
	8–20	7/10/13		18–	8/10/14
	21–23	7/10/14			
	24	8/10/14	8	3	10/13/–
				4	9/12/–
2	3–4	7/9/–		5–6	9/12/15
	5	7/10/–		7–14	8/11/15
	6–18	7/10/13		15–24	8/11/14
	19–21	7/10/14		25–	8/10/14
	22–	8/10/14			
			9	3	10/13/–
3	3–5	7/10/–		4	10/13/16
	6–14	7/10/13		5–7	9/12/16
	15–17	7/10/14		8	8/12/15
	18–	8/10/14		9–18	8/11/15
				19–31	8/11/14
4	3	8/–/–		32–	8/10/14
	4–7	8/10/13			
	8–	8/10/14	10	3	11/14/–
				4	10/13/17
5	3–4	9/11/–		5	9/13/17
	5–6	8/11/14		6	9/13/16
	7–	8/10/14		7–9	9/12/16

SOURCE. Adapted from Tukey (1959).

[a] The entries of this table are the critical values of the Tukey quick test statistics T_1 and T_2 for $\alpha = .025$, .005, and .005 in the one-sided tests, and T_3 for $\alpha = .05$, .01, and .001 in the two-sided test. Reject H_0 if the test statistic is *greater than or equal to* the appropriate critical value, which is obtained from the table by letting N_1 be the smaller of the two sample sizes and N_2 be the larger. Because of the discrete nature of the test statistics, the actual level of significance will be less than or equal to the apparent α used in the test.

Table 21   (CONTINUED)

$N_2 - N_1$	N_1	2.5%/.5%/.05% (one-sided) 5%/1%/.1% (two-sided)	$N_2 - N_1$	N_1	2.5%/.5%/.05% 5%/1%/.1%
10	10	8/12/15	14	2	13/–/–
	11–22	8/11/15		3	13/17/–
	23–42	8/11/14		4	11/16/19
	43–	8/10/14		5	11/15/19
				6	10/14/19
11	2	12/–/–		7	10/14/18
	3	11/15/–		8	9/13/18
	4	10/14/18		9–10	9/13/17
	5	10/13/17		11–12	9/12/17
	6–7	9/12/17		13	9/12/16
	8–10	9/12/16		14–16	8/12/16
	11–12	8/12/16			
	13–19	8/11/15	15	2	14/–/–
				3	13/18/–
12	2	12/–/–		4	12/16/20
	3	12/15/–		5	11/15/20
	4	11/15/18		6	10/15/19
	5	10/14/18		7	10/14/19
	6	10/13/17		8	10/14/18
	7–8	9/12/17		9	9/13/18
	9–10	9/12/16		10–11	9/13/17
	11–13	8/12/16		12–13	9/12/17
	14	8/11/16		14	9/12/16
	15–18	8/11/15		15	8/12/16
13	2	13/–/–	16	2	16/–/–
	3	12/16/–		3	13/18/–
	4	11/15/19		4	12/17/–
	5–6	10/14/18		5	11/16/20
	7	9/13/18		6	10/15/20
	8–9	9/13/17		7–8	10/14/19
	10	9/12/17		9	9/14/18
	11	9/12/16		10–11	9/13/18
	12–15	8/12/16		12	9/13/17
	16–17	8/11/16		13–14	9/12/17

Table 21 (Concluded)

$N_2 - N_1$	N_1	2.5%/.5%/.05% (one-sided) 5%/1%/.1% (two-sided)	$N_2 - N_1$	N_1	2.5%/.5%/.05% 5%/1%/.1%
17	2	16/–/–	19	2	17/–/–
	3	14/19/–		3	14/20/–
	4	12/18/–		4	13/19/23
	5	11/16/21		5	12/17/22
	6	11/16/20		6	11/16/22
	7	10/15/20		7	11/16/21
	8–9	10/14/19		8	10/15/20
	10–12	9/13/18		9	10/14/20
	13	9/13/17		10	10/14/19
				11	9/14/19
18	2	17/–/–	20	2	18/–/–
	3	14/20/–		3	15/21/–
	4	13/18/–		4	13/19/24
	5	11/17/22		5	12/18/23
	6	11/16/21		6	11/17/22
	7–8	10/15/20		7	11/16/21
	9	10/14/19		8	10/15/21
	10	9/14/19		9	10/15/20
	11–12	9/13/18		10	10/14/20

For sample sizes outside the range of this table, use the asymptotic approximation

$$\text{Prob}\,(T_3 \geq h) \cong \frac{2\lambda}{\lambda^2 - 1}\left(\frac{\lambda^h - 1}{(\lambda + 1)^h}\right) \qquad \text{where } \lambda = \frac{N_2}{N_1},$$

for the two-tailed test, with half that probability for the one-tailed tests.

Table 22 QUANTILES OF THE k-SAMPLE SLIPPAGE TEST STATISTIC[a]

	$k = 2$					$k = 3$					$k = 4$				
	$p=.80$	.90	.95	.98	.99	$p=.80$	.90	.95	.98	.99	$p=.80$	.90	.95	.98	.99
$n=3$	2	2				2	3					3			
4	2	3	3			3	3	4	4			4	4		
5	2	3	3	4		3	4	4	5	5	3	4	4	5	5
6	2	3	4	4	4	3	4	4	5	5	3	4	5	5	6
7	2	3	4	5	5	3	4	4	5	6	3	4	5	5	6
8	2	3	4	5	5	3	4	4	5	6	3	4	5	5	6
9	3	3	4	5	5	3	4	5	6	6	3	4	5	6	6
10	3	3	4	5	6	3	4	5	6	6	4	4	5	6	6
12	3	3	4	5	6	3	4	5	6	6	4	4	5	6	7
14	3	3	4	5	6	3	4	5	6	7	4	5	5	6	7
16	3	4	4	5	6	3	4	5	6	7	4	5	5	6	7
18	3	4	4	5	6	3	4	5	6	7	4	5	6	6	7
20	3	4	4	6	6	3	4	5	6	7	4	5	6	7	7
25	3	4	5	6	6	3	4	5	6	7	4	5	6	7	8
30	3	4	5	6	7	3	4	5	7	7	4	5	6	7	8
35	3	4	5	6	7	3	4	5	7	7	4	5	6	7	8
40	3	4	5	6	7	3	4	5	7	8	4	5	6	7	8
Approximation for $n > 40$	3	4	5	6	7	4	5	6	7	8	4	5	6	8	9

[a] The entries in this table are p quantiles w_p, for selected values of p, of the slippage test statistic as defined in Section 7.2. To obtain the p quantile, enter the portion of the table for the correct number of samples k, read across the row corresponding to the size n of each sample, to the entry in the column headed with the desired value of $p = 1 - \alpha$. Reject H_0 at the level α if the test statistic exceeds w_p.

411

Table 22 (CONTINUED)

	k = 5					k = 6					k = 7				
n	p = .80	.90	.95	.98	.99	p = .80	.90	.95	.98	.99	p = .80	.90	.95	.98	.99
n = 4	3	3				3	3				3	3			
5	3	4	4			3	4	4			3	4	4	5	
6	3	4	4	5	5	4	4	5	5	5	4	4	5	5	6
7	3	4	5	5	6	4	4	5	5	6	4	4	5	6	6
8	4	4	5	6	6	4	4	5	6	6	4	5	5	6	7
9	4	4	5	6	6	4	5	5	6	7	4	5	5	6	7
10	4	5	5	6	7	4	5	5	6	7	4	5	6	6	7
12	4	5	5	6	7	4	5	6	7	7	4	5	6	7	8
14	4	5	6	7	7	4	5	6	7	7	4	5	6	7	8
16	4	5	6	7	7	4	5	6	7	8	4	5	6	7	8
18	4	5	6	7	8	4	5	6	7	8	4	5	6	7	8
20	4	5	6	7	8	4	5	6	7	8	4	5	6	7	8
25	4	5	6	7	8	4	5	6	7	8	5	6	6	8	8
30	4	5	6	7	8	4	5	6	8	8	5	6	6	8	9
35	4	5	6	7	8	4	5	6	8	9	5	6	7	8	9
40	4	5	6	7	8	4	6	7	8	9	5	6	7	8	9
Approximation for n > 40	4	6	7	8	9	5	6	7	8	9	5	6	7	9	10

Table 22 (CONCLUDED)

	k = 8					k = 9					k = 10				
	p = .80	.90	.95	.98	.99	p = .80	.90	.95	.98	.99	p = .80	.90	.95	.98	.99
n = 4	3	4	4			3	4	4				4	4		
5	3	4	5	5		4	4	5	5		3	4	5	5	
6	4	5	5	6	6	4	5	5	6	6	4	5	5	6	6
7	4	5	5	6	6	4	5	5	6	6	4	5	5	6	7
8	4	5	5	6	7	4	5	5	6	7	4	5	6	7	7
9	4	5	6	6	7	4	5	6	6	7	4	5	6	7	7
10	4	5	6	6	7	4	5	6	7	7	4	5	6	7	7
12	4	5	6	7	8	5	5	6	7	8	5	5	6	7	8
14	4	5	6	7	8	5	6	6	7	8	5	6	6	7	8
16	4	5	6	7	8	5	6	6	7	8	5	6	6	7	8
18	5	6	6	7	8	5	6	6	8	8	5	6	7	8	8
20	5	6	7	8	9	5	6	7	8	9	5	6	7	8	9
25	5	6	7	8	9	5	6	7	8	9	5	6	7	8	9
30	5	6	7	8	9	5	6	7	8	9	5	6	7	8	9
35	5	6	7	8	9	5	6	7	8	9	5	6	7	8	9
40	5	6	7	8	9	5	6	7	8	9	5	6	7	8	9
Approximation for n > 40	5	6	7	9	10	5	6	8	9	10	5	7	8	9	10

Table 23 QUANTILES OF THE WALD-WOLFOWITZ TOTAL NUMBER OF RUNS STATISTIC[a]

N_1	N_2	$w_{.005}$	$w_{.01}$	$w_{.025}$	$w_{.05}$	$w_{.10}$	$w_{.90}$	$w_{.95}$	$w_{.975}$	$w_{.9}$	$w_{.995}$
2	5	—	—	—	—	3	—	—	—	—	—
	8	—	—	—	3	3	—	—	—	—	—
	11	—	—	—	3	3	—	—	—	—	—
	14	—	—	3	3	3	—	—	—	—	—
	17	—	—	3	3	3	—	—	—	—	—
	20	—	3	3	3	4	—	—	—	—	—
5	5	—	3	3	4	4	8	8	9	9	—
	8	3	3	4	4	5	9	10	10	—	—
	11	4	4	5	5	6	10	—	—	—	—
	14	4	4	5	6	6	—	—	—	—	—
	17	4	5	5	6	7	—	—	—	—	—
	20	5	5	6	6	7	—	—	—	—	—
8	8	4	5	5	6	6	12	12	13	13	14
	11	5	6	6	7	8	13	14	14	15	15
	14	6	6	7	8	8	14	15	15	16	16
	17	6	7	8	8	9	15	15	16	—	—
	20	7	7	8	9	10	15	16	16	—	—
11	11	6	7	8	8	9	15	16	16	17	18
	14	7	8	9	9	10	16	17	18	19	19
	17	8	9	10	10	11	17	18	19	20	21
	20	9	9	10	11	12	18	19	20	21	21
14	14	8	9	10	11	12	18	19	20	21	22
	17	9	10	11	12	13	20	21	22	23	23
	20	10	11	12	13	14	21	22	23	24	24
17	17	11	11	12	13	14	22	23	24	25	25
	20	12	12	14	14	16	23	24	25	26	27
20	20	13	14	15	16	17	25	26	27	28	29

For n or m greater than 20, the quantile w_p of T may be approximated by

$$w_p = \frac{2mn}{m + n} + 1 + x_p \sqrt{\frac{2mn(2mn - m - n)}{(m + n)^2(m + n - 1)}}$$

where x_p is the p quantile of a standard normal random variable, obtained from Table 1.

SOURCE. Adapted from Swed and Eisenhart (1943).

[a] The entries in this table are quantiles w_p of the Wald-Wolfowitz test statistic T. To enter the table let N_1 be the smaller sample size and N_2 the larger. If the exact values of N_1 and N_2 are not listed below, use the nearest values given as an approximation. This approximation will be exact in most cases. Reject H_0 at the level α if T is less than w_α (or greater than $w_{1-\alpha}$) for the one-tailed test, or, in the two-tailed test, if either $T < w_{\alpha/2}$ or $T > w_{1-\alpha/2}$. The test statistic is discrete, so the exact α will be less than or equal to the apparent α used in the test. For sample sizes greater than 20, the normal approximation given at the end of the tables may be used.

Table 24 TABLE OF SQUARES AND SQUARE ROOTS

Number	Square	Square Root	Number	Square	Square Root
1	1	1.0000	41	16 81	6.4031
2	4	1.4142	42	17 64	6.4807
3	9	1.7321	43	18 49	6.5574
4	16	2.0000	44	19 36	6.6332
5	25	2.2361	45	20 25	6.7082
6	36	2.4495	46	21 16	6.7823
7	49	2.6458	47	22 09	6.8557
8	64	2.8284	48	23 04	6.9282
9	81	3.0000	49	24 01	7.0000
10	1 00	3.1623	50	25 00	7.0711
11	1 21	3.3166	51	26 01	7.1414
12	1 44	3.4641	52	27 04	7.2111
13	1 69	3.6056	53	28 09	7.2801
14	1 96	3.7417	54	29 16	7.3485
15	2 25	3.8730	55	30 25	7.4162
16	2 56	4.0000	56	31 36	7.4833
17	2 89	4.1231	57	32 49	7.5498
18	3 24	4.2426	58	33 64	7.6158
19	3 61	4.3589	59	34 81	7.6811
20	4 00	4.4721	60	36 00	7.7460
21	4 41	4.5826	61	37 21	7.8102
22	4 84	4.6904	62	38 44	7.8740
23	5 29	4.7958	63	39 69	7.9373
24	5 76	4.8990	64	40 96	8.0000
25	6 25	5.0000	65	42 25	8.0623
26	6 76	5.0990	66	43 56	8.1240
27	7 29	5.1962	67	44 89	8.1854
28	7 84	5.2915	68	46 24	8.2462
29	8 41	5.3852	69	47 61	8.3066
30	9 00	5.4772	70	49 00	8.3666
31	9 61	5.5678	71	50 41	8.4261
32	10 24	5.6569	72	51 84	8.4853
33	10 89	5.7446	73	53 29	8.5440
34	11 56	5.8310	74	54 76	8.6023
35	12 25	5.9161	75	56 25	8.6603
36	12 96	6.0000	76	57 76	8.7178
37	13 69	6.0828	77	59 29	8.7750
38	14 44	6.1644	78	60 84	8.8318
39	15 21	6.2450	79	62 41	8.8882
40	16 00	6.3246	80	64 00	8.9443

Table 24 (CONTINUED)

Number	Square	Square Root	Number	Square	Square Root
81	65 61	9.0000	121	1 46 41	11.0000
82	67 24	9.0554	122	1 48 84	11.0454
83	68 89	9.1104	123	1 51 29	11.0905
84	70 56	9.1652	124	1 53 76	11.1355
85	72 25	9.2195	125	1 56 25	11.1803
86	73 96	9.2736	126	1 58 76	11.2250
87	75 69	9.3274	127	1 61 29	11.2694
88	77 44	9.3808	128	1 63 84	11.3137
89	79 21	9.4340	129	1 66 41	11.3578
90	81 00	9.4868	130	1 69 00	11.4018
91	82 81	9.5394	131	1 71 61	11.4455
92	84 64	9.5917	132	1 74 24	11.4891
93	86 49	9.6437	133	1 76 89	11.5326
94	88 36	9.6954	134	1 79 56	11.5758
95	90 25	9.7468	135	1 82 25	11.6190
96	92 16	9.7980	136	1 84 96	11.6619
97	94 09	9.8489	137	1 87 69	11.7047
98	96 04	9.8995	138	1 90 44	11.7473
99	98 01	9.9499	139	1 93 21	11.7898
100	1 00 00	10.0000	140	1 96 00	11.8322
101	1 02 01	10.0499	141	1 98 81	11.8743
102	1 04 04	10.0995	142	2 01 64	11.9164
103	1 06 09	10.1489	143	2 04 49	11.9583
104	1 08 16	10.1980	144	2 07 36	12.0000
105	1 10 25	10.2470	145	2 10 25	12.0416
106	1 12 36	10.2956	146	2 13 16	12.0830
107	1 14 49	10.3441	147	2 17 09	12.1244
108	1 16 64	10.3923	148	2 19 04	12.1655
109	1 18 81	10.4403	149	2 22 01	12.2066
110	1 21 00	10.4881	150	2 25 00	12.2474
111	1 23 21	10.5357	151	2 28 01	12.2882
112	1 25 44	10.5830	152	2 31 04	12.3288
113	1 27 69	10.6301	153	2 34 09	12.3693
114	1 29 96	10.6771	154	2 37 16	12.4097
115	1 32 25	10.7238	155	2 40 25	12.4499
116	1 34 56	10.7703	156	2 43 36	12.4900
117	1 36 89	10.8167	157	2 46 49	12.5300
118	1 38 24	10.8628	158	2 49 64	12.5698
119	1 41 61	10.9087	159	2 52 81	12.6095
120	1 44 00	10.9545	160	2 56 00	12.6491

Table 24 (Continued)

Number	Square	Square Root	Number	Square	Square Root
161	2 59 21	12.6886	201	4 04 01	14.1774
162	2 62 44	12.7279	202	4 08 04	14.2127
163	2 65 69	12.7671	203	4 12 09	14.2478
164	2 68 96	12.8062	204	4 16 16	14.2829
165	2 72 25	12.8452	205	4 20 25	14.3178
166	2 75 56	12.8841	206	4 24 36	14.3527
167	2 78 89	12.9228	207	4 28 49	14.3875
168	2 82 24	12.9615	208	4 32 64	14.4222
169	2 85 61	13.0000	209	4 36 81	14.4568
170	2 89 00	13.0384	210	4 41 00	14.4914
171	2 92 41	13.0767	211	4 45 21	14.5258
172	2 95 84	13.1149	212	4 49 44	14.5602
173	2 99 29	13.1529	213	4 53 69	14.5945
174	3 02 76	13.1909	214	4 57 96	14.6287
175	3 06 25	13.2288	215	4 62 25	14.6629
176	3 09 76	13.2665	216	4 66 56	14.6969
177	3 13 29	13.3041	217	4 70 89	14.7309
178	3 16 84	13.3417	218	4 75 24	14.7648
179	3 20 41	13.3791	219	4 79 61	14.7986
180	3 24 00	13.4164	220	4 84 00	14.8324
181	3 27 61	13.4536	221	4 88 41	14.8661
182	3 31 24	13.4907	222	4 92 84	14.8997
183	3 34 89	13.5277	223	4 97 29	14.9332
184	3 38 56	13.5647	224	5 01 76	14.9666
185	3 42 25	13.6015	225	5 06 25	15.0000
186	3 45 96	13.6382	226	5 10 76	15.0333
187	3 49 69	13.6748	227	5 15 29	15.0665
188	3 53 44	13.7113	228	5 19 84	15.0997
189	3 57 21	13.7477	229	5 24 41	15.1327
190	3 61 00	13.7840	230	5 29 00	15.1658
191	3 64 81	13.8203	231	5 33 61	15.1987
192	3 68 64	13.8564	232	5 38 24	15.2315
193	3 72 49	13.8924	233	5 42 89	15.2643
194	3 76 36	13.9284	234	5 47 56	15.2971
195	3 80 25	13.9642	535	5 52 25	15.3297
196	3 84 16	14.0000	236	5 56 96	15.3623
197	3 88 09	14.0357	237	5 61 69	15.3948
198	3 92 04	14.0712	238	5 66 44	15.4272
199	3 96 01	14.1067	239	5 71 21	15.4596
200	4 00 00	14.1421	240	5 76 00	15.4919

Table 24 (CONTINUED)

Number	Square	Square Root	Number	Square	Square Root
241	5 80 81	15.5242	281	7 89 61	16.7631
242	5 85 64	15.5563	282	7 95 24	16.7929
243	5 90 49	15.5885	283	8 00 89	16.8226
244	5 95 36	15.6205	284	8 06 56	16.8523
245	6 00 25	15.6525	285	8 12 25	16.8819
246	6 05 16	15.6844	286	8 17 96	16.9115
247	6 10 09	15.7162	287	8 23 69	16.9411
248	6 15 04	15.7480	288	8 29 44	16.9706
249	6 20 01	15.7797	289	8 35 21	17.0000
250	6 25 00	15.8114	290	8 41 00	17.0294
251	6 30 01	15.8430	291	8 46 81	17.0587
252	6 35 04	15.8745	292	8 52 64	17.0880
253	6 40 09	15.9060	293	8 58 49	17.1172
254	6 45 16	15.9374	294	8 64 36	17.1464
255	6 50 25	15.9687	295	8 70 25	17.1756
256	6 55 36	16.0000	296	8 76 16	17.2047
257	6 60 49	16.0312	297	8 82 09	17.2337
258	6 65 64	16.0624	298	8 88 04	17.2627
259	6 70 81	16.0935	299	8 94 01	17.2916
260	6 76 00	16.1245	300	9 00 00	17.3205
261	6 81 21	16.1555	301	9 06 01	17.3494
262	6 86 44	16.1864	302	9 12 04	17.3781
263	6 91 69	16.2173	303	9 18 09	17.4069
264	6 96 96	16.2481	304	9 24 16	17.4356
265	7 02 25	16.2788	305	9 30 25	17.4642
266	7 07 56	16.3085	306	9 36 36	17.4929
267	7 12 89	16.3401	307	9 42 49	17.5214
268	7 18 24	16.3707	308	9 48 64	17.5499
269	7 23 61	16.4012	309	9 54 81	17.5784
270	7 29 00	16.4317	310	9 61 00	17.6068
271	7 34 41	16.4621	311	9 67 21	17.6352
272	7 39 84	16.4924	312	9 73 44	17.6635
273	7 45 29	16.5227	313	9 79 69	17.6918
274	7 50 76	16.5529	314	9 85 96	17.7200
275	7 56 25	16.5831	315	9 92 25	17.7482
276	7 61 76	16.6132	316	9 98 56	17.7764
277	7 67 29	16.6433	317	10 04 89	17.8045
278	7 72 84	16.6733	318	10 11 24	17.8326
279	7 78 41	16.7033	319	10 17 61	17.8606
280	7 84 00	16.7322	320	10 24 00	17.8885

Table **24** (CONTINUED)

Number	Square	Square Root	Number	Square	Square Root
321	10 30 41	17.9165	361	13 03 21	19.0000
322	10 36 84	17.9444	362	13 10 44	19.0263
323	10 43 29	17.9722	363	13 17 69	19.0526
324	10 49 76	18.0000	364	13 24 96	19.0788
325	10 56 25	18.0278	365	13 32 25	19.1050
326	10 62 76	18.0555	366	13 39 56	19.1311
327	10 69 29	18.0831	367	13 46 89	19.1572
328	10 75 84	18.1108	368	13 54 24	19.1833
329	10 82 41	18.1384	369	13 61 61	19.2094
330	10 89 00	18.1659	370	13 69 00	19.2354
331	10 95 61	18.1934	371	13 76 41	19.2614
332	11 02 24	18.2209	372	13 83 84	19.2873
333	11 08 89	18.2483	373	13 91 29	19.3132
334	11 15 56	18.2757	374	13 98 76	19.3391
335	11 22 25	18.3030	375	14 06 25	19.3649
336	11 28 96	18.3303	376	14 13 76	19.3907
337	11 35 69	18.3576	377	14 21 29	19.4165
338	11 42 44	18.3848	378	14 28 84	19.4422
339	11 49 21	18.4120	379	14 36 41	19.4679
340	11 56 00	18.4391	380	14 44 00	19.4936
341	11 62 81	18.4662	381	14 51 61	19.5192
342	11 69 64	18.4932	382	14 59 24	19.5448
343	11 76 49	18.5203	383	14 66 89	19.5704
344	11 83 36	18.5472	384	14 74 56	19.5959
345	11 90 25	18.5742	385	14 82 25	19.6214
346	11 97 16	18.6011	386	14 89 96	19.6469
347	12 04 09	18.6279	387	14 97 69	19.6273
348	12 11 04	18.6548	388	15 05 44	19.6977
349	12 18 01	18.6815	389	15 13 21	19.7231
350	12 25 00	18.7083	390	15 21 00	19.7484
351	12 32 01	18.7350	391	15 28 81	19.7737
352	12 39 04	18.7617	392	15 36 64	19.7990
353	12 46 09	18.7883	393	15 44 49	19.8242
354	12 53 16	18.8149	394	15 52 36	19.8494
355	12 60 25	18.8414	395	15 60 25	19.8746
356	12 67 36	18.8680	396	15 68 16	19.8997
357	12 74 49	18.8944	397	15 76 09	19.9249
358	12 81 64	18.9209	398	15 84 04	19.9499
359	12 88 81	18.9473	399	15 92 01	19.9750
360	12 96 00	18.9737	400	16 00 00	20.0000

Table 24 (CONTINUED)

Number	Square	Square Root	Number	Square	Square Root
401	16 08 01	20.0250	441	19 44 81	21.0000
402	16 16 04	20.0499	442	19 53 64	21.0238
403	16 24 09	20.0749	443	19 62 49	21.0476
404	16 32 16	20.0998	444	19 71 36	21.0713
405	16 40 25	20.1246	445	19 80 25	21.0950
406	16 48 36	20.1494	446	19 89 16	21.1187
407	16 56 49	20.1742	447	19 98 09	21.1424
408	16 64 64	20.1990	448	20 07 04	21.1660
409	16 72 81	20.2237	449	20 16 01	21.1896
410	16 81 00	20.2485	450	20 25 00	21.2132
411	16 89 21	20.2731	451	20 34 01	21.2368
412	16 97 44	20.2978	452	20 43 04	21.2603
413	17 05 69	20.3224	453	20 52 09	21.2838
414	17 13 96	20.3470	454	20 61 16	21.3073
415	17 22 25	20.3715	455	20 70 25	21.3307
416	17 30 56	20.3961	456	20 79 36	21.3542
417	17 38 89	20.4206	457	20 88 49	21.3776
418	17 47 24	20.4450	458	20 97 64	21.4009
419	17 55 61	20.4695	459	21 06 81	21.4243
420	17 64 00	20.4939	460	21 16 00	21.4476
421	17 72 41	20.5183	461	21 25 21	21.4709
422	17 80 84	20.5426	462	21 34 44	21.4942
423	17 89 29	20.5670	463	21 43 69	21.5174
424	17 97 76	20.5913	464	21 52 96	21.5407
425	18 06 25	20.6155	465	21 62 25	21.5639
426	18 14 76	20.6398	466	21 71 56	21.5870
427	18 23 29	20.6640	467	21 80 89	21.6102
428	18 31 84	20.6882	468	21 90 24	21.6333
429	18 40 41	20.7123	469	21 99 61	21.6564
430	18 49 00	20.7364	470	22 09 00	21.6795
431	18 57 61	20.7605	471	22 18 41	21.7025
432	18 66 24	20.7846	472	22 27 84	21.7256
433	18 74 89	20.8087	473	22 37 29	21.7486
434	18 83 56	22.8327	474	22 46 76	21.7715
435	18 92 25	20.8567	475	22 56 25	21.7945
436	19 00 96	20.8806	476	22 65 76	21.8174
437	19 09 69	20.9045	477	22 75 29	21.8403
438	19 18 44	20.9284	478	22 84 84	21.8632
439	19 27 21	20.9523	479	22 94 41	21.8861
440	19 36 00	20.9762	480	23 04 00	21.9089

Table 24 (Continued)

Number	Square	Square Root	Number	Square	Square Root
481	23 13 61	21.9317	521	27 14 41	22.8254
482	23 23 24	21.9545	522	27 24 84	22.8473
483	23 32 89	21.9773	523	27 35 29	22.8692
484	23 42 56	22.0000	524	27 45 76	22.8910
485	23 52 25	22.0227	525	27 56 25	22.9129
486	23 61 96	22.0454	526	27 66 76	22.9347
487	23 71 69	22.0681	527	27 77 29	22.9565
488	23 81 44	22.0907	528	27 87 84	22.9783
489	23 91 21	22.1133	529	27 98 41	23.0000
490	24 01 00	22.1359	530	28 09 00	23.0217
491	24 10 81	22.1585	531	28 19 61	23.0434
492	24 20 64	22.1811	532	28 30 24	23.0651
493	24 30 49	22.2036	533	28 40 89	23.0868
494	24 40 36	22.2261	534	28 51 56	23.1084
495	24 50 25	22.2486	535	28 62 25	23.1301
496	24 60 16	22.2711	536	28 72 96	23.1517
497	24 70 09	22.2925	537	28 83 69	23.1733
498	24 80 04	22.3159	438	28 94 44	23.1948
499	24 90 01	22.3383	539	29 05 21	23.2164
500	25 00 00	22.3607	540	29 16 00	23.2379
501	25 10 01	22.3830	541	29 26 81	23.2594
502	25 20 04	22.4054	542	29 37 64	23.2809
503	25 30 09	22.4277	543	29 48 49	23.3024
504	25 40 16	22.4499	544	29 59 36	23.3238
505	25 50 25	22.4722	545	29 70 25	23.3452
506	25 60 36	22.4944	546	29 81 16	23.3666
507	25 70 49	22.5167	547	29 92 09	23.3880
508	25 80 64	22.5389	548	30 03 04	23.4094
509	25 90 81	22.5610	549	30 14 01	23.4307
510	26 01 00	22.5832	550	30 25 00	23.4521
511	26 11 21	22.6053	551	30 36 01	23.4734
512	26 21 44	22.6274	552	30 47 04	23.4947
513	26 31 69	22.6495	553	30 58 09	23.5160
514	26 41 96	22.6716	554	30 69 16	23.5372
515	26 52 25	22.6936	555	30 80 25	23.5584
516	26 62 56	22.7156	556	30 91 36	23.5797
517	26 72 89	22.7376	557	31 02 49	23.6008
518	26 83 24	22.7596	558	31 13 64	23·6220
519	26 93 61	22.7816	559	31 24 81	23.6432
520	27 04 00	22.8035	560	31 36 00	23.6643

Table 24 (Continued)

Number	Square	Square Root	Number	Square	Square Root
561	31 47 21	23.6854	601	36 12 01	24.5153
562	31 58 44	23.7065	602	36 24 04	24.5357
563	31 69 69	23.7276	603	36 36 09	24.5561
564	31 80 96	23.7487	604	36 48 16	24.5764
565	31 92 25	23.7697	605	36 60 25	24.5967
566	32 03 56	23.7908	606	36 72 36	24.6171
567	32 14 89	23.8118	607	36 84 49	24.6374
568	32 26 24	23.8238	608	36 96 64	24.6577
569	32 37 61	23.8537	609	37 08 81	24.6779
570	32 49 00	23.8747	610	37 21 00	24.6982
571	32 60 41	23.8956	611	37 33 21	24.7184
572	32 71 84	23.9165	612	37 45 44	24.7385
573	32 83 29	23.9374	613	37 57 69	24.7588
574	32 94 76	23.9583	614	37 69 96	24.7790
575	33 06 25	23.9792	615	37 82 25	24.7992
576	33 17 76	24.0000	616	37 94 56	24.8193
577	33 29 29	24.0208	617	38 06 89	24.8395
578	33 40 84	24.0416	618	38 19 24	24.8596
579	33 52 41	24.0624	619	38 31 61	24.8797
580	33 64 00	24.0832	620	38 44 00	24.8998
581	33 75 61	24.1039	621	38 56 41	24.9199
582	33 87 24	24.1247	622	38 68 84	24.9399
583	33 98 89	24.1454	623	38 81 29	24.9600
584	34 10 56	24.1661	624	38 93 76	24.9800
585	34 22 25	24.1868	625	39 06 25	25.0000
586	34 33 96	24.2074	626	39 18 76	25.0200
587	34 45 69	24.2281	627	39 31 29	25.0400
588	34 57 44	24.2487	628	39 43 84	25.0599
589	34 69 21	24.2693	629	39 56 41	25.0799
590	34 81 00	24.2899	630	39 69 00	25.0998
591	34 92 81	24.3105	631	39 81 61	25.1197
592	35 04 64	24.3311	632	39 94 24	25.1396
593	35 16 49	24.3516	633	40 06 89	25.1595
594	35 28 36	24.3721	634	40 19 56	25.1794
595	35 40 25	24.3926	635	40 32 25	25.1992
596	35 52 16	24.4131	636	40 44 96	25.2190
597	35 64 09	24.4336	637	40 57 69	25.2389
598	35 76 04	24.4540	638	40 70 44	25.2587
599	35 88 01	24.4745	639	40 83 21	25.2784
600	36 00 00	24.4949	640	40 96 00	25.2982

Table 24 (Continued)

Number	Square	Square Root	Number	Square	Square Root
641	41 08 81	25.3180	681	46 37 61	26.0960
642	41 21 64	25.3377	682	46 51 24	26.1151
643	41 34 49	25.3574	683	46 64 89	26.1343
644	41 47 36	25.3772	684	46 78 56	26.1534
645	41 60 25	25.3969	685	46 92 25	26.1725
646	41 73 16	25.4165	686	47 05 96	26.1916
647	41 86 09	25.4362	687	47 19 69	26.2107
648	41 99 04	25.4558	688	47 33 44	26.2298
649	42 12 01	25.4755	689	47 47 21	26.2488
650	42 25 00	25.4951	690	47 61 00	26.2679
651	42 38 01	25.4147	691	47 74 81	26.2869
652	42 51 04	25.5343	692	47 88 64	26.3059
653	42 64 09	25.5539	693	48 02 49	26.3249
654	42 77 16	25.5734	694	48 16 36	26.3439
655	42 90 25	25.5930	695	48 30 25	26.3629
656	43 03 36	25.6125	696	48 44 16	26.3818
657	43 16 49	25.6320	697	48 58 09	26.4008
658	43 29 64	25.6515	698	48 72 04	26.4197
659	43 42 81	25.6710	699	48 86 01	26.4386
660	43 56 00	25.6905	700	49 00 00	26.4575
661	43 69 21	25.7099	701	49 14 01	26.4764
662	43 82 44	25.7294	702	49 28 04	26.4953
663	43 95 69	25.7488	703	49 42 09	26.5141
664	44 08 96	25.7682	704	49 56 16	26.5330
665	44 22 25	25.7876	705	49 70 25	26.5518
666	44 35 56	25.8070	706	49 84 36	26.5707
667	44 48 89	25.8263	707	49 98 49	26.5895
668	44 62 24	25.8457	708	50 12 64	26.6083
669	44 75 61	25.8650	709	50 26 81	26.6271
670	44 89 00	25.8844	710	50 41 00	26.6458
671	45 02 41	25.9037	711	50 55 21	26.6646
672	45 15 84	25.9230	712	50 69 44	26.6833
673	45 29 29	25.9422	713	50 83 59	26.7021
674	45 42 76	25.9615	714	50 97 96	26.7208
675	45 56 25	25.9808	715	51 12 25	26.7395
676	45 69 76	26.0000	716	51 26 56	26.7582
677	45 83 29	26.0192	717	51 40 89	26.7769
678	45 96 84	26.0384	718	51 55 24	26.7955
679	46 10 41	26.0576	719	51 69 61	26.8142
680	46 24 00	26.0768	720	51 84 00	26.8328

Table 24 (Continued)

Number	Square	Square Root	Number	Square	Square Root
721	51 98 41	26.8514	761	57 91 21	27.5862
722	52 12 84	26.8701	762	58 06 44	27.6043
723	52 27 29	26.8887	763	58 21 69	27.6225
724	52 41 76	26.9072	764	58 36 96	27.6405
725	52 56 25	26.9258	765	58 52 25	27.6586
726	52 70 76	26.9444	766	58 67 56	27.6767
727	52 85 29	26.9629	767	58 82 89	27.6948
728	52 99 84	26.9815	768	58 98 24	27.7128
729	53 14 41	27.0000	769	59 13 61	27.7308
730	53 29 00	27.0185	770	59 29 00	27.7489
731	53 43 61	27.0370	771	59 44 41	27.7669
732	53 58 24	27.0555	772	59 59 84	27.7849
733	53 72 89	27.0740	773	59 75 29	27.8029
734	53 87 56	27.0924	774	59 90 76	27.8209
735	54 02 25	27.1109	775	60 06 25	27.8388
736	54 16 96	27.1293	776	60 21 76	27.8568
737	54 31 69	27.1477	777	60 37 29	27.8747
738	54 46 44	27.1662	778	60 52 84	27.8927
739	54 61 27	27.1846	779	60 68 41	27.9106
740	54 76 00	27.2029	780	60 84 00	27.9285
741	54 90 81	27.2213	781	60 99 61	27.9464
742	55 05 64	27.2397	782	61 15 24	27.9643
743	55 20 49	27.2580	783	61 30 89	27.9821
744	55 35 36	27.2764	784	61 46 56	28.0000
745	55 50 25	27.2947	785	61 62 25	28.0179
746	55 65 16	27.3130	786	61 77 96	28.0357
747	55 80 09	27.3313	787	61 93 69	28.0535
748	55 95 04	27.3496	788	62 09 44	28.0713
749	56 10 01	27.3679	789	62 25 21	28.0891
750	56 25 00	27.3861	790	62 41 00	28.1069
751	56 40 01	27.4044	791	62 56 81	28.1247
752	56 55 04	27.4226	792	62 72 64	28.1425
753	56 70 09	27.4408	793	62 88 49	28.1603
754	56 85 16	27.4591	794	63 04 36	28.1780
755	57 00 25	27.4773	795	63 20 25	28.1957
756	57 15 36	27.4955	796	63 36 16	28.2135
757	57 30 49	27.5136	797	63 52 09	28.2312
758	57 45 64	27.5318	798	63 68 04	28.2489
759	57 60 81	27.5500	799	63 84 01	28.2666
760	57 76 00	27.5681	800	64 00 00	28.2843

Table 24 (CONTINUED)

Number	Square	Square Root	Number	Square	Square Root
801	64 16 01	28.3019	841	70 72 81	29.0000
802	64 32 04	28.3196	842	70 89 64	29.0172
803	64 48 09	28.3373	843	71 06 49	29.0345
804	64 64 16	28.3549	844	71 23 36	29.0517
805	64 80 25	28.3725	845	71 40 25	29.0689
806	64 96 36	28.3901	846	71 57 16	29.0861
807	65 12 49	28.4077	847	71 74 09	29.1033
808	65 28 64	28.4253	848	71 91 04	29.1204
809	65 44 81	28.4429	849	72 08 01	29.1376
810	65 61 00	28.4605	850	72 25 00	29.1548
811	65 77 21	28.4781	851	72 42 01	29.1719
812	65 93 44	28.4956	852	72 59 04	29.1890
813	66 09 69	28.5132	853	72 76 09	29.2062
814	66 25 96	28.5307	854	72 93 16	29.2233
815	66 42 25	28.5482	855	73 10 25	29.2404
816	66 58 56	28.5657	846	73 27 36	29.2575
817	66 74 89	28.5832	857	73 44 49	29.2746
818	66 91 24	28.6007	858	73 61 64	29.2916
819	67 07 61	28.6082	859	73 78 81	29.3087
820·	67 24 00	28.6356	860	73 96 00	29.3258
821	67 40 41	28.6531	861	74 13 21	29.3428
822	67 56 84	28.6705	862	74 30 44	29.3598
823	67 73 29	28.6880	863	74 47 69	29.3769
824	67 89 76	28.7054	864	74 64 96	29.3939
825	68 06 25	28.7228	865	74 82 25	29.4109
826	68 22 76	28.7402	866	74 99 56	29.4279
827	68 39 29	28.7576	867	75 16 89	29.4449
828	68 55 84	28.7750	868	75 34 24	29.4618
829	68 72 41	28.7924	869	75 51 61	29.4788
830	68 89 00	28.8097	870	75 69 00	29.4958
831	69 05 61	28.8271	871	75 86 41	29.5127
832	69 22 24	28.8444	872	76 03 84	29.5296
833	69 38 89	28.8617	873	76 21 29	29.5466
834	69 55 56	28.8791	874	76 38 76	29.5635
835	69 72 25	28.8964	875	76 56 25	29.5804
836	69 88 96	28.9137	876	76 73 76	29.5973
837	70 05 69	28.9310	877	76 91 29	29.6142
838	70 22 44	28.9482	878	77 08 84	29.6311
839	70 39 21	28.9655	879	77 26 41	29.6479
840	70 56 00	28.9828	880	77 44 00	29.6648

Table 24 (CONTINUED)

Number	Square	Square Root	Number	Square	Square Root
881	77 61 61	29.6816	921	84 82 41	30.3480
882	77 79 24	29.6985	922	85 00 84	30.3645
883	77 96 89	29.7153	923	85 19 29	30.3809
884	78 14 46	29.7321	924	85 37 76	30.3974
885	78 32 25	29.7489	925	85 56 25	30.4138
886	78 49 96	29.7658	926	85 74 76	30.4302
887	78 67 69	29.7825	927	85 93 29	30.4467
888	78 85 44	29.7993	928	86 11 84	30.4631
889	79 03 21	29.8161	929	86 30 41	30.4795
890	79 21 00	39.8329	930	86 49 00	30.4959
891	79 38 81	29.8496	931	86 67 61	30.5123
892	79 56 64	29.8664	932	86 86 24	30.5287
893	79 74 49	29.8831	933	87 04 89	30.5450
894	79 92 36	29.8998	934	87 23 56	30.5614
895	80 10 24	29.9166	935	87 42 25	30.5778
896	80 28 16	29.9333	936	87 60 96	30.5941
897	80 46 09	29.9500	937	87 79 69	30.6105
898	80 64 04	29.9666	938	87 98 44	30.6268
899	80 82 01	39.9833	939	88 17 21	30.6431
900	81 00 00	30.0000	940	88 36 00	30.6594
901	81 18 01	30.0167	941	88 54 81	30.6757
902	81 36 04	30.0333	942	88 73 64	30.6920
903	81 54 09	30.0500	943	88 92 49	30.7083
904	81 72 16	30.0666	944	89 11 36	30.7246
905	81 90 25	30.0832	945	89 30 25	30.7409
906	82 08 36	30.0998	946	89 49 16	30.7571
907	82 26 49	30.1164	947	89 68 09	30.7734
908	82 44 64	30.1330	948	89 87 04	30.7796
909	82 62 81	30.1496	949	90 06 01	30.8058
910	82 81 00	30.1662	950	90 25 00	30.8221
911	82 99 21	30.1828	951	90 44 01	30.8383
912	83 17 44	30.1993	952	90 63 04	30.8545
913	83 35 69	30.2159	953	90 82 09	30.8707
914	83 53 96	30.2324	954	91 01 16	30.8869
915	83 72 25	30.2490	955	91 20 25	20.9031
916	83 90 56	30.2655	956	91 39 36	30.9192
917	94 08 89	30.2820	957	91 58 49	30.9354
918	84 27 24	30.2986	958	91 77 64	30.9516
919	84 45 61	30.3150	959	91 96 81	30.8677
920	84 64 00	30.3315	960	92 16 00	30.9839

TABLE 24   (CONTINUED)

Number	Square	Square Root	Number	Square	Square Root
961	92 35 21	31.0000	981	96 23 61	31.3209
962	92 54 44	31.0161	982	96 43 24	31.3369
963	92 73 69	31.0322	983	96 62 89	31.3528
964	92 92 96	31.0483	984	96 82 56	31.3688
965	93 12 25	31.0644	985	97 02 25	31.3847
966	93 31 56	31.0805	986	97 21 96	31.4006
967	93 50 89	31.0966	987	97 41 69	31.4166
968	93 70 24	31.1127	988	97 61 44	31.4325
969	93 89 61	31.1288	989	97 81 21	31.4484
970	94 09 00	31.1448	990	98 01 00	31.4643
971	94 28 41	31.1609	991	98 20 81	31.4802
972	94 47 84	31.1769	992	98 40 64	31.4960
973	94 67 29	31.1929	993	98 60 49	31.5119
974	94 86 76	31.2090	994	98 80 36	31.5278
975	95 06 25	31.2250	995	99 00 25	31.5436
976	95 25 76	31.2410	996	99 20 16	31.5595
977	95 45 29	31.2570	997	99 40 09	31.5753
978	95 64 84	31.2730	998	99 60 04	31.5911
979	95 84 41	31.2890	999	99 80 01	31.6070
980	96 04 00	31.3050	1000	100 00 00	31.6228

References

Abrahamson, I. G. (1967). Exact Bahadur efficiencies for the Kolmogorov-Smirnov and Kuiper one- and two-sample statistics. *The Annals of Mathematical Statistics*, **38**, 1475–1490 (6.1).

Adichie, J. N. (1967a). Asymptotic efficiency of a class of nonparametric tests for regression parameters. *The Annals of Mathematical Statistics*, **38**, 884–893 (5.5).

Adichie, J. N. (1967b). Estimates of regression parameters based on rank tests. *The Annals of Mathematical Statistics*, **38**, 894–904 (5.5).

Aitkin, M. A. and Hume, M. W. (1965). Correlation in a singly truncated bivariate normal distribution. II. Rank correlation. *Biometrika*, **52**, 639–643 (5.5).

Alling, D. W. (1963). Early decision in the Wilcoxon two-sample test. *Journal of the American Statistical Association*, **58**, 713–720 (5.3).

Andel, J. (1967). Local asymptotic power and efficiency of tests of Kolmogorov-Smirnov type. *The Annals of Mathematical Statistics*, **38**, 1705–1725 (6.2).

Anderson, T. W. (1962). On the distribution of the two-sample Cramér-von Mises criterion. *The Annals of Mathematical Statistics*, **33**, 1148–1159 (6.2).

Anderson, T. W. and Burstein, H. (1967). Approximating the upper binomial confidence limit. *Journal of the American Statistical Association*, **62**, 857–861 (3.1).

Anderson, T. W. and Burstein, H. (1968). Approximating the lower binomial confidence limit. *Journal of the American Statistical Association*, **63**, 1413–1415. Errata: Ibid. (1969), 669 (3.1).

Anderson, T. W. and Darling, D. A. (1952). Asymptotic theory of certain "goodness of fit" criteria based on stochastic processes. *The Annals of Mathematical Statistics*, **23**, 193–212 (6.1, 6.2).

Ansari, A. R. and Bradley, R. A. (1960). Rank-sum tests for dispersion. *The Annals of Mathematical Statistics*, **31**, 1174–1189 (5.3).

Arbuthnott, J. (1710). An argument for divine providence, taken from the constant regularity observed in the births of both sexes. *Philosophical Transactions*, **27**, 186–190 (3.4).

Arnold, H. J. (1965). Small sample power for the one sample Wilcoxon test for non-normal shift alternatives. *The Annals of Mathematical Statistics*, **36**, 1767–1778 (5.1).

Asano, C. (1965). Runs test for a circular distribution and a table of probabilities. *Annals of the Institute of Statistical Mathematics*, **17**, 331–346 (7.3).

Barlow, R. E. and Gupta, S. S. (1966). Distribution-free life test sampling plans. *Technometrics*, **8**, 591–614 (3.2).

Bartlett, N. S. and Govindarajulu, Z. (1968). Some distribution-free statistics and their application to the selection problem. *Annals of the Institute of Statistical Mathematics*, **20**, 79–98 (5.9).

Barton, D. E. (1967). Query: Completed runs of length k above and below median. *Technometrics*, **9**, 682–694 (7.3).

Basu, A. P. (1967a). On the large sample properties of a generalized Wilcoxon-Mann-Whitney statistic. *The Annals of Mathematical Statistics*, **38**, 905–915 (5.3).

Basu, A. P. (1967b). On two k-sample rank tests for censored data. *The Annals of Mathematical Statistics*, **38**, 1520–1535 (5.6).

Basu, A. P. (1968). On a generalized Savage statistic with applications to life testing. *The Annals of Mathematical Statistics*, **39**, 1591–1604 (5.3).

Basu, A. P. and Woodworth, G. (1967). A note on nonparametric tests for scale. *The Annals of Mathematical Statistics*, **38**, 274–277 (5.3).

Batschelet, E. (1965). *Statistical Methods for the Analysis of Problems in Animal Orientation and Certain Biological Rhythms*. The American Institute of Biological Sciences. (5.1, 5.3).

Bell, C. B. (1964). Some basic theorems of distribution-free statistics. *The Annals of Mathematical Statistics*, **35**, 150–156 (2.5).

Bell, C. B. and Doksum, K. A. (1965). Some new distribution-free statistics. *The Annals of Mathematical Statistics*, **36**, 203–214 (5.9).

Bell, C. B. and Doksum, K. A. (1967). Distribution-free tests of independence. *The Annals of Mathematical Statistics*, **38**, 429–446 (5.5).

Bell, C. B. and Haller, H. S. (1969). Bivariate symmetry tests: Parametric and nonparametric. *The Annals of Mathematical Statistics*, **40**, 259–269 (5.1).

Benard, A. and van Elteren, P. (1953). A generalization of the method of m rankings. *Proceedings Koninklijke Nederlandse Akademie van Wetenschappen* (*A*), **56** (Indagationes Mathematicae 15), 358–369 (5.8).

Bennett, B. M. (1965). On multivariate signed rank tests. *Annals of the Institute of Statistical Mathematics*, **17**, 55–61 (5.1).

Bennett, B. M. and Nakamura, E. (1963). Tables for testing significance in a 2 × 3 contingency table. *Technometrics*, **5**, 501–511 (4.2).

Bennett, B. M. and Nakamura, E. (1964). The power function of the exact test for the 2 × 3 contingency table. *Technometrics*, **6**, 439–458 (4.2).

Beran, R. J. (1969). The derivation of nonparametric two-sample tests from tests for uniformity of a circular distribution. *Biometrika*, **56**, 561–570 (5.3).

Bhapkar, V. P. and Deshpande, J. V. (1968). Some nonparametric tests for multi-sample problems. *Technometrics*, **10**, 578–585 (5.6).

Bhapkar, V. P. and Koch, G. G. (1968). Hypotheses of "no interaction" in multi-dimensional contingency tables. *Technometrics*, **10**, 107–124 (4.2).

Bhattacharyya, G. K. (1967). Asymptotic efficiency of multivariate normal score test. *The Annals of Mathematical Statistics*, **38**, 1753–1758 (5.9).

Bhattacharyya, G. K. and Johnson, R. A. (1968). Nonparametric tests for shift at unknown time point. *The Annals of Mathematical Statistics*, **39**, 1731–1743 (5.3).

Bickel, P. J. (1969). A distribution free version of the Smirnov two sample test in the *p*-variate case. *The Annals of Mathematical Statistics*, **40**, 1–23 (6.2).

Birnbaum, Z. W. (1953). On the power of a one-sided test of fit for continuous probability functions. *The Annals of Mathematical Statistics*, **24**, 484–489 (6.1).

Birnbaum, Z. W. (1962). *Introduction to Probability and Mathematical Statistics*. Harper, New York (4.5).

Birnbaum, Z. W. and Hall, R. A. (1960). Small sample distributions for multi-sample statistics of the Smirnov type. *The Annals of Mathematical Statistics*, **31**, 710–720 (6.3).

Birnbaum, Z. W. and Tingey, F. H. (1951). One-sided confidence contours for probability distribution functions. *The Annals of Mathematical Statistics*, **22**, 592–596 (6.1).

Birnbaum, Z. W. and Zuckerman, H. S. (1949). A graphical determination of sample size for Wilks' tolerance limits. *The Annals of Mathematical Statistics*, **20**, 313–317 (3.3).

Blomqvist, N. (1951). Some tests based on dichotomization. *The Annals of Mathematical Statistics*, **22**, 362–371 (4.6).

Blum, J. R. and Fattu, N. A. (1964). Nonparametric methods. *Review of Educational Research*, **24:5**, 467–487 (2.5).

Bofinger, E. and Bofinger, V. J. (1961). A runs test for sequences of random digits. *The Australian Journal of Statistics*, **3**, 37–41 (7.3).

Bofinger, V. J. (1965). The *k*-sample slippage problem. *Australian Journal of Statistics*, **7**, 20–31 (7.2).

Bohrer, R. (1968). A note on tolerance limits with type I censoring. *Technometrics*, **10**, 392 (3.3).

Bowden, D. C. (1968). Query: Tolerance interval in regression. *Technometrics*, **10**, 207–210 (3.3).

Boyd, W. C. (1965). A nomogram for chi-square. *Journal of the American Statistical Association*, **60**, 344–346 (1.6).

Bradley, J. V. (1968). *Distribution-free statistical tests*. Prentice-Hall, Englewood Cliffs, N.J. (5.9, 7.3).

Bradley, R. A., Martin, D. C., and Wilcoxon, F. (1965). Sequential rank tests I. Monte carlo studies of the two-sample procedure. *Technometrics*, **7**, 463–483 (5.3).

Bradley, R. A., Merchant, S. D., and Wilcoxon, F. (1966). Sequential rank tests II. Modified two-sample procedures. *Technometrics*, **8**, 615–624 (5.3).

Buckle, N., Kraft, C. H., and van Eeden, C. (1969). An approximation to the Wilcoxon-Mann-Whitney distribution. *Journal of the American Statistical Association*, **64**, 591–599 (5.3).

Burr, E. J. (1963). Distribution of the two-sample Cramér-von Mises criterion for small equal samples. *The Annals of Mathematical Statistics*, **34**, 95–101 (6.2).

Burr, E. J. (1964). Small-sample distributions of the two-sample Cramér-von Mises' W^2 and Watson's U^2. *The Annals of Mathematical Statistics*, **35**, 1091–1098 (6.2).

Capon, J. (1965). On the asymptotic efficiency of the Kolmogorov-Smirnov test. *Journal of the American Statistical Association*, **60**, 843–853 (6.2).

Cash, W. S. (1967). The power of Conover's k-sample slippage test. Unpublished master's report, Kansas State University (7.2).

Chacko, V. J. (1966). Modified chi-square test for ordered alternatives. *Sankhya (B)*, **28**, 185–190 (4.2).

Chanda, K. C. (1963). On the efficiency of two-sample Mann-Whitney test for discrete populations. *The Annals of Mathematical Statistics*, **34**, 612–617 (5.3).

Chapman, D. G. and Meng, R. C. (1966). The power of chi-square tests for contingency tables. *Journal of the American Statistical Association*, **61**, 965–975 (4.2).

Chatterjee, S. K. (1966). A bivariate sign test for location. *The Annals of Mathematical Statistics*, **37**, 1771–1782 (3.4).

Chernoff, H. (1967). Query: Degrees of freedom for chi-square. *Technometrics*, **9**, 489–490 (4.5).

Chernoff, H. and Lehmann, E. L. (1954). The use of maximum likelihood estimates in χ^2 tests for goodness of fit. *The Annals of Mathematical Statistics*, **25**, 579–686 (4.5).

Chung, J. H. and Fraser, D. A. S. (1958). Randomization tests for a multivariate two-sample problem. *Journal of the American Statistical Association*, **53**, 729–735 (7.4).

Claringbold, P. J. (1961). The use of orthogonal polynomials in the partition of chi-square. *The Australian Journal of Statistics*, **3**, 48–63 (4.2).

Cleroux, R. (1969). First and second moments of the randomization test in two-associate PBIB designs. *Journal of the American Statistical Association*, **64**, 1424–1433 (7.4).

432   References

Clopper, C. J. and Pearson, E. S. (1934). The use of confidence or fiducial limits illustrated in the case of the binomial. *Biometrika*, **26**, 404–413 (3.1).

Cochran, W. G. (1937). The efficiencies of the binomial series tests of significance of a mean and of a correlation coefficient. *Journal of the Royal Statistical Society*, **100**, 69–73 (3.4).

Cochran, W. G. (1950). The comparison of percentages in matched samples. *Biometrika*, **37**, 256–266 (4.6).

Cochran, W. G. (1963). *Sampling Techniques*, 2nd ed. John Wiley, New York (2.1).

Conover, W. J. (1965). Several k-sample Kolmogorov-Smirnov tests. *The Annals of Mathematical Statistics*, **36**, 1019–1026 (6.3).

Conover, W. J. (1967a). The distribution functions of Tsao's truncated Smirnov statistics. *The Annals of Mathematical Statistics*, **38**, 1208–1215 (6.2).

Conover, W. J. (1967b). A k-sample extension of the one-sided two-sample Smirnov test statistic. *The Annals of Mathematical Statistics*, **38**, 1726–1730 (6.3).

Conover, W. J. (1968). Two k-sample slippage tests. *Journal of the American Statistical Association*, **63**, 614–626 (7.2).

Cox, D. R. and Stuart, A. (1955). Some quick tests for trend in location and dispersion. *Biometrika*, **42**, 80–95 (3.4).

Cramér, H. (1928). On the composition of elementary errors. *Skandinavisk Aktuarietidskrift* **11**, 13–74 and 141–180 (6.1).

Cramér, H. (1946). *Mathematical Methods of Statistics*. Princeton Univ. Press, Princeton, N.J. (4.2, 4.4, 4.5).

Cronholm, J. N. (1968). Two tables connected with goodness-of-fit tests for equiprobable alternatives. *Biometrika*, **55**, 441 (6.1).

Crouse, C. F. (1966). Distribution-free tests based on the sample distribution function. *Biometrika*, **53**, 99–108 (5.6).

Csörgö, M. (1965). Some Smirnov type theoerms of probability. *The Annals of Mathematical Statistics*, **36**, 1113–1119 (6.2).

Cureton, E. E. (1967). The normal approximation to the signed rank sampling distribution when zero differences are present. *Journal of the American Statistical Association*, **62**, 1068–1069 (5.1).

Daniels, H. E. (1950). Rank correlation and population models. *Journal of the Royal Statistical Society* (B), **12**, 171–181 (5.5).

Danziger, L. and Davis, S. A. (1964). Tables of distribution-free tolerance limits. *The Annals of Mathematical Statistics*, **35**, 1361–1365 (3.3).

Davis, J. A. (1967). A partial coefficient for Goodman and Kruskal's gamma. *Journal of the American Statistical Association*, **62**, 189–193 (4.4).

Dempster, A. P. and Schatzoff, M. (1965). Expected significance level as a sensitivity index for test statistics. *Journal of the American Statistical Association*, **60**, 420–436 (2.3).

Deshpandé, J. V. (1970). A class of multisample distribution-free tests. *The Annals of Mathematical Statistics*, **41**, 227–236 (5.6).

Diamond, E. L. (1963). The limiting power of categorical data chi-square tests analogous to normal analysis of variance. *The Annals of Mathematical Statistics*, **34**, 1432–1441 (4.2).

Dixon, W. J. (1953). Power functions of the sign test and power efficiency for normal alternatives. *The Annals of Mathematical Statistics*, **24**, 467–473 (2.4, 3.4).

Dixon, W. J. (1954). Power under normality of several nonparametric tests. *The Annals of Mathematical Statistics*, **25**, 610–614 (7.3).

Doksum, K. (1967). Robust procedures for some linear models with one observation per cell. *The Annals of Mathematical Statistics*, **38**, 878–883 (5.7).

Dunn, J. E. (1969). A compound multiple runs distribution. *Journal of the American Statistical Association*, **64**, 1415–1423 (7.3).

Dunn, O. J. (1964). Multiple comparisons using rank sums. *Technometrics*, **6**, 241–252 (5.7).

Duran, B. S. and Mielke, P. W. Jr. (1968). Robustness of the sum of squared ranks test. *Journal of the American Statistical Association*, **63**, 338–344 (5.3).

Durbin, J. (1951). Incomplete blocks in ranking experiments. *British Journal of Psychology* (*Statistical Section*), **4**, 85–90 (5.8).

Durbin, J. (1961). Some methods of constructing exact tests. *Biometrika*, **48**, 41–55 (5.9).

Durbin, J. (1968). The probability that the sample distribution function lies between two parallel straight lines. *The Annals of Mathematical Statistics*, **39**, 398–411 (6.1).

van Eeden, C. (1964). Note on the consistency of some distribution-free tests for dispersion. *Journal of the American Statistical Association*, **59**, 105–119 (5.3).

Ehrenberg, A. S. C. (1951). Note on normal transformations of ranks. *British Journal of Psychology* (*Statistical Section*), **4**, 133–134 (5.9).

Erdös, P., and Renyi, A. (1959). On the central limit theorem for samples from a finite population. *Matem. Kutato Intezet Kolzem.*, vol. 4, p. 49 (1.6).

Federer, W. T. (1963). *Experimental Design*. Macmillan, New York (5.8).

Feller, W. (1968). *An Introduction to Probability Theory and Its Applications*, vol. I, 3rd ed. John Wiley, New York (1.1, 1.2).

Ferguson, T. S. and Kraft, C. H. (1955). A run test of the hypothesis that the median of a stochastic process is constant (abstract). *The Annals of Mathematical Statistics*, **26**, 770 (7.3).

Festinger, L. (1946). The significance of difference between means without reference to the frequency distribution function. *Psychometrika*, **11**, 97–105 (5.3).

Fine, T. (1966). On the Hodges and Lehmann shift estimator in the two sample problem. *The Annals of Mathematical Statistics*, **37**, 1814–1818 (6.2).

Finney, D. J. (1948). The Fisher-Yates test of significance in 2×2 contingency tables. *Biometrika*, **35**, 145–156 (4.2).

Fisher, R. A. (1935). *The Design of Experiments*. Oliver & Boyd, Edinburgh-London (1st ed. 1935, 7th ed. 1960) (7.4).

Fisher, R. A. and Yates, F. (1957). *Statistical Tables for Biological, Agricultural, and Medical Research*, 5th ed. Oliver and Boyd, Edinburgh (5.9).

Fisz, M. (1960). On a result by M. Rosenblatt concerning the von Mises-Smirnov test. *The Annals of Mathematical Statistics*, **31**, 427–429 (6.2).

Fisz, M. (1963). *Probability Theory and Mathematical Statistics*, 3rd ed. John Wiley, New York (1.6, 6.2).

Fleiss, J. L. (1965). A note on Cochran's *Q* test. *Biometrics*, **21**, 1008–1010 (4.6).

Fraser, D. A. S. (1957). *Nonparametric Methods in Statistics*. John Wiley, New York (5.9).

Freund, J. E. (1962). *Mathematical Statistics*. Prentice-Hall, Englewood Cliffs, N.J. (1.6).

Freund, J. E. and Ansari, A. R. (1957). *Two-way Rank Sum Test for Variances.* Technical Report No. 34, Virginia Polytechnic Institute, Blacksburg (5.3).

Friedman, M. (1937). The use of ranks to avoid the assumption of normality implicit in the analysis of variance. *Journal of the American Statistical Association*, **32**, 675–701 (5.7).

Gabriel, K. R. (1966). Simultaneous test procedures for multiple comparisons on categorical data. *Journal of the American Statistical Association*, **61**, 1081–1096 (4.3).

Gabriel, K. R. and Lachenbruch, P. A. (1969). Non-parametric ANOVA in small samples: A monte carlo study of the adequacy of the asymptotic approximation. *Biometrics*, **25**, 593–596 (5.6).

Gart, J. J. (1966). Alternative analyses of contingency tables. *Journal of the Royal Statistical Society (B)*, **28**, 164–179 (4.2).

Gastwirth, J. L. (1965a). Asymptotically most powerful rank tests for the two-sample problem with censored data. *The Annals of Mathematical Statistics*, **36**, 1243–1248 (5.3).

Gastwirth, J. L. (1965b). Percentile modifications of two sample rank tests. *Journal of the American Statistical Association*, **60**, 1127–1141 (5.3).

Gehan, E. A. (1965a). A generalized Wilcoxon test for comparing arbitrarily singly censored samples. *Biometrika*, **52**, 203–224 (5.3).

Gehan, E. A. (1965b). A generalized two-sample Wilcoxon test for doubly censored data. *Biometrika*, **52**, 650–653 (5.3).

Gehan, E. A. and Thomas, D. G. (1969). The performance of some two-sample tests in small samples with and without censoring. *Biometrika*, **56**, 127–132 (5.3).

Gelzer, J. and Pyke, R. (1965). The asymptotic relative efficiency of goodness-of-fit tests against scalar alternatives. *Journal of the American Statistical Association*, **60**, 410–419 (6.1).

Gerig, T. M. (1969). A multivariate extension of Freidman's χ_r^2 test. *Journal of the American Statistical Association*, **64**, 1595–1608 (5.7).

Gibbons, J. D. (1964). Effect of non-normality on the power function of the sign test. *Journal of the American Statistical Association*, **59**, 142–148 (3.4).

Gibbons, J. D. (1967). Correlation coefficients between nonparametric tests for location and scale. *Annals of the Institute of Statistical Mathematics*, **19**, 519–526 (5.3).

Gibbons, J. D. and Gastwirth, J. L. (1970). Properties of the percentile modified rank tests. *Annals of the Institute of Statistical Mathematics*, supplement 6, 95–114 (5.3).

Glasser, G. J. and Winter, R. F. (1961). Critical values of the coefficient of rank correlation for testing the hypothesis of independence. *Biometrika*, **48**, 444–448.

Gnedenko, B. V. and Korolyuk, V. S. (1951). On the maximum discrepancy between two empirical distributions. (Russian.) *Doklady Akademii* Nauk SSSR (N.S.), **80**, 525–528. English translation in *IMS and American Mathematical Society* (1961) (6.2).

Gokhale, D. V. (1968). On asymptotic relative efficiencies of a class of rank tests for independence of two variables. *Annals of the Institute of Statistical Mathematics*, **20**, 255–261 (5.5, 5.9).

Goodman, L. A. (1957). Runs tests and likelihood ratio tests for Markov chains (abstract). *The Annals of Mathematical Statistics*, **28**, 1072 (7.3).

Goodman, L. A. (1958). Simplified runs tests and likelihood ratio tests for Markoff chains. *Biometrika*, **45**, 181–197 (7.3).

Goodman, L. A. (1964). Simple methods for analyzing three-factor interaction in contingency tables. *Journal of the American Statistical Association*, **59**, 319–352 (4.2).

Goodman, L. A. (1965). On simultaneous confidence intervals for multinomial proportions. *Technometrics*, **7**, 247–254 (3.1).

Goodman, L. A. (1968). The analysis of cross-classified data: Independence, quasi-independence, and interactions in contingency tables with or without missing entries. *Journal of the American Statistical Association*, **63**, 1091–1113 (4.2).

Goodman, L. A. (1970). The multivariate analysis of qualitative data: Interactions among multiple classifications. *Journal of the American Statistical Association*, **65**, 226–256 (4.2).

Goodman, L. A. and Grunfeld, Y. (1961). Some nonparametric tests for comovements between time series. *Journal of the American Statistical Association*, **56**, 11–26 (7.3).

Goodman, L. A. and Kruskal, W. H. (1954). Measures of association for cross-classifications. *Journal of the American Statistical Association*, **49**, 732–764. Errata: ibid. (1957), 578 (4.4).

Goodman, L. A. and Kruskal, W. H. (1959). Measures of association for cross-classifications. II: Further discussion and references. *Journal of the American Statistical Association*, **54**, 123–163 (4.4).

Goodman, L. A. and Kruskal, W. H. (1963). Measures of association for cross-classifications. III: Approximate sample theory. *Journal of the American Statistical Association*, **58**, 310–364 (4.4).

Goodman, L. A. and Madansky, A. (1962). Parameter-free and nonparametric tolerance limits: The exponential case. *Technometrics*, **4**, 75–95 (3.3).

Govindarajulu, Z. (1968). Distribution-free confidence bounds for $P(X < Y)$. *Annals of the Institute of Statistical Mathematics*, **20**, 229–238 (5.4).

Gregory, G. (1961). Contingency tables with a dependent classification. *The Australian Journal of Statistics*, **3**, 42–47 (4.2).

Grizzle, J. E. (1967). Continuity correction in the χ^2-test for 2×2 tables. *The American Statistician*, **21:4**, 28–32 (4.1).

Haga, T. (1960). A two-sample rank test on location. *Annals of the Institute of Statistical Mathematics*, **11**, 211–219 (5.3).

Hájek, J. and Šidák, Z. (1967). *Theory of Rank Tests*. Academic Press, New York (5.9).

Hanson, D. L. and Owen, D. B. (1963). Distribution-free tolerance limits, elimination of the requirement that cumulative distribution functions be continuous. *Technometrics*, **5**, 518–522 (3.3).

Harkness, W. L. and Katz, L. (1964). Comparison of the power functions for the test of independence in 2×2 contingency tables. *The Annals of Mathematical Statistics*, **35**, 1115–1127 (4.1).

Harter, H. L. (1964). A new table of percentage points of the chi-square distribution. *Biometrika*, **51**, 231–240 (1.6).

Haynam, G. E. and Govindarajulu, Z. (1966). Exact power of the Mann-Whitney test for exponential and rectangular alternatives. *The Annals of Mathematical Statistics*, **37**, 945–953 (5.3).

Haynam, G. E. and Leone, F. C. (1965). Analysis of categorical data. *Biometrika*, **52**, 654–660 (4.2).

Healy, M. J. R. (1969). Exact tests of significance in contingency tables. *Technometrics*, **11**, 393–395 (4.2).

Hemelrijk, J. (1952). A theorem on the sign test when ties are present. *Proceedings Koninklijke Nederlandse Akademie van Wetenschappen (A)*, **55**, 322–326 (3.4).

Hettmansperger, T. P. (1968). On the trimmed Mann-Whitney statistic. *The Annals of Mathematical Statistics*, **39**, 1610–1614 (5.3).

Hodges, J. L. Jr. and Lehmann, E. (1956). The efficiency of some nonparametric competitors of the *t*-test. *The Annals of Mathematical Statistics*, **27**, 324–335 (3.4, 5.3, 5.6).

Hodges, J. L. Jr. and Lehmann, E. L. (1963). Estimates of location based on rank tests. *The Annals of Mathematical Statistics*, **34**, 598–611 (5.4).

Hoeffding, W. (1952). The large-sample power of tests based on permutations of observations. *The Annals of Mathematical Statistics*, **23**, 169–192 (7.4).

Hoeffding, W. (1965). Asymptotically optimal tests for multinomial distributions. *The Annals of Mathematical Statistics*, **36**, 369–400 (4.2).

Hollander, M. (1963). A nonparametric test for the two-sample problem. *Psychometrika*, **28**, 395–403 (5.3).

Hollander, M. (1967a). Asymptotic efficiency of two nonparametric competitors of Wilcoxon's two sample test. *Journal of the American Statistical Association*, **62**, 939–949 (5.3).

Hollander, M. (1967b). Rank tests for randomized blocks when the alternatives have an *a priori* ordering. *The Annals of Mathematical Statistics*, **38**, 867–877 (5.7).

Hollander, M. (1968). Certain uncorrelated nonparametric test statistics. *Journal of the American Statistical Association*, **63**, 707–714 (5.3).

Hollander, M. (1970). A distribution-free test for parallelism. *Journal of the American Statistical Association*, **65**, 387–394 (5.1).

Hotelling, H. and Pabst, M. R. (1936). Rank correlation and tests of significance involving no assumption of normality. *The Annals of Mathematical Statistics*, **7**, 29–43 (5.5).

Høyland, A. (1965). Robustness of the Hodges-Lehmann estimates for shift. *The Annals of Mathematical Statistics*, **36**, 174–197 (5.3).

Høyland, A. (1968). Robustness of the Wilcoxon estimate of location against a certain dependence. *The Annals of Mathematical Statistics*, **39**, 1196–1201 (5.2).

Hudimoto, H. (1959). On a two-sample nonparametric test in the case that ties are present. *Annals of the Institute of Statistical Mathematics*, **11**, 113–120 (5.3).

Ireland, C. T., Ku, H. H. and Kullback, S. (1969). Symmetry, and marginal homogeneity of an *r* × *r* contingency table. *Journal of the American Statistical Association*, **64**, 1323–1341 (4.2).

Ireland, C. T. and Kullback, S. (1968). Contingency tables with given marginals. *Biometrika*, **55**, 179–188 (4.2).

Ishii, G. (1960). Intraclass contingency tables. *Annals of the Institute of Statistical Mathematics*, **12**, 161–207. Errata: Ibid. (1960), 279 (4.2).

Ising, E. (1925). Beitrag zur Theorie des Ferromagnetismus. *Zeitschrift für Physik*, **31**, 253–258 (7.3).

Ives, K. H. and Gibbons, J. D. (1967). A correlation measure for nominal data. *The American Statistician*, **21:5**, 16–17 (4.4).

Jacobson, J. E. (1963). The Wilcoxon two-sample statistic: Tables and bibliography. *Journal of the American Statistical Association*, **58**, 1086–1103 (5.3).

Jogdeo, K. (1966). On randomized rank score procedures of Bell and Doksum. *The Annals of Mathematical Statistics*, **37**, 1697–1703 (5.9).

Kaarsemaker, L. and van Wijngaarden, A. (1953). Tables for use in rank correlation. *Statistica Neerlandica*, **7**, 41–54.

Kempthorne, O. (1955). The randomization theory of experimental inference. *Journal of the American Statistical Association*, **50**, 946–967 (7.4).

Kempthorne, O. and Doerfler, T. E. (1969). The behavior of some significance tests under experimental randomization. *Biometrika*, **56**, 231–248 (7.4).

Kendall, M. G. (1938). A new measure of rank correlation. *Biometrika*, **30**, 81–93 (5.5).

Kendall, M. G. (1942). Partial rank correlation. *Biometrika*, **32**, 277–283 (5.5).

Kendall, M. G. (1955). *Rank Correlation Methods*, 2nd ed. Hafner, New York (5.5).

Kendall, M. G. and Babington-Smith, B. (1939). The problem of m rankings. *The Annals of Mathematical Statistics*, **10**, 275–287 (5.7).

Kendall, M. G. and Sundrum, R. M. (1953). Distribution-free methods and order properties. *Review of the International Statistical Institute*, **21:3**, 124–134 (2.5).

Kim, P. J. (1969). On the exact and approximate sampling distribution of the two sample Kolmogorov-Smirnov criterion D_{mn} $m \leq n$. *Journal of the American Statistical Association*, **64**, 1625–1637 (6.2).

Klotz, J. (1962). Nonparametric tests for scale. *The Annals of Mathematical Statistics*, **33**, 498–512 (5.9).

Klotz, J. (1963). Small sample power and efficiency for the one sample Wilcoxon and normal scores tests. *The Annals of Mathematical Statistics*, **34**, 624–632 (5.1).

Klotz, J. (1965). Alternative efficiencies for signed rank tests. *The Annals of Mathematical Statistics*, **36**, 1759–1766 (5.1).

Klotz, J. (1966). The Wilcoxon, ties, and the computer. *Journal of the American Statistical Association*, **61**, 772–787 (5.3).

Klotz, J. (1967). Asymptotic efficiency of the two sample Kolmogorov-Smirnov test. *Journal of the American Statistical Association*, **62**, 932–938 (6.2).

Knight, W. R. (1966). A computer method for calculating Kendall's tau with ungrouped data. *Journal of the American Statistical Association*, **61**, 436–439 (5.5).

Koch, G. G. (1970). The use of non-parametric methods in the statistical analysis of a complex split plot experiment. *Biometrics*, **26**, 105–128 (5.7).

Kolmogorov, A. N. (1933). Sulla determinazione empirica di una legge di distribuzione. *Giornale dell' Istituto Italiano degli Attuari*, **4**, 83–91 (6.1).

Konijn, H. S. (1961). Non-parametric, robust and short-cut methods in regression and structural analysis. *The Australian Journal of Statistics*, **3**, 77–86 (5.5).

Kraft, C. H. and van Eeden, C. (1968). *A Nonparametric Introduction to Statistics*. Macmillan, New York.

Kruskal, W. H. (1952). A nonparametric test for the several sample problem. *The Annals of Mathematical Statistics*, **23**, 525–540 (5.6, 7.3).

Kruskal, W. H. (1958). Ordinal measures of association. *Journal of the American Statistical Association*, **53**, 814–861 (5.5).

Kruskal, W. H. and Wallis, W. A. (1952). Use of ranks on one-criterion variance analysis. *Journal of the American Statistical Association*, **47**, 583–621. Addendum: Ibid. (1953), 907–911 (5.6).

Ku, H. H. (1963). A note on contingency tables involving zero frequencies and the 2I test. *Technometrics*, **5**, 398–400 (4.2).

Kullback, S., Kupperman, M., and Ku, H. H. (1962). Tests for contingency tables and Markov chains. *Technometrics*, **4**, 573–608 (4.2).

Lapunov, A. M. (1901). Nouvelle forme du theoreme sur la limite de probabilite. *Mem. Acad. Sci. St. Petersburg*, **12**:5 (1.6).

Laubscher, N. F., Steffens, F. E., and DeLange, E. M. (1968). Exact critical values for Mood's distribution-free test statistic for dispersion and its normal approximation. *Technometrics*, **10**, 497–508 (5.3).

Lee, S. W. (1966). The power of the one-sided and one-sample Kolmogorov-Smirnov test. Unpublished master's report, Kansas State Univ. (6.1).

Lehmann, E. L. (1951). Consistency and unbiasedness of certain nonparametric tests. *The Annals of Mathematical Statistics*, **22**, 165–179 (6.2).

Lehmann, E. L. (1959). *Testing Statistical Hypotheses*. John Wiley, New York (2.4).

Lehmann, E. L. (1966). Some concepts of dependence. *The Annals of Mathematical Statistics*, **37**, 1137–1153 (5.5).

Lehmann, E. L. and Stein, C. (1949). On the theory of some nonparametric hypotheses. *The Annals of Mathematical Statistics*, **20**, 28–45 (7.4).

Levy, P. (1925). *Calcul des Probabilities*. Gauthiers-Villars, Paris (1.6).

Lilliefors, H. W. (1967). On the Kolmogorov-Smirnov test for normality with mean and variance unknown. *Journal of the American Statistical Association*, **62**, 399–402 (6.1).

Lilliefors, H. W. (1969). On the Kolmogorov-Smirnov test for the exponential distribution with mean unknown. *Journal of the American Statistical Association*, **64**, 387–389 (6.1).

Lindeberg, J. W. (1922). Eine neve Herleitung des Exponentialgesetzes. *Math. Zeitschrift*, **15**, 221 (1.6).

Lohrding, R. K. (1969a). Likelihood ratio tests of equal means when the variances are heterogeneous. Unpublished doctoral dissertation, Kansas State Univ. (7.2).

Lohrding, R. K. (1969b). A test of equality of two normal populations assuming homogeneous coefficients of variation. *The Annals of Mathematical Statistics*, **40**, 1374–1385 (7.2).

Maag, U. R. (1966). A k-sample analogue of Watson's U^2 statistic. *Biometrika*, **53**, 579–584 (6.2).

Maag, U. R. and Stephens, M. A. (1968). The V_{NM} two-sample test. *The Annals of Mathematical Statistics*, **39**, 923–935 (6.2).

Mack, C. (1969). Query: Tolerance limits for a binomial distribution. *Technometrics*, **11**, 201 (3.3).

MacKinnon, W. J. (1964). Table for both the sign test and distribution-free confidence intervals of the median for sample sizes to 1,000. *Journal of the American Statistical Association*, **59**, 935–956 (3.4).

Mann, H. B. and Whitney, D. R. (1947). On a test of whether one of two random variables is stochastically larger than the other. *The Annals of Mathematical Statistics*, **18**, 50–60 (5.3).

Mansfield, E. (1962). Power functions for Cox's test of randomness against trend. *Technometrics*, **4**, 430–432 (3.4).

Mantel, N. and Greenhouse, S. W. (1968). What is the continuity correction? *The American Statistician*, **22:5,** 27–30 (4.1).

Mardia, K. V. (1967a). A non-parametric test for the bivariate two-sample location problem. *Journal of the Royal Statistical Society (B)*, **29,** 320–342 (5.3).

Mardia, K. V. (1967b). Some contributions to contingency-type bivariate distributions. *Biometrika*, **54,** 235–249 (4.2).

Mardia, K. V. (1968). Small sample power of a nonparametric test for the bivariate two-sample location problem in the normal case. *Journal of the Royal Statistical Society ⟨B⟩*, **30,** 83–92 (5.3).

Marsaglia, G. (1968). Query: Psuedo random normal numbers. *Technometrics*, **10,** 401–402 (5.9).

Massey, F. J. (1950a). A note on the estimation of a distribution function by confidence limits. *The Annals of Mathematical Statistics*, **21,** 116–119 (6.1).

Massey, F. J. (1950b). A note on the power of a nonparametric test. *The Annals of Mathematical Statistics*, **21,** 440–443. Errata: Ibid. (1952), 637–638 (6.1).

Massey, F. J. (1952). Distribution table for the deviation between two sample cumulatives. *The Annals of Mathematical Statistics*, **23,** 435–441. Errata: Davis, L. S. (1958). *Mathematical Tables and Other Aids to Computation*, **12,** 262–263 (6.2).

Matthes, T. K. and Truax, D. R. (1965). Optimal invariant rank tests for the *k*-sample problem. *The Annals of Mathematical Statistics*, **36,** 1207–1222 (5.6).

Maxwell, A. E. (1961). *Analyzing Qualitative Data*. John Wiley, New York (4.2).

McCarthy, P. J. (1965). Stratified sampling and distribution-free confidence intervals for a median. *Journal of the American Statistical Association*, **60,** 772–783 (5.2).

McCornack, R. L. (1965). Extended tables of the Wilcoxon matched pairs signed rank statistics. *Journal of the American Statistical Association*, **60,** 864–871 (5.1).

McDonald, B. J. and Thompson, W. A. (1967). Rank sum multiple comparisons in one- and two-way classifications. *Biometrika*, **54,** 487–498 (5.6, 5.7).

McNeil, D. R. (1967). Efficiency loss due to grouping in distribution-free tests. *Journal of the American Statistical Association*, **62,** 954–965 (5.3).

McNemar, Q. (1962). *Psychological Statistics*, 3rd ed. John Wiley, New York (4.4).

Mehra, K. L. and Sarangi, J. (1967). Asymptotic efficiency of certain rank tests for comparative experiments. *The Annals of Mathematical Statistics*, **38,** 90–107 (5.7).

Mielke, P. W. Jr. (1967). Note on some squared rank tests with existing ties. *Technometrics*, **9,** 312–314 (5.3).

Mielke, P. W. Jr. and Siddiqui, M. M. (1965). A combinatorial test for independence of dichotomous responses. *Journal of the American Statistical Association*, **60,** 437–441 (4.2).

Mikulski, P. W. (1963). On the efficiency of optimal nonparametric procedures in the two sample case. *The Annals of Mathematical Statistics*, **34,** 22–32 (5.3).

Miller, L. H. (1956). Table of percentage points of Kolmogorov statistics. *Journal of the American Statistical Association*, **51**, 111–121.

Miller, R. G. Jr. (1966). *Simultaneous Statistical Inference*. McGraw-Hill, New York (5.7).

Milton, R. C. (1964). An extended table of critical values for the Mann-Whitney (Wilcoxon) two-sample statistic. *Journal of the American Statistical Association*, **59**, 925–934 (5.3).

von Mises, R. (1931). *Wahrscheinlichkeitsrechnung und ihre Anwendung in der Statistik und theoretischen Physik*. F. Deuticke, Leipzig (6.1).

Mood, A. M. (1940). The distribution theory of runs. *The Annals of Mathematical Statistics*, **11**, 367–392 (7.3).

Mood, A. M. (1954). On the asymptotic efficiency of certain nonparametric two-sample tests. *The Annals of Mathematical Statistics*, **25**, 514–522 (5.3, 7.3).

Moore, G. H. and Wallis, W. A. (1943). Time series significance test based on signs of differences. *Journal of the American Statistical Association*, **38**, 153–164 (7.3).

Moses, L. E. (1952). Nonparametric statistics for psychological research. *Psychol. Bull.*, **49**, 122–143 (7.4).

Moses, L. E. (1963). Rank tests of dispersion. *The Annals of Mathematical Statistics*, **34**, 973–983 (5.3).

Moses, L. E. (1965). Query: Confidence limits from rank tests. *Technometrics*, **7**, 257–260 (5.2, 5.4).

Mosteller, F. (1941). Note on an application of runs to quality control charts. *The Annals of Mathematical Statistics*, **12**, 228–232 (7.3).

Mosteller, F. (1948). A *k*-sample slippage test for an extreme population. *The Annals of Mathematical Statistics*, **19**, 58–65 (7.2).

Mosteller, F. (1968). Association and estimation in contingency tables. *Journal of the American Statistical Association*, **63**, 1–29 (4.2).

Mosteller, F., Rourke, R. E. K., and Thomas, G. B. Jr. (1961). *Probability: A First Course*. Addison-Wesley, Palo Alto, Calif. (1.1).

Mote, V. L. and Anderson, R. L. (1965). An investigation of the effect of misclassification on the properties of χ^2-tests in the analysis of categorical data. *Biometrika*, **52**, 95–110 (4.2).

Murphy, R. B. (1948). Nonparametric tolerance limits. *The Annals of Mathematical Statistics*, **19**, 581–588 (3.3).

Neave, H. R. (1966). A development of Tukey's quick test of location. *Journal of the American Statistical Association*, **61**, 949–964 (7.1).

Neave, H. R. and Granger, W. J. (1968). A monte carlo study comparing various two-sample tests for differences in mean. *Technometrics*, **10**, 509–522 (7.1).

Nelson, L. S. (1966). Query: Combining values of observed χ^2's. *Technometrics*, **8**, 709 (1.6, 4.1).

Nemenyi, P. (1969). Variances: An elementary proof and a nearly distribution-free test. *The American Statistician*, **23:5**, 35–37 (5.3).

Noé, M. and Vandewiele, G. (1968). The calculation of distributions of Kolmogorov-Smirnov type statistics including a table of significance points for a particular case, *The Annals of Mathematical Statistics*, **39**, 233–241 (6.1).

Noether, G. E. (1963). Efficiency of the Wilcoxon two-sample statistic for randomized blocks. *Journal of the American Statistical Association*, **58**, 894–898 (5.3).

Noether, G. E. (1967a). *Elements of Nonparametric Statistics*. John Wiley, New York (2.4, 3.3, 5.1, 5.4, 5.8, 6.1, 6.2).

Noether, G. E. (1967b). Wilcoxon confidence intervals for location parameters in the discrete case. *Journal of the American Statistical Association*, **62**, 184–188 (5.2).

Odeh, R. E. (1967). The distribution of the maximum sum of ranks. *Technometrics*, **9**, 271–278 (5.6).

Olmstead, P. S. and Tukey, J. W. (1947). A corner test for association. *The Annals of Mathematical Statistics*, **18**, 495–513 (7.1).

Olshen, R. A. (1967). Sign and Wilcoxon tests for linearity. *The Annals of Mathematical Statistics*, **38**, 1759–1769 (3.4).

Ott, R. L. and Free, S. M. (1969). A short-cut rule for a one-sided test of hypothesis for qualitative data. *Technometrics*, **11**, 197–200 (4.1).

Owen, D. B. (1962). *Handbook of Statistical Tables*. Addison-Wesley, Reading, Mass (3.3).

Page, E. B. (1963). Ordered hypotheses for multiple treatments: A significance test for linear ranks. *Journal of the American Statistical Association*, **58**, 216–230 (5.7).

Pearson, E. S. (1947). The choice of statistical test illustrated on the interpretation of data classed in a 2 × 2 table. *Biometrika*, **34**, 139–167 (4.1).

Pearson, E. S. and Hartley, H. O. (1962). *Biometrika Tables for Statisticians*, vol. 1, 2nd ed. Cambridge University Press, Cambridge, England (5.8).

Pearson, K. (1900). On the criterion that a given system of deviations from the probable in the case of a correlated system of variables is such that it can reasonably be supposed to have arisen from random sampling. *Philosophical Magazine* (5), **50**, 157–175 (4.5).

Percus, O. E. and Percus, J. K. (1970). Extended criterion for comparison of empirical distributions. *Journal of Applied Probability*, **7**, 1–20 (6.2).

Pitman, E. J. G. (1937/38). Significance tests which may be applied to samples from any populations. I. *Suppl. Journal of the Royal Statistical Society*, **4** (1937), 119–130. II. The correlation coefficient test. *Suppl Journal of the Royal Statistical Society*, **4** (1937), 225–232. III. The analysis of variance test. *Biometrika*, **29** (1938), 322–335 (7.4).

Plackett, R. L. (1964). The continuity correction in 2 × 2 tables. *Biometrika*, **51**, 327–338 (4.1).

Plackett, R. L. (1965). A class of bivariate distributions. *Journal of the American Statistical Association*, **60**, 516–522 (4.2).

Potthoff, R. F. (1963). Use of the Wilcoxon statistic for a generalized Behrens-Fisher problem. *The Annals of Mathematical Statistics*, **34**, 1596–1599 (5.3).

Prairie, R. R., Zimmer, W. J., and Brookhouse, J. K. (1962). Some acceptance sampling plans based on the theory of runs. *Technometrics*, **4**, 177–185 (7.3).

Pratt, J. W. (1959). Remarks on zeros and ties in the Wilcoxon signed rank procedures. *Journal of the American Statistical Association*, **54**, 655–667 (5.1).

Puri, M. L. (1965). On some tests of homogeneity of variances. *Annals of the Institute of Statistical Mathematics*, **17**, 323–330 (5.3).

Puri, M. L. (1968). Multi-sample scale problem: Unknown location parameters. *Annals of the Institute of Statistical Mathematics*, **20**, 99–106 (5.3).

Puri, M. L. and Puri, P. S. (1969). Multiple decision procedures based on ranks for certain problems in analysis of variance. *The Annals of Mathematical Statistics*, **40**, 619–632 (5.6).

Puri, M. L. and Sen, P. K. (1967). On some optimum nonparametric procedures in two-way layouts. *Journal of the American Statistical Association*, **62**, 1214–1229 (5.7).

Puri, M. L. and Sen, P. K. (1968). Nonparametric confidence regions for some multivariate location problems. *Journal of the American Statistical Association*, **63**, 1373–1378 (5.2).

Puri, M. L. and Sen, P. K. (1969a). Analysis of covariance based on general rank scores. *The Annals of Mathematical Statistics*, **40**, 610–618 (5.6).

Puri, M. L. and Sen, P. K. (1969b). On the asymptotic theory of rank order tests for experiments involving paired comparisons. *Annals of the Institute of Statistical Mathematics*, **21**, 163–173 (5.8).

Putter, J. (1964). The χ^2 goodness-of-fit test for a class of cases of dependent observations. *Biometrika*, **51**, 250–252 (4.5).

Quade, D. (1965). On the asymptotic power of the one-sample Kolmogorov-Smirnov tests. *The Annals of Mathematical Statistics*, **36**, 1000–1018 (6.1).

Quade, D. (1966). On analysis of variance for the k-sample problem. *The Annals of Mathematical Statistics*, **37**, 1747–1758 (5.6).

Quade, D. (1967). Rank analysis of covariance. *Journal of the American Statistical Association*, **62**, 1187–1200 (5.6).

Quesenberry, C. P. and Gessaman, M. P. (1968). Nonparametric discrimination using tolerance regions. *The Annals of Mathematical Statistics*, **39**, 664–673 (3.3).

Quesenberry, C. P. and Hurst, D. C. (1964). Large sample simultaneous confidence intervals for multinomial proportions. *Technometrics*, **6**, 191–195 (3.1).

Radhakrishna, S. (1965). Combination of results from several 2 × 2 contingency tables. *Biometrics*, **21**, 86–99 (1.6, 4.1).

Raghavachari, M. (1965a). The two-sample scale problem when locations are unknown. *The Annals of Mathematical Statistics*, **36**, 1236–1242 (5.3).

Raghavachari, M. (1965b). On the efficiency of the normal scores test relative to the *F*-test. *The Annals of Mathematical Statistics*, **36**, 1306–1307 (5.9).

Ramachandramurty, P. V. (1966). On the Pitman efficiency of one-sided Kolmogorov and Smirnov tests for normal alternatives. *The Annals of Mathematical Statistics*, **37**, 940–944 (6.2).

Rand Corporation (1955). *A Million Random Digits with 100,000 Normal Deviates.* Free Press, Glencoe, Ill. (5.9).

Rao, T. S. (1968). A note on the asymptotic relative efficiencies of Cox and Stuart's tests for testing trend in dispersion of a *p*-dependent time series. *Biometrika*, **55**, 381–386 (3.4).

van der Reyden, D. (1952). A simple statistical significance test. *Rhodesia Agricultural Journal*, **49**, 96–104 (5.3).

Rhyne, A. L. and Steel, R. G. D. (1965). Tables for a treatments versus control multiple comparisons sign test. *Technometrics*, **7**, 293–306 (3.4).

Riedwyl, H. (1967). Goodness of fit. *Journal of the American Statistical Association*, **62**, 390–398 (6.1).

Rizvi, M. H. and Sobel, M. (1967). Nonparametric procedures for selecting a subset containing the population with the largest α-quantile. *The Annals of Mathematical Statistics*, **36**, 1788–1803 (5.6).

Rizvi, M. H., Sobel, M., and Woodworth, G. G. (1968). Nonparametric ranking procedures for comparison with a control. *The Annals of Mathematical Statistics*, **39**, 2075–2093 (5.6).

Robertson, W. H. (1960). Programming Fisher's exact method of comparing two percentages. *Technometrics*, **2**, 103–107 (4.2).

Rosenbaum, S. (1965). On some two-sample nonparametric tests. *Journal of the American Statistical Association*, **60**, 1118–1126 (7.1).

Rosenblatt, M. (1952). Limit theorems associated with variants of the von Mises statistic. *The Annals of Mathematical Statistics*, **23**, 617–623 (6.2).

Ruist, E. (1955). Comparison of tests for nonparametric hypotheses. *Arkiv For Matematik*, **3:2**, 133–163 (2.4).

Sandelius, M. (1968). A graphical version of Tukey's confidence interval for slippage. *Technometrics*, **10**, 193–194 (7.1).

Savage, I. R. (1962). *Bibliography of Nonparametric Statistics.* Harvard, Cambridge, Mass.

Saw, J. G. (1966). A nonparametric comparison of two samples, one of which is censored. *Biometrika*, **53**, 599–603 (5.3).

Schach, S. (1969a). Nonparametric symmetry tests for circular distributions. *Biometrika*, **56**; 571–577 (5.1).

Schach, S. (1969b). On a class of nonparametric two-sample tests for circular distributions. *The Annals of Mathematical Statistics*, **40**, 1791–1800 (5.3).

Scheffé, H. (1943). Statistical inference in the nonparametric case. *The Annals of Mathematical Statistics*, **14**, 305–332 (7.4).

Scheffé, H. and Tukey, J. W. (1944). A formula for sample sizes for population tolerance limits. *The Annals of Mathematical Statistics*, **15**, 217 (3.3).

Sen, P. K. (1962). On studentized non-parametric multi-sample location tests. *Annals of the Institute of Statistical Mathematics*, **14,** 119–131 (5.6).

Sen, P. K. (1963). On weighted rank-sum tests for dispersion. *Annals of the Institute of Statistical Mathematics*, **15,** 117–135 (5.3).

Sen, P. K. (1965). Some nonparametric tests for *m*-dependent time series. *Journal of the American Statistical Association*, **60,** 134–147 (7.3).

Sen, P. K. (1966). On nonparametric simultaneous confidence regions and tests for the one criterion analysis of variance problem. *Annals of the Institute of Statistical Mathematics*, **18,** 319–336 (5.6).

Sen, P. K. (1967a). A note on the asymptotic efficiency of Friedman's χ_r^2 test. *Biometrika*, **54,** 677–679 (5.7).

Sen, P. K. (1967b). On some multisample permutation tests based on a class of *U*-statistics. *Journal of the American Statistical Association*, **62,** 1201–1213 (7.4).

Sen, P. K. (1968a). Estimates of the regression coefficient based on Kendall's tau. *Journal of the American Statistical Association* **63,** 1379–1389 (5.5).

Sen, P. K. (1968b). On a class of aligned rank order tests in two-way layouts. *The Annals of Mathematical Statistics*, **39,** 1115–1124 (5.7).

Sen, P. K. and Govindarajulu, Z. (1966). On a class of *c*-sample weighted rank-sum tests for location and scale. *Annals of the Institute of Statistical Mathematics*, **18,** 87–105 (5.6).

Sen, P. K. and Puri, M. L. (1967). On the theory of rank order tests for location in the multivariate one sample problem. *The Annals of Mathematical Statistics*, **38,** 1216–1228 (5.1).

Serfling, R. J. (1968). The Wilcoxon two-sample statistic on strongly mixing processes. *The Annals of Mathematical Statistics*, **39,** 1202–1209 (5.3).

Shapiro, S. S., Wilk, M. B., and Chen, Mrs. H. J. (1968). A comparative study of various tests for normality. *Journal of the American Statistical Association*, **63,** 1343–1372 (6.1).

Sherman, E. (1965). A note on multiple comparisons using rank sums. *Technometrics*, **7,** 255–256 (5.6).

Shorack, G. R. (1967). Testing against ordered alternatives in model I analysis of variance: normal theory and nonparametric. *The Annals of Mathematical Statistics*, **38,** 1740–1752 (5.6).

Shorack, G. R. (1969). Testing and estimating ratios of scale parameters. *Journal of the American Statistical Association*, **64,** 999–1013 (5.3).

Shorack, R. A. (1967). Tables of the distribution of the Mann-Whitney-Wilcoxon *U*-statistic under Lehmann alternatives. *Technometrics*, **9,** 666–678 (5.3).

Shorack, R. A. (1968). Recursive generation of the distribution of several nonparametric test statistics under censoring. *Journal of the American Statistical Association*, **63,** 353–366 (5.3).

Siegel, S. (1956). *Nonparametric Statistics for the Behavioral Sciences*. McGraw-Hill, New York (4.4).

Siegel, S. and Tukey, J. W. (1960). A nonparametric sum of ranks procedure for relative spread in unpaired samples. *Journal of the American Statistical Association*, **55**, 429–444. Errata: Ibid. (1961) 1005 (5.3).

Slakter, M. J. (1965). A comparison of the Pearson chi-square and Kolmogorov goodness-of-fit tests with respect to validity. *Journal of the American Statistical Association*, **60**, 854–858 (6.1).

Slakter, M. J. (1966). Comparative validity of the chi-square and two modified chi-square goodness-of-fit tests for small but equal expected frequencies. *Biometrika*, **53**, 619–623 (4.5).

Slakter, M. J. (1968). Accuracy of an approximation to the power of the chi-square goodness of fit test with small but equal expected frequencies. *Journal of the American Statistical Association*, **63**, 912–918 (4.5).

Smirnov, N. V. (1936). Sur la distribution de ω^2 (Criterium de M. R. v. Mises). *Comptes Rendus* (Paris), **202**, 449–452 (6.1).

Smirnov, N. V. (1939). Estimate of deviation between empirical distribution functions in two independent samples. (Russian) *Bulletin Moscow Univ.*, **2:2**, 3–16 (6.1, 6.2).

Smirnov, N. V. (1948). Table for estimating goodness of fit of empirical distributions. *The Annals of Mathematical Statistics*, **19**, 279–281 (6.1).

Smith, K. (1953). Distribution-free statistical methods and the concept of power efficiency. In L. Festinger and D. Katz (eds.), *Research Methods in the Behavioral Sciences*. Dryden, New York, 536–577 (7.4).

Sobel, M. (1967). Nonparametric procedures for selecting the t populations with the largest α-quantile. *The Annals of Mathematical Statistics*, **38**, 1804–1816 (5.6).

Spearman, C. (1904). The proof and measurement of association between two things. *American Journal of Psychology*, **15**, 72–101 (5.5).

Steck, G. P. (1968). A note on contingency-type bivariate distributions. *Biometrika*, **55**, 262–264 (4.2).

Steck, G. P. (1969). The Smirnov two sample tests as rank tests. *The Annals of Mathematical Statistics*, **40**, 1449–1466 (6.2).

Steel, R. G. D. (1960). A rank sum test for comparing all pairs of treatments. *Technometrics*, **2**, 197–207 (5.6).

Stephens, M. A. (1964). The distribution of the goodness-of-fit statistic, U_N^2. II. *Biometrika*, **51**, 393–398 (6.1).

Stephens, M. A. (1965a). The goodness-of-fit statistic V_N: distribution and significant points. *Biometrika*, **52**, 309–322 (6.1).

Stephens, M. A. (1965b). Significance points for the two-sample statistic $U_{M,N}^2$. *Biometrika*, **52**, 661–663 (6.2).

Stephens, M. A. (1969). A goodness-of-fit statistic for the circle, with some comparisons. *Biometrika*, **56**, 161–168 (6.1).

Stephens, M. A. and Maag, U. R. (1968). Further percentage points for W_N^2. *Biometrika*, **55**, 428–430 (6.1).

Stevens, S. S. (1946). On the theory of scales of measurement. *Science*, **103**, 677–680 (2.1).

Stone, M. (1967). Extreme tail probabilities for the null distribution of the two-sample Wilcoxon statistic. *Biometrika*, **54**, 629–640 (5.3).

Stone, M. (1968). Extreme tail probabilities for sampling without replacement and exact Bahadur efficiency of the two-sample normal scores test. *Biometrika*, **55**, 371–376 (5.9).

Stuart, A. (1953). The estimation and comprison of strengths of association in contingency tables. *Biometrika*, **40**, 105–110 (4.4).

Stuart, A. (1954). Asymptotic relative efficiency of tests and the derivatives of their power functions. *Skandinavisk Aktaurietidskrift*, parts 3–4, pp. 163–169 (2.4, 5.5).

Stuart, A. (1956). The efficiencies of tests of randomness against normal regression. *Journal of the American Statistical Association*, **51**, 285–287 (3.4, 5.5).

Stuart, A. (1963). Calculation of Spearman's rho for ordered two-way classifications. *The American Statistician*, **17**:4, 23–24 (5.5).

Sugiura, N. and Otake, M. (1968). Numerical comparison of improved methods of testing in contingency tables with small frequencies. *Annals of the Institute of Statistical Mathematics*, **20**, 505–517 (4.2).

Suzuki, G. (1967). On exact probabilities of some generalized Kolmogorov's D-statistics. *Annals of the Institute of Statistical Mathematics*, **19**, 373–388 (6.1).

Suzuki, G. (1968). Kolmogorov-Smirnov tests of fit based on some general bounds. *Journal of the American Statistical Association*, **63**, 919–924 (6.1).

Swed, F. S., and Eisenhart, C. (1943). Tables for testing randomness of grouping in a sequence of alternatives. *The Annals of Mathematical Statistics*, **14**, 66–87 (7.3).

Tamura, R. (1963). On a modification of certain rank tests. *The Annals of Mathematical Statistics*, **34**, 1101–1103 (5.3).

Tate, M. W. and Brown, S. M. (1970). Note on the Cochran Q test. *Journal of the American Statistical Association*, **65**, 155–160 (4.6).

Teichroew, D. (1965). A history of distribution sampling prior to the era of the computer and its relevance to simulation. *Journal of the American Statistical Association*, **60**, 27–49 (6.1).

Tiku, M. L. (1965). Chi-square approximations for the distribution of goodness-of-fit statistics U_N^2 and W_N^2. *Biometrika*, **52**, 630–633 (6.1).

Thompson, R. (1966). Bias of the one-sample Cramér-von Mises test. *Journal of the American Statistical Association*, **61**, 246–247 (6.1).

Thompson, R., Govindarajulu, Z., and Doksum, K. A. (1967). Distribution and power of the absolute normal scores test. *Journal of the American Statistical Association*, **62**, 966–975 (5.9).

Tobach, E., Smith, M., Rose, G., and Richter, D. (1967). A table for rank sum multiple paired comparisons. *Technometrics*, **9**, 561–568 (5.6).

Tsao, C. K. (1954). An extension of Massey's distribution of the maximum deviation between two sample cumulative step functions. *The Annals of Mathematical Statistics*, **25**, 587–592 (6.2).

Tukey, J. W. (1949). The simplest signed-rank tests. *Mimeographed report no 17, Statistical Research Group, Princeton Univ.* (5.2).

Tukey, J. W. (1959). A quick, compact, two-sample test to Duckworth's specifications. *Technometrics*, **1**, 31–48 (7.1).

Ury, H. K. (1966). Large-sample sign tests for trend in dispersion. *Biometrika*, **53**, 289–291 (3.4).

Note. Surnames with "von" or "van" are alphabetized according to the principal word in the surname.

Verdooren, L. R. (1963). Extended tables of critical values for Wilcoxon's test statistic. *Biometrika*, **50**, 177–186 (5.3).

Vorlickova, D. (1970). Asymptotic properties of rank tests under discrete distributions. *Zeitschrift für Wahrscheinlichkeitstheorie*, **14**, 275–289 (5.1).

van der Waerden, B. L. (1952/53). Order tests for the two-sample problem and their power. *Proceedings Koninklijke Nederlandse Akademie van Wetenschappen (A)*, **55** (Indagationes Mathematical 14), 453–458, and **56** (Indagationes Mathematicae 15), 303–316. Errata: Ibid. (1953), 80 (5.9).

van der Waerden, B. L. (1953). Testing a distribution function. *Proceedings Koninklijke Nederlandse Akademie van Wetenschappen (A)*, **56** (Indagationes Mathematicae 15), 201–207 (6.1).

Wald, A. and Wolfowitz, J. (1939). Confidence limits for continuous distribution functions. *The Annals of Mathematical Statistics*, **10**, 105–118 (6.1).

Wald, A. and Wolfowitz, J. (1940). On a test whether two samples are from the same population. *The Annals of Mathematical Statistics*, **11**, 147–162 (7.3).

Walker, H. M. and Lev, J. (1953). *Statistical Inference*. Holt, New York (5.2, 5.4).

Wallis, W. A. (1939). The correlation ratio for ranked data. *Journal of the American Statistical Association*, **34**, 533–538 (5.7).

Wallis, W. A. and Moore, G. H. (1941). A significance test for time series. *Journal of the American Statistical Association*, **36**, 401–409 (7.3).

Walsh, J. E. (1949). Some significance tests for the median which are valid under very general conditions. *The Annals of Mathematical Statistics*, **20**, 64–81 (5.1).

Walsh, J. E. (1951). Some bounded significance level properties of the equal-tail sign test. *Annals of Mathematical Statistics*, **22**, 408–417 (3.4).

Walsh, J. E. (1960). Probabilities for Cramér-von Mises-Smirnov test using grouped data. *Annals of the Institute of Statistical Mathematics*, **12**, 143–145 (6.1).

Walsh, J. E. (1962). *Handbook of Nonparametric Statistics*. Van Nostrand, Princeton, N.J. (2.5).

Walsh, J. E. (1963). Bounded probability properties of Kolmogorov-Smirnov and similar statistics for discrete data. *Annals of the Institute of Statistical Mathematics*, **15**, 153–158 (6.1, 6.2).

Walsh, J. E. (1965). *Handbook of Nonparametric Statistics, II.* Van Nostrand, Princeton, N.J.

Weiss, L. (1960). Two-sample tests for multivariate distributions. *The Annals of Mathematical Statistics*, **31**, 159–164 (7.3).

Weiss, L. (1966). On the asymptotic power of Cramér-von Mises tests of fit. *Annals of the Institute of Statistical Mathematics*, **18**, 149–153 (6.1).

Welch, B. L. (1937). On the z-test in randomized blocks and Latin squares, *Biometrika*, **29**, 21–52 (7.4).

Wheeler, S. and Watson, G. S. (1964). A distribution-free two-sample test on a circle. *Biometrika*, **51**, 256–257 (5.3).

White, C. (1952). The use of ranks in a test of significance for comparing two treatments. *Biometrics*, **8**, 33–41 (5.3).

Wilcoxon, F. (1945). Individual comparisons by ranking methods. *Biometrics*, **1**, 80–83 (5.1, 5.3).

Wilcoxon, F. (1949). *Some Rapid Approximate Statistical Procedures.* American Cyanamid Co., Stamford Research Laboratories (5.1, 5.7).

Wilks, S. S. (1962). *Mathematical Statistics.* John Wiley, New York (3.3).

Woinsky, M. N. and Kurz, L. (1969). Sequential nonparametric two-way classification with prescribed maximum asymptotic error probability. *The Annals of Mathematical Statistics*, **40**, 445–455 (5.3).

Woodroofe, M. (1967). On the maximum deviation of the sample density. *The Annals of Mathematical Statistics*, **38**, 475–481 (6.1).

Yates, F. (1934). Contingency tables involving small numbers and the χ^2 test. *Journal of the Royal Statistical Society*, **1**, 217–235 (4.1).

Yule, G. U. and Kendall, M. G. (1950). *An Introduction to the Theory of Statistics*, 14th ed. Hafner, New York (4.4, 4.5).

Zaremba, S. K. (1965). Note on the Wilcoxon-Mann-Whitney statistic. *The Annals of Mathematical Statistics*, **36**, 1058–1060 (5.3).

Answers

CHAPTER 1

Sec. 1.2: 1. 3^{10}; 2. 10^4; 3. 5^4; 4. 26^4; 5. $26 \cdot 25 \cdot 24 \cdot 23$; 6. $26 \cdot 25 \cdot 25 \cdot 25$; 7. 220; 8. 20; 9. 105; 10. 21; 11. 420; 12. 14; 13. 15/4; 14. 232/729; 15. 74,250; 16. (a) $\binom{N_1}{n_1}\binom{N_2}{n_2} \cdots \binom{N_k}{n_k}$, (b) 0.

Sec. 1.3: 1. HHH, HHT, HTH, HTT, TTT, TTH, THT, THH; 2. (a) At least 2H and at least 2T, (b) At least 1T and at most 1H; 3. 255; 4. 2^n; 5. n; 6. $9/2^8$; 7. 1/8; 8. 7/8; 9. 1/7; 10. 1/4; 11. 511/512; 12. $735/2^{14}$; 13. $\binom{10}{4}\binom{8}{3}\bigg/\binom{18}{7}$; 14. $\sum_{i=0}^{8}\binom{10}{i}\binom{8}{i}\bigg/2^{18}$; 15. $\binom{10}{3}\bigg/2^{10}$.

Sec. 1.4: 1. (a) 1/729, (b) 64/729, (c) 0, (d) 496/729, (e) 0, (f) 1; 2. (a) 1/12, (b) 1/12, (c) 0, (d) 0, (e) 0, (f) 3/12, (g) 1, (h) 0; 3. (a) 1/128, (b) 4/128, (c) 0, (d) 4/128, (e) 0, (f) 1/128, (g) 8/128, (h) 99/128; 4. (a) 1, (b) 3/7, (c) 0, (d) 1/7, (e) 0; 5. $\sum_{i=1}^{5} f(x) = 5/6$ instead of 1; (6) (a) A, B, C, D, E, F, (b) $P(A) = 1/6$, $P(B) = 1/6, \ldots, P(F) = 1/6$, (c) $X(A) = 3$, $X(B) = 2$, $X(C) = 5$, $X(D) = 1$, $X(E) = 4$, $X(F) = 6$; 7. (a) $91/10^{10}$, (b) Yes, (c) All possible lists of cured and not cured for 10 people, (d) $(.1)^x$ $(.9)^{10-x}$. For x cures and $10 - x$ noncures in a specified order, (e) $X =$ no. of patients cured, (f) Binomial; 8. (a) All possible lists of pass or fail for 17 people, (b) 7/136, (c) Hypergeometric, (d) Same as (b).

Sec. 1.5: 1. (a) 7/6, (b) 41/36, (c) 29/6, (d) $X = 1$, (e) .5, (f) 1; 2. (a) 2, (b) 3/2, (c) −2, (d) 1.5, (e) 3, (f) 1; 3. (a) 1/2, (b) 1/2, (c) 1/4, (d) 1, (e) 0, (f) 1/2, (g) 1/2 (g) yes; 4. (a) 1/2, (b) 1/2, (c) 1/8, (d) 1/8 (e) −1/8 (f) 1/2, (g) No; 5. 2211; 6. 2535; 7. 9455; 8. (a) 7/2, (b) 35/12, (c) 56/3; 9. (a) 31, (b) 434/3.

Sec. 1.6: 1. .001; 2. Approx 1; 3. 207; 4. 03; 5. .06; 6. (a) 124.3, (b) 123.26, (c) 124.19, (d) 124.27; 7. (a) .970, (b) .97.

CHAPTER 2

Sec. 2.1: 1. (a) The nation's high schools, (b) High schools in Washington, D.C., 2. (a) Voting citizens in Kansas, (b) All people in viewing area who could call, (c) People who called, 372; people who said No, 164; people who said yes, 208, (d) Nomial and ratio; 3. (a) Ordinal, (b) Total number of points, used to determine winner; 4. Nominal; 5. (a) Interval, (b) Interval, (c) Ratio, 6. (a) 35, (b) 1/35, (c) 1, (d) 2/35; 7. No; 8. Yes

Sec. 2.2: 1. (b) 2150, (c) 4300, (d) 24,324,000, (e) $\sqrt{24{,}324{,}000} \approx 4932$ or 4933; 2. (b) 78, (c) 10, (d) 74; 3. $\dfrac{\text{Number of sample values} \le C}{n}$, 1/5; 4.
$X^{(n)} - X^{(1)}$, No, Yes, Yes; 5. (a) 6, (b) (4, 6), (4, 7), (4, 10), (6, 7), (6, 10), (7, 10), (c) 1/6, (d) 5, 5.5, 7, 6.5, 8, 8.5, (3) 1/6, (g) Same as for median.

Sec. 2.3: 1. H_0: The new method is no better than the existing.
H_1: The new method is better than the existing.
2. (a) The judge; (b) H_0: the defendant is innocent.
H_1: the defendant is guilty.
(c) The evidence and all true facts; (d) Level of significance; P(convicting and innocent); power; P(convicting the guilty) 3. (a) Fertilizer B is not as good as fertilizer A, (b) My opponent is cheating, (c) The occurrence of sun spots does affect the economic cycle; 4. (a) The subject does not have extrasensory perception, (b) The dowsing rod is not effective in finding water, (c) Our average yearly temperatures are not rising; 5. .08; 6. .10, .91; 7. 3/16 or .187; 8. (a) 220, (b) 11, (c) 1/220, (d) .05; 9. (a) 1320, (b) 66, (c) 1/1320, (d) .05; 10. (a) 35, (b) 128, (c) 10, (d) 1/128, (e) 5/68 or .078; 11. (a) 32/2187, (b) 128/2187, (c) 576/2187 or .26; 12. (a) Simple, (b) Simple, (c) Simple, (d) Simple; 13. (a) 1, (b) 1, (c) 2,

Sec. 2.4: 1. (a) $(1 - p)^7 + 5p(1 - p)^6 + 4p^2(1 - p)^5$, (b) No; 2. (a) Same as 1(a). (b) No; 3. (a) $p^3 + (1 - p)^3$, (b) .4375, (c) Yes; 4. Yes; 5. No; 6. Yes; 7. (a) .57, (b) 1.75; 8. (a) .5, (b) No; 9. 32.

CHAPTER 3

Sec. 3.1: 1. No, $P(0 \le p \le .19) = .95$; 2. $\hat\alpha = .0033$ Assumption Number 2, 3. (a) Yes, (b) $P(.15 \le p \le .28) = .95$, (c) $P(.15 \le p \le .27) = .95$; 4. $P(.17 \le p \le .61) = .90$; 5. $P(.06 \le p \le .44) = .95$; 6. No, $\hat\alpha = .01$.

Sec. 3.2: 1. $\hat\alpha = .1154$; 2. $\hat\alpha = .4148$; 3. $\hat\alpha = .2375$; 4. (a) $P(117 \le x_{.50} \le 137) = .8846$, (b) $P(117 \le x_{.05} \le 137) \approx .90$; 5. No; 6. $P(70.5 \le x_{.95} \le 76.0) = .95$.

Sec. 3.3: 1. (a) 77, (b) 76.35, $\therefore$ use 77; 2. (a) 29, (b) 28.45, $\therefore$ use 29; 3. (a) 15, (b) 14.198, $\therefore$ use 15, (c) 15, $\alpha = .0874$; 4. (a) 473, (b) 472.528, use 473; 5. 26.

Sec. 3.4:   1. $\hat{\alpha} = .1094$; 2. $\hat{\alpha} \approx .0001$; 3. $\hat{\alpha} = 0$; 4. $\hat{\alpha} = .3592$; 5. $\hat{\alpha} = .0078$;
6. $\hat{\alpha} = .0003$; 7. $\hat{\alpha} = .0037$; 8. $\hat{\alpha} = .001$; 9. $\hat{\alpha} = .0312$; 10. $\hat{\alpha} = .24$;
11. $\hat{\alpha} = .47$; 12. $\hat{\alpha} = .0312$; 13. $\hat{\alpha} = .0287$.

CHAPTER 4

Sec. 4.1:   1. $\hat{\alpha} = .427$;   2. $\hat{\alpha} = .087$;   3. $\hat{\alpha} = .355$;   4. $P(T = 0) = 2/32$, $P(T = 5/36) = 12/32$, $P(T = 5/6) = 6/32$, $P(T = 15/8) = 4/32$, $P(T = 20/9) = 6/32$, $P(T = 5) = 2/32$, $\alpha = 2/32$; 5. $\hat{\alpha} = .114$.
Sec. 4.2:   1. 2×3 $(R_1 = 8, R_2 = 10)$, $\hat{\alpha} = .005$; 2. $\hat{\alpha} = .111$, 3. $\hat{\alpha} = .00004$.
Sec. 4.3:   1. $\hat{\alpha} = .78$; 2. $\hat{\alpha} \ll .00001$; 3. $\hat{\alpha} = .3785$.
Sec. 4.4:   1. (a) 6.34, (b) .18, (c) .0317, (d) .244, (e) .0634, (f) .1780; 2. (a) 2.05, (b) .17, (c) .041, (d) .198, (e) .041, (f) .2023; 3. (a) $R_5 = .149$, $\hat{\alpha} = .12$, (b) .3636, (c) 0.
Sec. 4.5:   2. $\hat{\alpha} = .134$.
Sec. 4.6:   1. (a) $\hat{\alpha} = .025$, (b) $\hat{\alpha} = .02$, (c) $\hat{\alpha} = .04$; 2. $\hat{\alpha} = .018$; 3. $\hat{\alpha} = .005$.

CHAPTER 5

Sec. 5.1:   1. $\hat{\alpha} \cong .50$; 2. $P(T = x) = 1/32$ for $x = 0, 1, 2, 13, 14, 15$
$= 2/32$ for $x = 3, 4, 11, 12$
$= 3/32$ for $x = 5, 6, 7, 8, 9, 10$;
3. $\hat{\alpha} < .01$; 4. $\hat{\alpha} \cong .34$. Probably symmetry and continuity; possibly independence.
Sec. 5.2:   1. 5.5 to 7.45, 6.5 is middle dot; 2. 3.0 to 5.8.
Sec. 5.3:   1. $\hat{\alpha} < .001$; 2. $\hat{\alpha} = .0046$; 6. (a) $\hat{\alpha} > .20$, (b) $\hat{\alpha} \cong .38$, (c) $\hat{\alpha} = .54$, (d) $\hat{\alpha} = .50$.
Sec. 5.4:   1. 0 to 14; 2. -6 to 2.
Sec. 5.5:   1. (a) $-.60303$, (b) $-.51111$, (c) $\hat{\alpha} = .07$, (d) $\hat{\alpha} = .037$; 7. (b) .0435, (c) .503, (d) for ρ, $\hat{\alpha} \cong .015$; for T, $\hat{\alpha}$ is close to 1.0, (e) $R = 0$.
Sec. 5.6:   1. $\hat{\alpha} < .009$; 2. $\hat{\alpha} = .022$; 5. Exact tables are available for small sample sizes, and one-tailed tests are easier to perform.
Sec. 5.7:   1. $\hat{\alpha} = .004$; 2. $\hat{\alpha} < .001$ for lists $\hat{\alpha} < .001$ for students;

7. Let $T = \dfrac{k-1}{k} \sum_{j=1}^{k} \dfrac{[R_j - E(R_j)]^2}{\text{Var}\,(R_j)}$, where $E(R_j) = 3b(k-1)/k$ and

$\text{Var}\,(R_j) = b(5k - 9)/k^2$. Use Table 2, with $k - 1$ degrees of freedom.
Sec. 5.8:   1. $\hat{\alpha} = .016$.
Sec. 5.9:   Answers vary widely according to numbers selected.

CHAPTER 6

Sec. 6.1:   2. (a) $\hat{\alpha} > .20$, (b) $\hat{\alpha} = .3$; 3. $\hat{\alpha} > .20$.
Sec. 6.2:   1. $\hat{\alpha} = .05$; 2. $\hat{\alpha} = .05$; 3. $\hat{\alpha} = .04$; 4. $x_{.80} = 5/6$, $x_{.90} = 1$, $x_{.95} = 1$, $x_{.90}$, and $x_{.95}$ are not given in Table 17 because there is no corresponding critical region; 6. For $n = 30$ the exact $x_{.95}$, .6, is smaller than the approximation .608. For $n = 10$, 1/3 (exact) $< .350$ (approximate).

Sec. 6.3:   1. Birnbaum-Hall: $\hat{\alpha} = .10$, 2-sided Smirnov: $\hat{\alpha} = .05$;   2. $\hat{\alpha} = .10$;
3. $\hat{\alpha} = .005$.

CHAPTER 7

Sec. 7.1:   1. $\hat{\alpha} > .05$; 2. $\hat{\alpha} < .0005$; 3. $\hat{\alpha} < .025$; 4. $\hat{\alpha} < .001$.
Sec. 7.2:   1. $\hat{\alpha} \leq .02$, lighting schemes C and E appear to be better; 2. $\hat{\alpha} = .05$;
4. $\hat{\alpha} = .01$; Population 3 appears larger.
Sec. 7.3:   1. $\hat{\alpha} = .02$; 2. $\hat{\alpha} = .84$; 3. $\hat{\alpha} = .90$; 4. $\hat{\alpha} = .025$.
Sec. 7.4:   1. $\hat{\alpha} = .125$; 2. $\hat{\alpha} = 59/64$; 3. $\hat{\alpha} = .107$; 4. $\hat{\alpha} = 1/28$.
5. Neither procedure would affect the results;
6. Neither procedure would affect the results.

Index